THE ASCENT OF AFFECT

THE ASCENT OF AFFECT

GENEALOGY AND CRITIQUE

RUTH LEYS

THE UNIVERSITY OF CHICAGO PRESS

CHICAGO AND LONDON

The University of Chicago Press, Chicago 60637
The University of Chicago Press, Ltd., London

Published 2017
Printed in the United States of America

26 25 24 23 22 21 20 19 18 17   1 2 3 4 5

ISBN-13: 978-0-226-48842-4 (cloth)
ISBN-13: 978-0-226-48856-1 (paper)
ISBN-13: 978-0-226-48873-8 (e-book)
DOI: 10.7208/chicago/9780226488738.001.0001

Library of Congress Cataloging-in-Publication Data

Names: Leys, Ruth, author.
Title: The ascent of affect : genealogy and critique / Ruth Leys.
Description: Chicago ; London : The University of Chicago Press, 2017. | Includes index.
Identifiers: LCCN 2016046895 | ISBN 9780226488424 (cloth : alk. paper) | ISBN 9780226488561 (pbk. : alk. paper) | ISBN 9780226488738 (e-book)
Subjects: LCSH: Emotions—Research—History—20th century. | Emotions—Research—History—21st century. | Affect (Psychology) | Emotions and cognition. | Intentionality (Philosophy)
Classification: LCC BF531 .L465 2017 | DDC 152.4—dc23 LC record available at https://lccn.loc.gov/2016046895

♾ This paper meets the requirements of ANSI/NISO Z39.48-1992 (Permanence of Paper).

To Jennifer Ashton and Walter Benn Michaels

CONTENTS

[INTRODUCTION]

SETTING THE STAGE

Those who wish to inform themselves about the state of emotion research today are well advised to read the *Emotion Review*. The first journal of the International Society for Research on Emotion, *Emotion Review* was started in 2009, at a time when there was a groundswell of interest in affect among researchers in the United States, Europe, and elsewhere. From the beginning, the editors of *Emotion Review* conceived of the journal as a place where chiefly conceptual articles would be published in an effort to achieve an interdisciplinary, integrative view of the field. They opened its pages not only to the opinions of the best scientists working in the affective sciences but also to the ideas of historians, philosophers, and others with an interest in the emotions. As a result, the journal offers readers a highly accessible and informative picture of the state of play in contemporary emotion research.

But those same readers might be surprised or disconcerted to discover that there is no consensus regarding the science of emotion's most basic assumptions. On the contrary, articles that directly contradict each other appear with regularity, as if criticisms that have just been raised have no bearing whatsoever on the present author's claims and do not deserve even token acknowledgment. The impression left is of a scientific domain in stasis, one in which the majority of researchers cling to their contested positions and research strategies, leaving fundamental questions unresolved. This is all the more striking because the editors have repeatedly invited contributions and commentaries on pertinent selected themes, in which it might be expected that differences would be confronted and thrashed out. The situation is captured in James Gross's paper "The Future's So Bright, I Gotta Wear Shades" (2010), in which the author considers the future of emotion research. In a section that he describes as not for those readers who are "faint of heart," he reviews the "many daunting conceptual and empirical challenges" facing the field before going on to recommend the "incredible conceptual and empirical opportunities" for research in a discussion that he warns readers to avoid "if you are prone to getting dizzy."[1]

1. James J. Gross, "The Future's So Bright, I Gotta Wear Shades," *Emotion Review* 2 (3) (July 2010): 212.

I first turned my attention to the emotions around 2000, after completing a book on the genealogy of the concept of psychic trauma (*Trauma: A Genealogy* [Chicago, 2000]). I thought it was an interesting and important fact that in the third edition of the *Diagnostic and Statistical Manual of Mental Disorders* (1980) (DSM-III), when the American Psychiatric Association officially recognized Post-Traumatic Stress Disorder, survivor guilt was included as a criterion for the disorder. But in the subsequent 1987 revision of DSM-III (DSM-IIIR), this emotion was demoted to the position of merely an "associated" feature of the condition. In pursuing this topic, I came to realize that shame had taken the place previously attributed to survivor guilt, and the goal of my next book, subsequently published as *From Guilt to Shame: Auschwitz and After* (Princeton, 2007), was to explain why and in what terms the shift from guilt to shame had occurred. As a result, I began to immerse myself more generally in the literature of the emotion sciences.

One of the key figures in this development was the brilliant literary critic Eve Kosofsky Sedgwick who, starting in 1995, had launched the turn to affect in the humanities by drawing the attention of scholars to the work on shame of the American psychologist Silvan S. Tomkins (1911–1991). Sedgwick was convinced that Tomkins was a hitherto neglected but very significant thinker for those interested in theorizing the emotions.[2] Yet, the more I looked into Sedgwick's writings and Tomkins's work, the more dubious I became about the claims of both authors. Moreover, Sedgwick believed she was delivering news about the relevance of Tomkins's scientific ideas to contemporary debates about the emotions and selfhood, whereas in fact, as I soon discovered, those ideas had long been entrenched in American psychology and indeed had informed the chief paradigm in the emotion sciences, that proposed by Tomkins's admirers and supporters Paul Ekman and Carroll Izard.

On evolutionary and other grounds, Tomkins and his followers posited the existence of a limited number of discrete, primary, or "basic emotions" as part of a universal human nature, emotions that were held to be characterized by signature facial expressions and specific patterns of behavioral and autonomic responses. Ekman's assumptions and findings in particular dominated scientific research when I began investigating these topics. But I was not convinced that his claims were correct. On the contrary, I suspected that some of the iconic experiments Ekman had carried out were problematic, and I was interested to learn somewhat later that sev-

2. See especially Eve Kosofsky Sedgwick and Adam Frank, eds., *Shame and Its Sisters: A Silvan S. Tomkins Reader* (Durham, NC, 1995); and Eve Kosofsky Sedgwick, *Touching Feeling: Affect, Pedagogy, Performativity* (Durham, NC, 2003).

eral scientists had questioned their validity. In particular, I discovered that already in 1994 Ekman's former student Alan J. Fridlund and the psychologist James A. Russell had launched powerful critiques of the Tomkins–Ekman position, arguing that the experimental evidence cited in its support was inadequate and that the common interpretations of the results were unwarranted.[3] It took me some time, though, to find my way to these authors' publications, for the reason I mentioned at the start, namely, that scientists convinced of the validity of Ekman's views routinely failed to acknowledge these or related criticisms that had been raised against them.

Take for instance what has arguably been the most influential book on the affects in recent times, Paul Griffiths's *What Emotions Really Are: The Problem of Psychological Categories* (Chicago, 1997). In this supremely confident work, Griffiths—a philosopher with a special interest in the philosophy of biology—launched a set of related claims and contentions. He staged his book as an intervention in the long-standing debate between the cognitivists, who had emphasized the importance of cognition, intention, and meaning in the emotions, and the noncognitivists, who had denied that cognition is a necessary condition for many emotional responses. Criticizing the cognitivists as armchair philosophers who indulged in conceptual analyses while ignoring empirical findings, Griffiths championed instead the Tomkins–Ekman "affect program theory," or Basic Emotion Theory, as it has come to be called, as the correct scientific approach, at least for certain classes of emotional phenomena.[4] I came across Griffiths's book early in my research. To my mind, it was among the most illuminating works available at the time, steeped as it was in the latest results of various relevant sciences and committed as it also was to certain robustly defended arguments. Surprisingly, though, in his book Griffiths did not mention existing criticisms of Ekman's work, especially the critiques that had recently been published by Fridlund and Russell, with the result that his perspective on the field came to strike me as skewed. As we shall see, Griffiths later acknowledged the interest if not the validity of those critiques, in order to defuse them. But in an important sense the damage had been done, because his 1997 book was widely read as definitively answering the most pressing issues in emotion research.

3. Alan J. Fridlund, *Human Facial Expression: An Evolutionary View* (San Diego, CA, 1994). Fridlund's book included a reprint of Russell's critique of Ekman's and others' cross-cultural studies: "Is There Universal Expression of Emotion from Facial Expression? A Review of Cross-Cultural Studies," *Psychological Bulletin* 115 (1994): 102–41.

4. Paul E. Griffiths, *What Emotions Really Are: The Problem of Psychological Categories* (Chicago and London, 1997); hereafter abbreviated as *WERA*.

THE EMOTION PROBLEM, NATURALISM, AND THE AFFECT PROGRAM THEORY

But what were those pressing issues that had long divided the emotion field and that informed Griffiths's decision to throw his lot in with Ekman's position? As the philosopher Phil Hutchinson has recently emphasized, the fundamental "Emotion Problem" confronting scientists and philosophers alike has been the perceived need to resolve two apparently contradictory demands.[5] On the one hand, there has been the felt requirement to acknowledge the intentionality of the emotions, that is, the fact or idea that emotions are directed at cognitively apprehended objects and are sensitive to "reasons." Thus, when I am sad, I am sad about something—some object or situation—and the fact that my emotion is directed at such an object or situation is said, following the influential writings of the philosopher Franz Brentano, to give that emotion its intentionality or meaning.[6] On the other hand, emotions—or at least some of them—appear to be common to both humans and nonhuman animals. But nonhuman animals as typically conceived lack the cognitive capacity, specifically the linguistic ability, to make propositions on which the intentionality of emotions has seemed to depend.

Faced with that apparent dilemma, theorists and researchers have tended to line up on two opposing sides. On one side are the cognitivists—appraisal psychologists such as Richard Lazarus, social constructionists

5. Phil Hutchinson, *Shame and Philosophy* (Basingstoke and New York, 2008), 124–25.

6. "The following facts are commonly cited as examples of 'intentionality,'" the philosopher Hilary Putnam has observed in this regard: "(1) the fact that words, sentences, and other 'representations' have meaning; (2) the fact that representations may refer to (i.e., be true to) some actually existing thing or each of a number of actually existing things; (3) the fact that representations may be about something which does not exist; (4) the fact that a state of mind may have a 'state of affairs' as its object, as when someone says, 'she believes that he is trustworthy,' 'he hopes that his boss will get fired,' 'she fears that there won't be food in the house.'" Hilary Putnam, *Representation and Reality* (Cambridge, MA, 1988), 1. As Griffiths has commented, the intentionality of the emotions has been a central concern in the philosophy of affect since the 1960s, when Anthony Kenny, a student of Elizabeth Anscombe and Ludwig Wittgenstein, was the "most prominent of several philosophers to argue that emotions are subject to normative standards of 'fit' to the world—they can be appropriate, reasonable or unreasonable. This suggests that emotions have intentional objects—it is inappropriate and unreasonable to fear things that are not dangerous because fear represents the world as dangerous." Paul E. Griffiths, "Current Emotion Research in Philosophy," *Emotion Review* 5 (2) (2013): 216.

such as Rom Harré, and many philosophers, such as Anthony Kenny, Robert Solomon, Peter Goldie, and Martha Nussbaum—all of whom in the post–World War II period have stressed the intentionality of the emotions, but are thought to have trouble accommodating the existence of emotions in nonhuman animals (or in infants before they acquire language).[7] On the other side are the noncognitivists—postwar psychologists such as Tomkins, Ekman, and Izard, and philosophers such as Griffiths—who, often influenced by Darwinian considerations and the work of William James, have emphasized the importance of bodily changes and subpersonal processes in the emotions but are seen to have difficulty explaining how it is that emotions have meaning.

In attacking the cognitivist position, Griffiths—who already in 1989 had rejected it as a degenerative research program—in his book listed several problems with the cognitivists' views.[8] In particular, he identified two difficulties that have repeatedly surfaced in critiques of this kind: the problem of so-called *objectless emotions*, such as states of depression or anxiety that seem to occur in the absence of intentional objects or any accompanying beliefs; and the problem of so-called *reflex emotions* (or what Griffiths also named "the problem of emotional recalcitrance"), in the grip of which, for example, a person may rationally believe that it is safe to fly but nevertheless remains terrified of flying. In this last connection, Griffiths also cited as evidence against the cognitivists the existence of affective responses that, as claimed by the well-known researcher Robert Zajonc, appear to be triggered by subliminal stimuli without the individual's awareness or conscious cognition (*WERA*, 28–29).

Also central to Griffiths's approach was an argument from homology, according to which similarities among animals based on a shared evolutionary history reflect real correspondences in nature. This was an argument that, by appealing to phylogeny or the evolutionary history of the species, justified treating the so-called primary or basic emotions as the same or similar in humans and many nonhuman creatures. Griffiths reasoned that even if, owing to evolution, the function of fear has been "subtly altered" by the different meaning of "danger" for, say, humans and

7. For Lazarus's views see this book, chapters 3 and 4. See also Rom Harré, ed., *The Social Construction of Emotions* (Oxford, 1986); Anthony Kenny, *Action, Emotion, and Will* (London, 1963); Robert Solomon, *The Passions* (Garden City, NY, 1976); idem, *Not Passion's Slave* (Oxford, 2003); Peter Goldie, *The Emotions: A Philosophical Exploration* (Oxford, 2000); and Martha Nussbaum, *Upheavals of Thought* (Cambridge, 2001).

8. Paul E. Griffiths, "The Degeneration of the Cognitive Theory of Emotions," *Philosophical Psychology* 2 (3) (1989): 297–313.

chimps, the computational mechanisms used to process danger, along with the neural structures and systems implementing them, are very similar, thereby justifying the treatment of the basic emotions as alike in such animals (*WERA*, 213–16).[9] Griffiths's argument from homology relied in turn on a hierarchical view of brain function according to which the emotions shared by humans and nonhumans are controlled by phylogenetically old, subpersonal structures and processes. Those structures and processes are located in subcortical or "limbic system" brain modules that, according to his hypothesis, are informationally encapsulated in the sense that they are anatomically and functionally independent of the higher cognitive systems. Griffiths mentioned the amygdala—a small group of nuclei located subcortically in the brain—as one such limbic system module playing a crucial role in affective responses (*WERA*, 96). (The year before Griffiths published his book, on the basis of experiments on rats, the neuroscientist Joseph LeDoux had made the case for the amygdala as the seat of the fear response.)[10]

Each of Griffiths's assumptions had been challenged in the past. To take just one example, in a tradition going back to Wittgenstein and earlier, the argument had been made that a seemingly objectless or "undirected emotion," such as anxiety, is not without an object, it's just that the object, real or imaginary, is abstract, indefinite, diffuse, or vague.[11] For a Freudian, alternatively, the object of the emotion exists, but the subject is unaware of it because it is repressed. However, Griffiths clearly felt that he had the philosophical and empirical resources to answer such objections. He therefore proceeded as if the key issues could now be settled in favor of the noncognitivist position, at least as regards the so-called primary or basic emotions.

In particular, he was convinced that the scientific evidence was on his side. He was persuaded that Ekman had demonstrated the existence of a set of universal, hardwired, discrete emotions or "affect program phenomena," each characterized by signature facial expressions and distinct patterns of related behavioral, physiological, and autonomic nervous sys-

9. See also Paul E. Griffiths, "Is Emotion a Natural Kind?," in *Thinking About Feeling: Contemporary Philosophers on Emotion*, ed. Robert C. Solomon (Oxford, 2004), 237–38.

10. Joseph LeDoux, *The Emotional Brain: The Mysterious Underpinnings of Emotional Life* (New York, 1996).

11. See Michel Ter Hark, *Beyond the Inner and the Outer: Wittgenstein's Philosophy of Psychology* (Dordrecht, 1990), 219–20. Cf. Phil Hutchinson, "Facing Atrocity: Shame and Its Absence," *Passions in Context* 2 (1) (2011): 93–117.

tem processes. In a chapter on the psycho-evolutionary approach to the emotions, Griffiths cited Ekman's well-known cross-cultural judgment studies and laboratory experiments as proving that certain facial expressions of emotion, first discussed in the modern period by Darwin, are reliably associated with specific emotions in all human cultures (*WERA*, 51–55). That is, Griffiths treated Ekman's research as decisive scientific evidence in favor of the view that the affect program phenomena are pancultural behavioral-physiological responses that can be discharged in an automatic, involuntary, noncognitive fashion by unlearned triggers. Although he recognized that, according to Ekman, such affect program phenomena may also be activated by culturally learned stimuli and their manifestations controlled by cultural conventions or "display rules," Griffiths argued against the cognitivists by asserting that "The affect program phenomena are a standing example of the emotional and passionate. They are sources of motivation not integrated into the system of beliefs and desires. The characteristic properties of the affect program states, their informational encapsulation, and their involuntary triggering, necessitate a concept of mental state separate from the concept of belief and desire" (*WERA*, 243). Or, as he also observed: "The psychoevolved emotions occur in a particularly informationally encapsulated modular subsystem of the mind/brain. The processes that occur therein, the 'beliefs' of the system and the 'judgements' it makes, are not beliefs and judgements of the person in the traditional sense, any more than the 'beliefs' and 'judgements' of the balance mechanisms fed by the inner ear."[12]

Griffiths's endorsement of the affect program theory was closely linked to another claim, fundamental to his project, namely, that emotions do not form a single category or "natural kind." A natural kind approach takes it for granted that, as Robert Roberts has observed, the boundaries of categories, such as the affect program events, are "set by the nature of the phenomena under investigation" and not by the "limitations of scientific technique and interests," because nature is carved at its joints.[13] But Griffiths argued that it was necessary to assume a natural kind "pluralism," according to which there are several natural kind emotional phenomena: the affect program theory accounts for a discrete number of primary affects, whereas the so-called "higher" or "cognitive" emotions, such as guilt, envy, jealousy, and shame (and even some categories of anger and

12. Griffiths, "The Degeneration of the Cognitive Theory of Emotions," 298.

13. Robert C. Roberts, *Emotions: An Essay in Aid of Moral Psychology* (Cambridge, 2003), 23.

disgust), constitute a different "kind" or category of emotion, requiring an alternative explanation.[14]

Griffiths described those higher emotions as marked by certain features, such as their cultural variability, the fact that they are not brief but are sustained over time, their lack of stereotypical behavioral and physiological attributes, and above all their integration with complex, often conscious cognitive processes such as beliefs and desires "in a way quite alien to the affect program model" (*WERA*, 102). These "higher" emotions were of course the types of emotion that had been paradigmatically viewed as intentional states. But because, as Griffiths himself admitted, he did not have a firm explanation for those more complex or higher emotions, he left the impression that in his effort to explain them he had postponed rather than fully elucidated the problem of intentionality or meaning. Moreover, it was evident that a great deal hung on his commitment to Ekman's affect program theory: if the solidity of that theory was doubtful, as certain emotion scientists contended, then not only would Griffiths's position be undermined, but the very distinction between the basic affects and the cognitive emotions on which his scheme depended would be compromised.

THE PROBLEM OF INTENTIONALITY

How central the problem or question of intentionality has remained to the study of the emotions is evident in a more recent work by a philosopher who, convinced of the validity of Ekman's affect program theory, also takes aim at cognitivism. In his book *Passionate Engines: What Emotions Reveal About Mind and Artificial Intelligence* (Oxford, 2002), Craig DeLancey follows Griffiths's lead by proposing what he regards as a much-needed corrective to various cognitivists who fail to take advantage of the "best scientific understanding" that emotion research has to offer.[15] He aggressively denounces philosophers such as Donald Davidson and John McDowell for several sins (*PE*, 49–67, 142–47), especially for ignoring discoveries that he believes support Ekman's affect program theory, discoveries

14. In fact, in his book Griffiths divided the "higher cognitive emotions" into two discrete types, "irruptive motivations" (which he viewed as passive emotional states that overwhelm the subject) and "disclaimed" or "socially constructed" emotions (*WERA*, chapters 5 and 6). He left open the question of whether the latter constitute a discrete natural kind.

15. Craig DeLancey, *Passionate Engines: What Emotions Reveal About Mind and Artificial Intelligence* (Oxford, 2002); hereafter abbreviated as *PE*.

DeLancey then offers as a solution to the "emotion problem" along lines very close to those put forward by Griffiths.

We therefore find in DeLancey's book the same depreciation of the role of reason and intention in certain emotional responses; the same commitment to a pluralism of natural kinds, so that Ekman's affect program phenomena can be seen to form a class of emotions distinct from other kinds of affects; the same adherence to the idea of the homology between humans and nonhuman animals such that the basic emotions are said to be found across different species; the same hierarchical view of brain function, with the accompanying claim that the affective and cognitive systems are subserved by separate anatomical-neural structures or modules; and the same resort to subpersonal processes to explain the affect program phenomena.[16] Nowhere in his book, though, does DeLancey acknowledge any of the scientific criticisms of Ekman's affect program theory, criticisms that by 2002 even Griffiths had started to confront.

Nevertheless, DeLancey's work is of interest, in part because he tries to address that feature of cognitivism which, as he acknowledges, has made it so appealing, even convincing, to so many: its ability to explain the intentionality or meaningfulness of the emotions. As he puts it:

> [C]ognitive theories are perhaps most compelling when they are used to account for those features of emotion that ally them with what would normally be called cognitive features. These include the intentionality of emotions (the fact that they are often in some sense "about" something), their evaluative nature (they are often like judgments, which can be seen as evaluations made by the subject), and their interesting connections to rationality . . . These are all features for which any theory of emotion should account. (*PE*, 43–44)

What is noteworthy is that, in order to respond to the necessity of solving the problem of the intentionality of the emotions, DeLancey provocatively suggests that the affect program theory itself can clarify the issue even more effectively than cognitivist theories are able to do. Because Ekman and Tomkins had explicitly presented the affect program phenomena as

16. A cognitivist might reply to DeLancey that the latter begs questions when he offers certain problematic cases as counterexamples to the view that emotional responses are reducible to, or depend on, cognition. As one example of a response that he considers independent of belief and desire and hence a "purely expressive action," DeLancey mentions "jumping for joy upon getting some good news" (*PE*, 52). But a critic might object that a person won't jump for joy unless she understands the news, and if she understands the news she is already doing cognition.

independent of intention, DeLancey appears to break new ground. In a discussion that until that point had been relentlessly hostile to appeals to ordinary language usage in emotion theory as a form of empty conceptual talk, he considers "grammatical" evidence about how we normally speak about our emotions—as when we say we are angry at another person, or are afraid of a dog or snake—in order to introduce a terminological distinction that he believes allows him to perform the necessary work of analysis.

He observes in this regard that the word "object" can be ambiguous because it refers both to "whatever thing an intentional state is directed at, or to a (potentially) concrete thing in the world" (*PE*, 89). In order to resolve this supposed ambiguity, DeLancey distinguishes between the intentional object defined as "a kind of logical term," referring to "whatever thing to which the intentional state is directed," and the "*concretum*," defined as "a concrete object (e.g., a snake)"—a concrete object that, in conformity with Brentano's views, can be either actual or imagined. DeLancey therefore argues that "Objects therefore include not only concreta, but also events or states of affairs (that is, the referents of propositions), either real or possible or imagined" (*PE*, 89).

On the basis of this somewhat murky distinction, the author then proposes that Ekman's affect program phenomena can be defined as intentional simply because they are, as he puts it, "directed at" a concrete object or *concretum.* Thus, according to DeLancey, while certain instances of the basic emotions can be understood in terms of propositional mental content of the kind "Eric is afraid [because he believes] that the rattlesnake will bite him," some basic emotions "can occur in forms where their intentionality is only sufficiently described as concretum directed" (*PE*, 90). He gives the following example: if Tony is mad at Eric not for one particular act but for a long list of acts, yet Tony can't easily remember all the reasons he is mad at Eric, then in such a case the role of beliefs is so weak that "it starts to become not some identifiable thing that constitutes or causes the anger, but some vague collection with perhaps some extremely general shared properties (slights, inappropriate behavior toward the agent, etc.)." In short, DeLancey contends that if the only thing that is common to all the possible intentional contents in such a list of propositions is "the concretum Eric," then "why assume that any old proposition will do, as opposed to the concretum term 'Eric'? Thus, if this case is possible, it shows that the best explanation in such a case is that the basic emotion is directed at a concretum (Eric), not at some events or states of affairs" (*PE*, 91).

DeLancey adds that much scientific evidence is "clearly consistent" with a concretum-directed interpretation. Among the findings he cites are Zajonc's claims for the existence of affective preferences stemming from subliminal exposure to stimuli apparently independently of reportable memories of the event; the results of fear-conditioning experiments; the production of emotional responses by direct stimulation of subcortical systems in the brain; the behavior of nonhuman animals who lack propositional attitudes, as when a rat flees a cat; and, "perhaps most important," instances of basic emotions for which "no gloss into a propositional attitude is going to be able to capture the same content." He gives as an example of the latter the inherited human predisposition to fear snakes or spiders because in such cases "there may be fear of a concretum for which there is no appropriate kind of believed sentence" (*PE*, 93).

DeLancey does not deny that propositions may be an irreducible feature of even certain "basic emotion" responses in humans, because the sentences in which we describe our feelings can't be reliably translated into equivalent instances of concretum-directed intentional states (*PE*, 93). His aim, rather, is to propose the existence of a heterogeneous intentionality of basic emotions, grounded in an argument for the plurality of natural kinds, to the effect that some emotional responses in humans are intentional because they have propositional content, while other, more fundamental, emotional responses shared by humans and nonhuman animals are also intentional in the minimal sense of being "directed at" concrete objects. In this scenario, as Griffiths has recently summarized DeLancey's position, basic emotions "can be intentionally directed at a state of affairs, so their content is a proposition. For example, I may be afraid *that this dingo may bite me.*" But, as Griffiths goes on to remark:

> the very same emotion may be intentionally directed at what DeLancey calls a "concretum," meaning an object as such, rather than as an element in a proposition. For example, I may be afraid of *this dingo.* This is where affect-program theory comes in. Because emotions are intrinsically action-directing, an emotion whose content is a concretum can nevertheless explain action. I flee the dingo because I am afraid of it. In order to flee I do not need a proposition about the dingo, such as that it is dangerous or that it will bite me, combined with a desire to avoid danger or not to be bitten. I just need to be afraid, and for the target of my fear to be the dingo. What we have in common with other animals, DeLancey argues, is the ability to have emotions that are intentionally directed at a concre-

tum. What distinguishes us is the ability to have emotions intentionally directed at a proposition.[17]

Cognitivists might protest that DeLancey's argument amounts to nothing more than sleight of hand: he simply redefines as "intentional" what, under the terms of the affect program theory, had been considered causally induced, quasi-reflexive responses functioning independently of the beliefs, desires, and cognitions that, as he himself notes, are usually held to be intrinsic to intentionality. Critics might point out in this connection that the little words "directed at," as in the statement that the rat's fear is "directed at" the cat—words that are repeatedly deployed by DeLancey to describe such basic emotion responses—are here doing a lot of work: these words obscure from view, or appear to deny, what DeLancey elsewhere in his book quietly recognizes, namely, that according to the affect program theory, the objects or "concreta" he designates as the intentional objects are in fact simply the "mere stimuli" or "elicitors" that "trigger" the basic affect programs (*PE*, 92, 96, 99, 214). In short, critics might argue that DeLancey simply reverses the relationship between stimulus and response posited by the affect program theory by treating the cause of the emotional response as its supposed intentional object. The result is that he illegitimately applies the notion of intentionality to actions that lack the attributes of intentionality since they are, by definition, subpersonal, quasi-reflexive emotional reactions that function independently of cognition, beliefs, and desires.

17. Griffiths, "Current Research in Philosophy," 216–17. Griffiths's statement is strange because it seems to suggest that propositions are external to and independent of the emotions themselves, indeed that propositions are in some sense the objects at which the emotions are "directed." This contrasts with Wittgenstein's claim that beliefs and cognitions are constitutive of the emotions and that it belongs to the "grammar" of emotions that this is the case. As Ter Hark has stated of Wittgenstein's position in this regard, emotions are "constituted by beliefs, and via those beliefs emotions are essentially related to an object" (Ter Hark, *Beyond the Inner and the Outer*, 197). Or, as Ter Hark has also remarked of Wittgenstein's views: "The language-game 'I am afraid' already contains the object . . . The concept of emotion is internally related to an object at which the emotion is directed. If the relation is internal, this entails that the description of an emotion is *at the same time* the description of the object of emotion—not the cause . . . This is not to reject a causal explanation, but merely to put it in its place: an inventory of causes is not constitutive of the language-game of emotions. We could, for instance, describe all the causal happenings when we cry without mentioning what we are crying about . . . But as long as we have not mentioned what we are crying about, we have not described an emotion" (217–18).

AN EMBODIED WORLD-TAKING COGNITIVISM

There is more to be said about these matters. Throughout their work, both Griffiths and DeLancey equate cognitivism with the human capacity to speak and make propositions, in order to reject it (*WERA*, 2). Nor are they wrong to suggest that many cognitivists do tend to equate cognition with the human capacity for speech and the making of propositions.[18] But that is not the only kind of cognitivism on offer. In particular, the philosopher Phil Hutchinson has recently proposed a framework for understanding emotions, which does not assume that cognitions must have propositional content. He calls his position an "embodied world-taking cognitivism." It is a cognitivism that accepts as valid the neo-Kantian position of the philosopher John McDowell who, rejecting the idea of nonconceptual, subpersonal mental content, maintains instead that perceptual experience is conceptual through and through (*Shame and Philosophy*, 102–8). McDowell's is arguably the most influential attempt in recent years to come to grips with the problem of intentionality and mind-world relations by returning to, and building on, Kant's and Hegel's treatments of these topics (Wittgenstein is also an important influence).[19] Hutchinson's is a version of McDowell's "cognitivism" that, expanding on McDowell's recent revision of his own tendency to equate conceptuality and intention with the human linguistic ability to state propositions, no longer equates the capacity for conceptuality with the human capacity for discourse. Hutchinson's position thus acknowledges that both humans and some nonhuman animals enjoy forms of cognition and conceptuality, even though nonhuman animal cognitions and concepts are different from those of humans because they are nonpropositional (*Shame and Philosophy*, 130–36).[20]

18. Although, as Hutchinson has observed, even a "'pure cognitivist'" such as Robert Solomon does not hold that emotions necessarily have propositional content, suggesting that Griffiths runs the risk of erecting as his nemesis "no more than a straw man" (*Shame and Philosophy*, 87).

19. John McDowell, *Mind and World* (Cambridge, 1994).

20. In *Mind and World,* McDowell took his argument to imply the existence of a fundamental difference between humans and nonhuman animals based on the equation between human conceptuality and propositional attitudes (or the human capacity for language). But, persuaded especially by some arguments by the philosopher Charles Travis, he has recently modified his position. McDowell now suggests that perceptual conceptual content is not propositional but "intuitional" in Kant's sense. See McDowell, "Avoiding the Myth of the Given," in *John McDowell: Experience, Norm,*

McDowell is one of DeLancey's targets in his critique of cognitivism. But Hutchinson's main McDowellian objection to the views of Griffiths and other like-minded authors (although he does not mention DeLancey) is that they are in the grip of the wrong picture of how the mind relates to the world. Griffiths and DeLancey assume that the mind is linked to the world in an external fashion, in the sense that the intentional states held to characterize the mind are understood as added on to, or derived from, a nature that is defined in modern, post-Cartesian terms as intrinsically meaningless. The challenge for these authors is therefore to show how intentionally can be explained naturalistically by locating its origin in, or reducing it to, non-semantic, physical properties. As Jerry Fodor, one of the most influential post–World War II architects of naturalism in the philosophy of mind, has stated of the goals of this project: "It's hard to see . . . how one can be a Realist about intentionality without also being, to some extent or other, a Reductionist. If the semantic and the intentional are real properties of things, it must be in virtue of their identity with (or maybe of their supervenience on?) properties that are themselves *neither* intentional *nor* semantic. If aboutness is real, it must be really something else." Or, as he has also observed: "Here, then, are the ground rules. I want a *naturalized* theory of meaning: a theory that articulates in non-semantic and non-intentional terms, sufficient conditions for one bit of the world to *be about* (to express, represent, or be true of) another bit."[21]

This was precisely the program of the cognitive science revolution in the post–World War II period. Indeed, Fodor was one of the architects of cognitive psychology's goal to naturalize mind and meaning by positing the existence of mental representations and processes "in the head" and by making explicit the causal laws that were held to govern those representations. It was widely accepted by those committed to the cognitive science program that the computer provided the best model for such a reductive science of the mind by explaining intentionality as the effect of the computational processing of the hypothesized mental representations.

---

and Nature (Oxford, 2008), 1–14; Charles Travis, "Reason's Reach," in ibid., 176–99; and McDowell's response to Travis, in ibid., 234–35. Meanwhile, influenced by McDowell, Travis, and others, as I have observed, Hutchinson has made the case for a cognitivism that is not committed to a propositional attitude account of mental content. For a succinct statement of his position, see also Hutchinson, "Emotion-Philosophy-Science," in *Emotions and Understanding: Wittgensteinian Perspectives*, eds. Ylva Gustafsson, Camilla Kronqvist, and Michael McEachrane (Basingstoke and New York, 2009), 60–80.

21. Jerry A. Fodor, *Psychosemantics* (Cambridge, MA, 1987), 97 and 98, cited by Mason Cash, "Normativity Is the Mother of Intention: Wittgenstein, Normative Practices and Neurological Representations," *New Ideas in Psychology* 27 (2009): 139.

As heirs to that same cognitive science program, Griffiths, DeLancey, and other like-minded emotion theorists therefore assume that the contents of intentional states ought to be explicable in terms of internal representations in the mind-brain, and that each representation's content must be reducible to physical properties and causal processes. How successful such materialist, reductionist projects have been is an open question.

But Hutchinson follows McDowell in arguing that we don't have to accept this account of how the mind relates to the world. As he puts it, the picture proposed by Griffiths and other like-minded theorists is one that

> presupposes a disenchanted and non-conceptual world, which stands external to our minds. This world is located outside the space of reasons, is governed by purely causal laws, and thus can *only* have causal impact upon us. It presupposes that our minds are located in our heads (and modelled on, or taken to be, the brain). The world only gains meaning, becomes enchanted, through our cognitively projecting values/meaning on to it, or more accurately, projecting values onto the impressions it causes on our senses. If we assume such a world to be our world . . . and if . . . *not all* emotions involve cognition, how then, indeed, how can *all* emotions be meaningful? (*Shame and Philosophy*, 118–19)

Rejecting such a picture of mind-world relations, Hutchinson argues instead that

> There are other ways to understand human cognitive powers, the human mind, and the world in which we live with others . . . We do not . . . situate the world in some brute Given realm (outside the space of reasons), thus creating for ourselves the problem as to how things caused by this Given realm can give rise to meanings in, and for, us minded animals, operating within the space of reasons. *Of course* if one characterizes things in *that* . . . way, we will have a pretty serious problem and pretty serious trouble in finding the resources to overcome the problem.
>
> We can . . . rather see the mind as a structured system of object-involving abilities; we can see our cognitive powers as our having grasped concepts, learnt how to employ them in contexts, through recognising their significance, for ourselves and for others, recognising their place in a life, and being able to respond to requests for reasons for our actions. We can understand the world and our mental capacities as internally related, rather than externally (causally) so. The world is our world; our minds reach out to meet it, or rather

> they think of it and act in it . . . I do not interpret sense data; I see the world. Furthermore, I convey what I see to you, employing concepts that we both have grasped. (*Shame and Philosophy*, 118–19)

In these passages, Hutchinson refers only to human capabilities, but as noted, his cognitivism extends to nonhuman animals. According to him, many of these, too, can be understood to have conceptual-intentional capabilities of various degrees of sophistication, even though the evolution of the human mind-brain has brought with it capacities for symbolic representation more complex than the conceptual abilities of even its closest nonhuman relatives.

It seems to me that Hutchinson's is the most interesting alternative to the noncognitivism of Griffiths and DeLancey and to various existing cognitive theories of emotion. He follows McDowell in rejecting as inadequate a "bald" or "restricted" naturalism of the kind adopted by Griffiths (and DeLancey), a naturalism that, in invoking Ekman's affect program theory and related ideas, marginalizes the intact person with his or her intentions and meanings and raises difficult questions about how conceptuality can be added on to nonconceptual mental contents and processes.[22] In fact, one of the objections raised by Hutchinson and

22. For a defense of McDowell's view on rationality and a critique of "tack-on" theories of conceptual content suggesting that concepts can be added on to a more primordial system governed by raw sensory inputs, see Matthew Boyle, "Essentially Rational Animals," in *Rethinking Epistemology*, eds. Abel Günter and James Conant, vol. 2 (Berlin and Boston, 2012), 395–427; and idem, "Additive Theories of Rationality: A Critique," *Journal of European Philosophy*, in press. In a paper on the importance of the distinction between personal and subpersonal mechanisms (or information-processing mechanisms) in the work of Daniel Dennett, McDowell remarks: "What could an internal information-processing device really tell an animal? To give a positive answer, we would need to deal satisfactorily with the question . . . about how to make sense of the frog's being on the receiving end of 'sub-personal' telling; but my point now is not that we have no inkling how that might be done. What could an information-processing device *really* tell *anything* (including another component in a sub-personal or 'sub-personal' informational system)? It is essential to realize that the answer to this question can be—in fact is—'Nothing,' without the slightest threat being posed to the utility, or even the theoretical indispensability, of cognitive science" (John McDowell, "The Content of Perceptual Experience," in *Mind, Value, and Reality* [Cambridge, MA, and London, 1998], 350). Hutchinson agrees with McDowell's answer to his own question, but disagrees with the conclusion McDowell draws regarding the lack of threat to cognitive science. In Hutchinson's view—and I agree with him on this point—McDowell's philosophical position does challenge the utility and theoretical indispensability of much of the work by Griffiths and others that has

other critics against Griffiths and like-minded theorists of emotion is that they often end up more or less surreptitiously transferring to subpersonal mechanisms the very powers of "cognition" and meaning-making they deny the person—as when DeLancey attributes to nonconceptual and noncognitive affect programs located subcortically in the brain the very "intentionality" he has denied to the individual. Bennett and Hacker have called this maneuver the "mereological fallacy," by which they mean the fallacy or mistake of referring to a part what properly belongs to the whole person or organism.[23]

In place of such a restricted naturalism, Hutchinson follows McDowell's lead in accepting a "liberal"—or what in the same spirit the philosopher Jennifer Hornsby calls a "naïve"—naturalism, a naturalism that takes in its stride the idea that forms of mindedness, such as intentional mental states, are through and through a part of the natural world and not alien or external to it, as post-Cartesian forms of naturalism, including modern cognitive psychology, assume.[24] Hutchinson's objections to the restricted naturalism of Griffiths and other noncognitivists resonate with the work of other scholars and critics, such as Vincent Descombes, Meredith Williams, Stuart Shanker, Jeff Coulter, and Wes Sharrock, all of whom on Wittgensteinian and related philosophical grounds have argued powerfully against the cognitive science thesis that the correct way to understand mental phenomena is by locating them inside the skull "in the internal flux of representations."[25]

---

been carried out on the emotions in the name of that science (*Shame and Philosophy*, 171–72, n. 14).

23. M. R. Bennett and P. M. S. Hacker, *Philosophical Foundations of Neuroscience* (Malden, MA, and Oxford, 2003), 71–107.

24. Jennifer Hornsby, *Simple Mindedness: In Defense of Naïve Naturalism in the Philosophy of Mind* (Cambridge, MA, and London, 1997). For Hornsby's valuable discussion of the subpersonal-personal distinction, see ibid., 157–84. I have also benefited greatly from the following works: Jason Bridges, "Teleofunctionalism and Psychological Explanation," *Pacific Philosophical Quarterly* 87 (2006): 403–21; idem, "Davidson's Transcendental Externalism," *Philosophy and Phenomenological Research* 73 (2006): 290–315; idem, "Does Informational Semantics Commit Euthyphro's Fallacy?," *Nous* 60 (2006): 522–47; David Finkelstein, *Expression and the Inner* (Cambridge, MA, 2003); idem, "Holism and Animal Minds," in *Wittgenstein and the Moral Life: Essays in Honor of Cora Diamond,* ed. Alice Crary (Cambridge, MA, 2007), 251–78.

25. The phrase "internal flux of representations" is found in Vincent Descombes, *The Mind's Provisions: A Critique of Cognitivism*, trans. Stephen Adam Schwartz (Princeton, 2001), 2. See also Jeff Coulter, *Rethinking Cognitive Theory* (New York, 1983); Graham Button and Jeff Coulter, *Computers, Mind, and Conduct* (Cambridge, 1995); Stuart Shan-

Rather, such critics locate mindedness in the public expressions and exchanges between persons or between nonhuman animals. They reject as misguided the idea that mental phenomena must be understood in terms of raw sensory Givens and internal representations defined as computational or non-semantic processes to which meaning is somehow tacked on. We might put it that for these critics the emphasis falls instead on the "ecological" or "ethological" determination of the behavior and interactions of intact animals in their natural or ordinary settings and contexts, which is to say, in the determination and understanding of human and nonhuman animal "forms of life." What the implications of such a relaxed or naïve naturalism might be for the possibilities of a science of emotion are unclear, which is to say that it remains to be seen what empirical research projects comport with the kind of embodied world-taking cognitivism advocated by Hutchinson—a topic on which I hope to throw some light in this book.

A word here about the term "cognitive." It should be apparent by now that the term has several meanings. In particular, it can suggest a concern with what is known as "cognitive psychology," which is very far from the meaning attached to the term by Hutchinson and other theorists who reject the assumptions of "cognitive science," above all the reductionist assumption that intentionality can be derived from nonconceptual and non-semantic processes.[26] It is also worth remarking that the kind of cognitivism advocated by Hutchinson and others is sometimes reproached for offering too intellectualist or disembodied an account of emotional behavior, as if emotions can be explained by appeal to a disembodied mind or set of beliefs. But there is nothing about the "world-taking cognitivism" position that is opposed to the idea that humans and nonhuman animals are embodied creatures and that this fact is of the highest importance.

Hutchinson's goal is to analyze the conceptual-philosophical problems inherent in recent approaches to the emotions, not to examine the genealogy of scientific research on the affects. Yet the history of developments in the science of emotion reflects several of his main concerns. As we have seen, the anti-intentionalism so pervasive today in emotion theory

---

ker, *Wittgenstein's Remarks on the Foundations of AI* (London and New York, 1998); Meredith Williams, *Wittgenstein, Mind, and Meaning: Towards a Social Conception of Mind* (New York, 1999); Jeff Coulter and Wes Sharrock, *Brain, Mind, and Human Behavior in Contemporary Cognitive Science: Critical Assessments of the Philosophy of Psychology* (Lewiston, NY, 2007); and Alan Leudar and Alan Costall, eds., *Against Theory of Mind* (Basingstoke, 2009).

26. For a useful discussion of the various "ideal types" of cognitivism, see Hutchinson, *Shame and Philosophy*, 97–108.

and research has a genealogy that for our purposes can be traced back to developments in the psychological sciences beginning in the early 1960s. At that time two very different scientific approaches to the emotions were simultaneously proposed. One approach, first published in 1962, was associated with the work of Tomkins, who argued that the affects and cognition constituted two entirely separate systems, and that accordingly the emotions should be theorized in anti-intentionalist terms. His views were taken up and elaborated in active research programs by Paul Ekman and Carroll Izard. The other approach, associated with a famous (if problematic) experiment by Stanley Schachter and Jerome Singer, published in 1962, claimed to demonstrate that cognition and a neutrally valenced physiological arousal mutually constitute emotion.

At first the Schachter–Singer "cognitive" model of the emotions or other variants of cognitivism prevailed. Thus, during the 1960s, psychologist Richard S. Lazarus demonstrated in a series of artfully designed experiments that viewing stressful films could induce powerful emotional and physiological responses that depended crucially on the viewer's appraisals, beliefs, and coping styles. Such findings led Lazarus in the 1980s to take a prominent role in defending the cognitivist position when it was disputed in a well-known debate with Robert Zajonc. During these same years, social constructionists and other appraisal theorists also supported the cognitivist position. But for various reasons, including especially Ekman's advocacy and the results of his highly productive laboratory research program, Tomkins's approach began to displace the cognitivist model, with the result that starting in the 1980s his had become the mainstream position. This meant that when in 1994 Fridlund published his critique of the theoretical and empirical claims underpinning the Tomkins–Ekman paradigm, he was taking on a highly entrenched position. The same year, in a superb analysis of the cross-cultural judgment studies that had been reported by Ekman and his colleagues, Russell showed that the results were artifactual, depending on forced-choice response formats and other problematic methods that begged the questions to be proved in ways that fundamentally undermined Ekman's claims for the universality of the basic emotions.

Fridlund went on to propose that facial movements should not be viewed as expressions of hardwired, discrete internal emotions leaking out into the external world, as Tomkins and Ekman claimed, but as intentional behaviors that have evolved in order to communicate motives in an ongoing interpersonal or interindividual context or transaction. From this perspective, one that linked up with a "new ethology" that likewise emphasized the communicative value of nonhuman animal displays, facial displays are relational signals that take other (real or imagined) organisms into account. According to Fridlund, humans and nonhuman animals produce facial behaviors or displays when it is strategically advantageous for them to do so and not at other times, because displays are dynamic and often highly plastic social and communicative signals. In short, Fridlund made the question of intentionality—including nonhuman animal intentionality—central to his account of the emotions. Russell, too, as we have seen, as well as scientists such as José-Miguel Fernández-Dols, Lisa Feldman Barrett, and Brian Parkinson, have all raised objections to the Ekman affect program theory and have proposed alternative approaches that challenge the disjunction between emotion and meaning on which the highly successful Tomkins–Ekman paradigm rests.

The present situation therefore offers to the historian the engrossing phenomenon of an ongoing clash between competing ways of thinking about the emotions. What is especially striking is that some of the researchers who have been formed by and trained in Ekman's presuppositions and methods have voiced misgivings about the anti-intentionalist position. But as powerful and even intellectually decisive as these scientists' objections may be, for reasons I will be examining, it will not be easy for them to overthrow Ekman's views. How long this state of affairs will prevail is an open question. Nor will it be simple for critics to come up with viable alternative models of the emotions, not only because intentionality is an intrinsically difficult topic and theorizing its role a challenging task, but it is not clear how best to operationalize intentionalist views in terms compatible with scientific requirements—more on this later.

A GENEALOGY OF RESEARCH ON THE EMOTIONS

In the following book, I offer a history of post–World War II theoretical and experimental approaches to the emotions against the background of the conceptual and epistemological issues and concerns I have just sketched. Although numerous books on the emotions have appeared in recent years, none has attempted to understand the present quandary in the emotion sciences from a genealogical perspective.[27] Of course, brief histories of emotion research are included in many surveys of psychology

27. However, I draw attention here to Jan Plamper's important book *The History of Emotions: An Introduction* (Oxford, 2015; German edition, 2012), which as a self-described "synthesis" of the current state of knowledge of the emotions also offers an "intervention" in current debates in terms that are generally compatible with my analysis and critique.

written by psychologists as introductory texts. Moreover, as a result of the burgeoning interest in the emotions by historians, there now exist several valuable studies that shed light on the nature of various emotional communities or modes of feeling in the past.[28] What has been missing is an investigation that takes the recent emotion sciences as its principal object of inquiry, and that is what I intend to provide.[29] I do not aim at comprehensiveness: the literature on the emotions is so vast and the proliferation of new articles so rapid that it would be an impossible task. Instead, in what is a work of intellectual history, I have focused on a select number of exemplary figures and episodes in the postwar history of approaches to the affects in order to throw light on the empirical, conceptual, and epistemological questions that have repeatedly surfaced.

In many respects, my choices regarding what to include are not controversial. For the most part, I have selected for discussion what are widely considered the chief episodes or figures in the past fifty or more years of research. But there are some important topics I have had to leave out. For instance, precisely because I am chiefly interested in laboratory and experimental contributions, I have touched only lightly and in passing on the large literature of social constructionism. Luckily this is a topic that has been ably discussed by Jan Plamper in his recent introduction to the history of the emotions.[30] There are also some matters I have chosen

28. There now exist centers for the study of the history of emotion in Berlin, London, and elsewhere. In a rapidly growing literature on the history of emotions, see especially William M. Reddy, *The Navigation of Feeling: A Framework for the History of Emotions* (Cambridge, 2001); Thomas Dixon, *From Passions to Emotions: The Creation of a Secular Psychological Category* (Cambridge, 2003); Joanne Bourke, *Fear: A Cultural History* (London, 2005); Jan Plamper, "The History of Emotions: An Interview with William Reddy, Barbara H. Rosenwein, and Peter Stearns," *History and Theory* 49 (2) (2010): 237–65; Ute Frevert, *Emotions in History: Lost and Found* (Budapest, 2011); Ute Frevert, Monique Scheer, Anne Schmidt, Pascal Eitler, Bettina Hitzer, Nina Verheyen, Benno Gammerl, Christian Bailey, and Margrit Pernau, *Emotional Lexicons: Continuity and Change in the Vocabulary of Feeling 1700–2000* (Oxford, 2014); and Jan Plamper and Benjamin Lazier, eds., *Fear: Across the Disciplines* (Pittsburgh, 2012).

29. But see Otniel Dror, "Modernity and the Scientific Study of Emotions, 1880–1950," PhD dissertation, Princeton University, 1998; idem, "The Affect of Experiment: The Turn to Emotions in Anglo-American Physiology, 1880–1940," *Isis* 90 (2) (1999): 205–37; idem, "Counting the Affects: Discoursing in Numbers," *Social Research* 68 (2) (2001): 357–78; idem, "Techniques of the Brain and the Paradox of Emotions, 1880–1930," *Science in Context* 14 (4) (2001): 643–60; idem, "Afterword: A Reflection on Feelings and the History of Science," *Isis* 100 (4) (2009): 848–51; F. Biess and Daniel M. Gross, eds., *Science and Emotions After 1945: A Transatlantic Perspective* (Chicago, 2014).

30. Jan Plamper, *The History of Emotions: An Introduction*, trans. Keith Tribe (Oxford, 2015), 75–146.

to highlight in ways others might not have done. For example, I have devoted two chapters to Fridlund's work and responses to it, both because, as a former student of Ekman, his attempt to dismantle the latter's affect program theory from the "inside" is significant, and because his analyses raise instructive questions about what kind of science of the emotions is possible, once the presuppositions animating Ekman's paradigm are abandoned. In other words, I have operated on the assumption that Fridlund is "good to think with" when it comes to assessing debates over the nature of emotion. I not only regard his research and writings as important contributions to the recent history of empirical work on the affects, I treat the various responses to his work as a kind of litmus test of how scientists are handling the conceptual and empirical issues raised by his critique.

My book is divided into the following chapters:

In chapter 1, I discuss the work of Silvan S. Tomkins, who helped reinvigorate the study of the emotions in the 1960s by proposing a new theory of affect. I show how, through the twin influences of Darwin and the new science of cybernetics, Tomkins proposed the existence of a limited number of universal "affect programs" or basic emotions, which he theorized as innately triggered responses that functioned independently of objects or cognitions. Through his influence especially on Ekman and Izard, Tomkins set the stage for the emergence of an influential noncognitive paradigm of the emotions.

In chapter 2, I continue the story of the affect program theory by focusing on the work of Paul Ekman, arguably the most influential figure in emotion science to this day. I explain that according to Ekman's "neurocultural" version of Tomkins's theory, we expand through socialization the range of stimuli that can elicit the affect programs and can learn voluntarily to moderate our facial movements according to the conventions of "display rules." But under certain conditions the underlying affect program phenomena will betray themselves, sometimes in micro movements of the face discernible only to the trained expert. In order to clarify the methodological issues at stake in Ekman's research, I discuss some of the problems raised by his use of photographs of posed facial expressions as an experimental tool. I explore the implications of the fundamental physiognomic assumption underlying Ekman's work in this regard, namely, the idea that a distinction can be strictly maintained between authentic and artificial expressions of emotion based on differences between the faces we make when we are alone and those we make when we are with others. Throughout my discussion, I aim to bring out some of the tensions and contradictions inherent in Ekman's affect program model. I conclude the chapter with a discussion of the influential neuroscientist Antonio Damasio's related claims about the emotions.

Richard Lazarus played a very important role in the post–World War II history of research on the emotions, and for this reason I have devoted two chapters to his work. In chapter 3, I confront head-on the challenges Lazarus faced in his attempts to provide a cognitive account of the emotions. I show that Lazarus's ideas about the role of "appraisal" in emotion were often tentative and confused, in part because of the difficulty he had in deciding what kind of claim it is that emotions are intentional states or actions. Is the claim fundamentally a constitutive-conceptual one, according to which it belongs to the very "grammar" of the emotions that they are intentional states? Or is it a causal claim about how emotions are aroused? Are those two kinds of claims incompatible, or can one adopt both a conceptual grammatical and a causal explanation of the affects? Lazarus did not find it easy to answer these questions, even as he pursued a major research program designed to do so. The aim of this chapter is to examine Lazarus's experiments on the emotions and appraisal in light of these difficulties.

In chapter 4, I examine the famous 1980s debate between Lazarus and Robert Zajonc over the role of cognition in emotion. I emphasize the role of information-processing theories of the mind in enabling Zajonc to propose a noncognitive theory of the affects in terms not unlike those of Tomkins. In this connection, I discuss James S. Uleman and John A. Bargh's *Unintended Thought* (1989) for the light it throws on the pervasiveness of information-processing ideas about the automaticity and nonintentionality of many mental functions. And I analyze Lazarus's several attempts to rebut Zajonc, based in part on his criticisms of computer models of mentation, models to which, somewhat incoherently, his own thought also fell captive. Throughout this chapter, I hope to make clear the difficulties facing Lazarus in his attempt to incorporate intentionality and meaning into a science of the emotions.

In chapter 5, I focus on Fridlund's critique of Ekman's Basic Emotion Theory. I suggest that one reason for the success of Ekman's theory is that it appears to solve the problem of deception in everyday life by suggesting that expressions have evolved to convey accurate information to others about our internal emotional states. On this model, although we are able to disguise our feelings through the voluntary management of our facial signals, under the right conditions the emotional truth of our inner states will betray itself. It is on the basis of this model that Ekman has played an influential role in federally funded post-9/11 surveillance research designed to find ways to identify terrorists before they can act. Ekman's goal is to ameliorate fears about our own tendencies to dissimulate, by providing a technological means by which authentic facial expressions can be reliably distinguished from false ones, the genuine from the feigned.

But on evolutionary, ethological, and other grounds, Fridlund has challenged that assumption. The purpose of chapter 5 is thus to lay out the essence of Fridlund's arguments and to examine the evidence he cites in support of his views. I also analyze the details of Fridlund's dismantling of one of Ekman's famous experiments on emotion, a by-now-iconic experiment on which so much of the validity of the affect program theory has been seen to depend. At the end of the chapter, I discuss Ekman's various attempts to respond to Fridlund's criticisms, attempts that reveal the contradictions and slippages inherent in the former's version of Tomkins's affect program theory.

In chapter 6, I address the question of whether Fridlund's interventions, as well as the contributions of others who have recently criticized Ekman's model of the emotions, herald a paradigm change in the emotion field or whether instead it is business as usual, in the sense that the affect program theory continues to dominate the field. In order to understand the current situation, I define certain features of Fridlund's position that, in his own words, make it a "tough sell" even if, in my view, those features reflect something incontestably right about his arguments. In order to throw further light on the recalcitrance of those committed to Ekman's views, I discuss in detail the recent writings of the author Paul Griffiths. Along with his colleague Andrea Scarantino, and in response to the critiques offered by Fridlund and others, these authors have offered a spirited defense of the basic emotion approach in psychology. The writings of Griffiths and Scarantino are especially interesting because they reveal how fundamental is the need to deflate claims about the intentionality of the emotions by retheorizing the intentional behaviors emphasized by Fridlund in non-intentional terms. How successful these authors are in this effort at retheorization is another matter.

In recent years, the topic of emotion has attracted more and more attention from cultural critics and theorists, and the purpose of my final chapter is to explore the terms in which this turn to affect has been taking place. In particular, I wish to understand why the anti-intentionalism that has informed so much of the work on the emotions in the psychological sciences now exerts such a fascination for scholars in the humanities and social sciences, whether they borrow from Tomkins's or Ekman's emotion theories or from ideas about affect originating in the work of Gilles Deleuze and related figures. I offer my analysis in the spirit of a "history of the present," that is, as an attempt to understand the rise of a non-intentionalist "affect theory" in the light of the genealogy I have charted and to explain why I think the views being forwarded are a mistake. I examine some of the scientific research cited by such affect theorists in support of their ideas in order to see how well that research stands up to scrutiny, and I draw out the aesthetic, political, and philosophical implications of their views.

Finally, I bring the different threads of my argument together in a short epilogue. In 2015, just as I was finalizing the draft of this book, some of the chief scientists in the long-standing dispute over the nature of the emotions—Fridlund, Russell, and Dacher Keltner and Daniel Cordaro (the last are followers of Ekman's basic emotions model)—were invited by Scarantino, editor of *Emotion Researcher*, the newsletter of the International Society for Research on Emotion, to publish online statements concerning their respective views. In addition, Scarantino posed questions to these scientists, and the statements and the exchanges that ensued were also posted online. If Scarantino hoped that the debate would generate a consensus about the nature of emotion, he must have been disappointed, since no such consensus emerged. In my final comments, I take a brief look at this debate and say why I think this outcome was inevitable.

[CHAPTER ONE]

SILVAN S. TOMKINS'S AFFECT THEORY

Cognition without affect is weak; affect without cognition is blind.

—Silvan S. Tomkins, 1981[1]

The last text Foucault published before his death had at its center the concept of life. In his teacher Georges Canguilhem's greatest book, *Le Normal et Le Pathologique* (*The Normal and the Pathological*, [1943] 1966), Foucault suggested, the problem of the specificity of life found itself inflected in the direction of certain crucial problems:

> At the center of these problems is that of error. For at life's most basic level, the play of code and decoding leaves room for a kind of chance that, before being disease, deficit or monstrosity, is something like a perturbation in the information system, something like a "mistake." At the limit, we might say that life—and this is its radical character—is what is capable of error. And perhaps we have to ask whether it is because of this given, or rather this fundamental eventuality, that the question of anomaly pervades the whole of biology. Of this eventuality also we have to ask for the explanation of mutations and the evolutionary processes they induce. Equally this eventuality must be interrogated about this unique yet hereditary error which produces in man a living being who never feels completely in his place, a living being who is destined "to err" and "to be deceived" . . . [Canguilhem], himself so "rationalist," is a philosopher of error: I mean to say that it is in starting from error that he poses philosophical problems, more exactly, the problem of truth and life . . . Does not the entire theory of the subject have to be reformulated once knowledge, instead of opening onto the truth of the world, is rooted in the "errors" of life?[2]

1. Silvan S. Tomkins, "The Rise, Fall, and Resurrection of the Study of Personality," *Journal of Mind and Behavior* 4 (2) (1981): 448.

2. Michel Foucault, "La vie: Experience et Science," *Revue de Metaphysique et de Morale* (January–March 1985): 13. In my translation, I have chosen to bring out the idea of "feeling" in the phrase "*un vivant qui ne se trouve jamais tout à fait a sa place,*" and the idea of "deception" in the phrase "*un vivant qui est voué a 'errer' et à 'se tromper.'*" My thanks to Jacques Neefs for help in translating Foucault's difficult, nuanced French.

The philosopher Giorgio Agamben has recently stressed what he sees as the anti-intentionalist implications of Foucault's remarks by observing that

> [W]hat is at issue here is . . . something like a new experience that necessitates a general reformulation of the relations between truth and the subject and that, nevertheless, concerns the specific area of Foucault's research. Tearing the subject from the terrain of the *cogito* and consciousness, this experience roots it in life. But insofar as this life is essentially errancy, it exceeds the lived experiences and intentionality of phenomenology . . . What is the nature of a knowledge that has as its correlate no longer the opening to a world and to truth, but only to life and its errancy? . . . It is clear that what is at issue in Foucault is not simply an epistemological adjustment but, rather, another dislocation of the theory of knowledge, one that opens onto entirely unexplored terrain.[3]

Foucault's impressive discussion of Canguilhem, along with Agamben's comments, can serve as a frame for my analysis of a moment in the history of the human sciences when the idea of life's essential errancy informed an explicitly anti-intentionalist account of the affects. My goal in this chapter is to analyze the emergence in the United States in the post–World War II period of a new anti-intentionalist paradigm of the affects based on the role of error. That new paradigm eventually displaced alternative ways of thinking about the emotions. Indeed, its success has been so great that today it informs the work not only of the majority of neuroscientists and psychologists but also, increasingly, that of literary critics and political theorists. (There is also a reaction against it—we shall see.)

My story begins in the early 1960s, when the American psychologist Silvan S. Tomkins (1911–1991) published a new theory of the affects in two volumes of what would become a massive four-volume study of the emotions.[4] It was not his first publication on the topic. He had already sketched his ideas in an essay in French in a volume edited by the French psychoanalyst Jacques Lacan.[5] But Tomkins's orientation was radically

3. Giorgio Agamben, "Absolute Immanence," in Jean Khalfa, *Introduction to the Philosophy of Gilles Deleuze* (Bath, 1999), 151–52.

4. Silvan S. Tomkins, *Affect Imagery Consciousness: The Complete Edition* (New York, 2008), 2 vols. The first volume of the Complete Edition reprints vol. 1 and 2 of Tomkins's original (1962, 1963); the second volume reprints vol. 3 and 4 (1991, 1992); hereafter abbreviated as *AIC*, vol. 1–2, or *AIC*, vol. 3–4.

5. Silvan S. Tomkins, "La Conscience et L'Inconscient Représentés Dans Un Modèle de L'Être Humain," *Psychanalyse* 1 (1956): 275–86. See Eve Kosofsky Sedgwick and Adam

different from that of psychoanalysis. His work was strongly marked instead by the new science of cybernetics, which in the 1960s was exerting a powerful metadisciplinary influence. Many postwar researchers in America were also looking for ways to bring purpose or teleology back into the psychological sciences, from which it had been banished by the behaviorist movement. Cybernetics was attractive in this context because it seemed to suggest that intentionality could be theorized, in effect recast, using strictly mechanical assumptions. (At roughly the same moment, Canguilhem was following analogous trends by redescribing genetic mutation and genetic transmission in terms borrowed from information theory.)[6]

A trio of influential thinkers, Arturo Rosenblueth, Norbert Wiener, and Julian Bigelow, had put forward the crucial arguments for such a view in a famous article of 1943. There they had proposed that all purposive animal behaviors could be regarded as "servo-mechanisms" requiring negative feedback; the model was that of a mechanical device, such as a thermostat, operating on the feedback principle.[7] For Rosenblueth, Wiener, and Bigelow, there were no significant qualitative differences between living organisms and machines in this regard. In spite of some sharp philosophical criticisms, Rosenblueth, Wiener, and Bigelow's claims were widely embraced.[8] In a book titled *Plans and the Structure of Behavior* (1960), which everyone in American psychology was reading in the early 1960s, George A. Miller, Eugene Galanter, and Karl Pribram—all leading figures in the field—claimed that the "cybernetic hypothesis" or feedback loop was the fundamental building-block of the nervous system, and that the living organism could be conceptualized as a set of hierarchically organized

---

Frank, eds., *Shame and Its Sisters: A Silvan Tomkins Reader* (Durham and London, 1995), 6.

6. Georges Canguilhem, *The Normal and the Pathological*, with an introduction by Michel Foucault, transl. Carolyn Fawcett in collaboration with Robert S. Cohen (1966; New York, 1991), 276–77.

7. Arturo Rosenblueth, Norbert Wiener, and Julian Bigelow, "Behavior, Purpose and Teleology," *Philosophy of Science* 10 (1943): 23–24.

8. For a philosophical critique accusing Rosenblueth, Wiener, and Bigelow of "gross confusion" in their treatment of the concepts of purpose and teleology, see Richard Taylor, "Comments on a Mechanistic Conception of Purposefulness," *Philosophy of Science* 17 (1950): 310–17. For Rosenblueth and Wiener's reply, see "Purposeful and Non-Purposeful Behavior," ibid., 318–26, and for Taylor's rejoinder, see "Purposeful and Non-Purposeful Behavior: A Rejoinder," ibid., 327–32. N. Katherine Hayles has discussed the Rosenblueth–Wiener–Taylor exchange in *How We Became Posthuman: Virtual Bodies in Cybernetics, Literature, and Informatics* (Chicago, 1999), 93–97. See also Peter Galison, "The Ontology of the Enemy: Norbert Wiener and the Cybernetic Vision," *Critical Inquiry* 21 (1) (Autumn, 1994): 228–66.

"plans" or "programs." They asserted that in this way intentionality, even unconscious intentionality, could be brought back into the human sciences without stirring up the ghosts of nineteenth- or twentieth-century vitalisms of the kind associated with the work of Hans Dreisch, Henri Bergson, Walter M. Elsasser, and others.[9]

Tomkins was among those who were influenced by these cybernetic developments. But in a conference he organized in 1962 on the "Computer Simulation of Personality," he rightly observed that one critical factor was missing from existing proposals to simulate mechanically the behavior of living organisms, especially human beings. What was lacking was any discussion of the role of motivation—especially the role of emotions. As he put it: "The automaton must be motivated."[10] The trouble, Tomkins went on to assert, was that philosophers and psychologists from Plato to Freud had gotten motivation wrong. This was because they had subordinated the affects to the drives. Tomkins invited his audience to consider the need for air. It might seem that a momentary interruption of the air supply immediately creates an urgent need to gasp for breath. But Tomkins argued that what is primarily observed in the response to air deprivation is not the *drive* for air, but the rapidly mounting *emotional panic* that amplifies the drive signal. That the drive signal itself is not necessarily strong or insistent unless appropriately amplified by emotion could be seen, he suggested, when pilots in World War II suffered a gradual loss of air at 35,000 feet but refused to wear their oxygen masks. Rather than panic, they experienced the euphoria that accompanies slow anoxic deprivation. As Tomkins reported, "some of these men died with a smile on their lips" (*CSP*, 13).

Tomkins therefore suggested that the drives and the affects are two completely separate systems, and that it is the affects, rather than the drives, that constitute the primary motives. Moreover, he declared that what makes the affect system different from the drive system, and what

9. George A. Miller, Eugene Galanter, and Karl H. Pribram, *Plans and the Structure of Behavior* (New York, 1960). Gardner notes the "tremendous impact" of this book on psychology and allied fields in Howard Gardner, *The Mind's New Science: A History of the Cognitive Revolution* (New York, 1985), 32. See also Jean-Pierre Dupuy, *The Mechanization of the Mind: On the Origins of Cognitive Science*, trans. M. B. DeBevoise (Princeton, 2000), where the author characterizes the theoretical power of cybernetics as involving the "capacity of certain complex physical systems, through their behavior, to mimic—to *simulate*—the manifestations of what in everyday language, unpurified by scientific rigor, we call purposes and ends, even intention and finality" (9).

10. Silvan S. Tomkins and Samuel Messick, *Computer Simulation of Personality: Frontier of Psychological Theory* (New York and London, 1963), 12; hereafter abbreviated as *CSP*. Gardner notes the de-emphasis on affect in *The Mind's New Science*, 41–42.

gives it its motivating power, is above all the fact that, unlike the drive system, which is inherently oriented to the objects that satisfy it, the affect system is only contingently connected to its triggering events, which indeed it may fail to identify correctly. In short, according to Tomkins, the affect system operates blindly and is marked by a tendency to error. Tomkins argued as follows.[11] He said that the drives are too narrowly constrained in their aims, time relations, and above all their objects to make them a suitable basis for human motivation, which requires a much higher degree of flexibility or freedom. As literary critic Eve Kosofsky Sedgwick has stated in endorsing Tomkins's position, "only a tiny subset of gases satisfy my need to breathe or of liquids my need to drink."[12] Tomkins believed this objection applied to the sexual drive as well. For anyone familiar with the work of psychoanalysts Jean Laplanche and J.-B. Pontalis, who argued that for Freud desire has extraordinary freedom with respect to objects precisely because it has no predetermined objects of its own, it may come as a surprise to learn from Tomkins that sexuality is constrained in its object orientation in much the same way that hunger is.[13] Nevertheless, for Tomkins what makes even the sexual drive an inadequate basis for human motivation is a trait it shares with the other drives, its *immediate instrumentality*, its "defining orientation toward a specified aim and end different from itself" (*TF*, 19).

Tomkins thought the affect system does not have this instrumental character. As he stated in his reflections on the requirements for constructing a human automaton (or what he called a "humanomaton"):

> There must be built into such a machine a number of responses which have self-rewarding and self-punishing characteristics. This means that these responses are inherently acceptable or inherently unacceptable. These are essentially aesthetic characteristics of the affective responses—and in one sense no further reducible. Just as the experience of redness could not be further described to a color-blind man, so the particular qualities of excitement, joy, fear, sadness, shame, and anger cannot be further described if one is missing the necessary effector and receptor apparatus . . . [I]t is . . . the phenomenological quality which, we are urging, has intrinsic

11. Parts of the next several passages are taken from my *From Guilt to Shame: Auschwitz and After* (Princeton, 2007).

12. Eve Kosofsky Sedgwick, *Touching Feeling: Affect, Pedagogy, Performativity* (Durham, 2003), 18; hereafter abbreviated as *TF*.

13. Jean Laplanche and J.-B. Pontalis, "Fantasy and the Origin of Sexuality," *International Journal of Psychoanalysis* 49 (1968): 1–17.

rewarding or punishing characteristics. If and when the humanomaton learns English, we would require a spontaneous reaction to joy or excitement of the sort "I like this," and to fear and shame and distress, "Whatever this is, I don't care for it." We cannot define this quality in terms of the instrumental behavioral responses to it, since it is the gap between these responses and instrumental responses which is necessary if the affective response is to function like a human motivational response. There must be introduced into the machine a critical gap between the conditions which instigate the self-rewarding or self-punishing responses, which maintain them, which turn them off, and the "knowledge" of these conditions, and the further response to the knowledge of these conditions. The machine initially would know only that it liked some of its own responses and disliked some of its own responses but not that they might be turned on, or off, and not how to turn them on, or off, or up, or down in intensity. (*CSP*, 18–19)

It is because of sexual desire's instrumental character—because like the other drives it is structured by lack and oriented toward an aim and object different from itself—that Tomkins demoted the role of sexuality in human motivation (and here he broke with Freud).

In contrast to the drives, Tomkins held that affects have far greater freedom with respect to objects. In a statement that mixes direct quotation and commentary, Sedgwick has observed in this connection:

"[A]ny affect may have any 'object.' This is the basic source of complexity of human motivation and behavior." The object of affects such as anger, enjoyment, excitement, or shame is not proper to the affects in the same way that air is the object proper to respiration. "There is literally no kind of object which has not historically been linked to one or other of the affects" . . . Affects can be, and are, attached to things, people, ideas, sensations, relations, activities, ambitions, institutions, and any number of other things, including other affects. Thus, one can be excited by anger, disgusted by shame, or surprised by joy. (*TF*, 19)

But in what I think is a questionable move, Tomkins then went on to argue that because the emotions are not tied to any one object but can be contingently attached to a vast range of objects, *they are intrinsically independent of all objects*. He therefore claimed that the multiplicity of objects of the affects means that the emotions are in principle objectless, and hence can be satisfied without the means-end logic of the drives. I consider this a mistake: it does not follow that because the affects can have a multiplicity—even a vast multiplicity—of objects, they are inherently without any relation to objects whatsoever. The mistake, in other words, is thinking that having multiple objects undoes objectality altogether. Put another way, for Tomkins the affects are non-intentional states.

Now, the idea that one or other of the emotions can be discharged in a self-rewarding or self-punishing fashion independently of any object whatsoever implies that the way to understand joy or anger or sadness is to say that these affects are elicited or activated by what we call the object, but the object is nothing more than a stimulus or trip wire for an inbuilt behavioral-physiological response. We might put it that in this account the object of the emotion is turned into the "trigger" of the reaction, with the result that the response is purged of instrumentality. Tomkins adopted such a trigger theory of the emotions. In his words:

> If the affects are our primary motives, what are they and where are they? Affects are sets of muscle, vascular, and glandular responses located in the face and also widely distributed through the body, which generate sensory feedback which is inherently either "acceptable" or "unacceptable." These organized sets of responses are triggered at subcortical centers where specific "programs" for each distinct affect are stored. These programs are innately endowed and have been genetically inherited. They are capable, when activated, of simultaneously capturing such widely distributed organs as the face, the heart, and the endocrines and imposing on them a specific pattern of correlated responses. One does not learn to be afraid or to cry or to be startled, any more than one learns to feel pain or to gasp for air . . . If we are happy when we smile and sad when we cry, why are we reluctant to agree that smiling or crying is primarily what it means to be happy or sad?[14]

14. Silvan Tomkins, "Affect as the Primary Motivational System," *Loyola Symposium on Feelings and Emotions 1968* (New York, 1970), 105–6. The term "program" was of course a term of art in cybernetics and the emerging field of cognitive science. As the philosopher Jerry Fodor explained and defended the position: "[T]he causal processes underlying such performances as, for example, tying one's shoes often consist of events of devising and executing plans, programs, and the like which are explicitly represented in the mind . . . That kind of theory (which Ryle 1949 explicitly stigmatized as 'intellectualist,' to say nothing of 'Cartesian') was both computationalist about mental processes *and* realist about mental representations." Jerry Fodor, *LOT2, The Language of Thought Revisited* (Oxford, 2008), 7, n. 9.

This passage brings out Tomkins's commitment to the idea that there exists a limited number of discrete emotions defined as pan-cultural or universal, inherited, and adaptive responses of the organism, an idea that in Tomkins's cybernetic-inspired model treats the emotions as distinct affect "programs" or "assemblies," which can and do combine in "central assemblies" with the purposive-cognitive and other systems, but from which they are in principle independent. As hardwired, reflex-like, subcortical, and hence noncognitive, species-typical genetic programs, behaviors, and physiological reactions, the affects have activators or triggers that are innate and hence independent of learning, although they can also be stimulated by the learned activators of memory, imagination, and thinking (*AIC*, vol. 1–2, 248). At first Tomkins thought there were eight different primary affects, but he later decided there were nine.[15] In short, according to him the emotions are "natural kinds," that is to say, categories that correspond to real divisions in nature.

Tomkins thus proposed a noncognitive, or non-intentionalist, account of the affects. What are the general implications of his account? What is interesting about his theory is the way it makes it a delusion to say you are happy because your child got a job, or sad because your mother died, for the simple reason that your child's getting a job, or your mother's death, is only a trigger for your happiness or sadness, emotions that could in principle be triggered by something else. In other words, Tomkins held that the affects are inherently objectless, because they are bodily responses, like a sneeze or an orgasm or an itch: I laugh when I am tickled, but I am not laughing at *you* (at anything).[16] As Donald Nathanson, a leading follower of Tomkins, recently declared in this regard, the affects "are completely free of inherent meaning or association to their triggering source. There is nothing about sobbing to tell us anything about the steady-state stimulus that has triggered it; sobbing itself has nothing to do with hunger or cold or loneliness. Only the fact that we grow up with an increasing experience of sobbing lets us form some idea about its meaning."[17]

For Tomkins, therefore, the paradigm of the affects is the miserable newborn baby who cries without knowing why or what can be done about

15. Tomkins's eight primary or innate affects were interest, surprise, joy, anger, fear, distress, disgust, and shame. Later he added contempt (or what he called "dissmell").

16. For the comparison between the affect response and a sneeze or orgasm, see Silvan S. Tomkins, "The Quest for Primary Motives: Biography and Autobiography of an Idea," *Journal of Personality and Social Psychology* 41 (1981): 323; hereafter abbreviated as "Q."

17. Donald Nathanson, *Shame and Pride: Affect, Sex, and the Birth of the Self* (New York, 1992), 66.

it. For him, free-floating distress or anxiety of the kind one might attribute to the wailing infant is a paradigm of the affects just *because* it is free floating and hence can be experienced without relation to an object or to cognition. Although we may search to provide the anxiety with an object, there is no object to which it inherently belongs. For Freud, free-floating anxiety is only apparently free from the object, because the latter is not absent, only unconscious repressed. For Tomkins, the anxiety really is free, and the attribution to it of an object always a potential illusion. The point for him was not to define the affects in terms of cognitively apprehended objects but as intentionless states. As I observed in the introduction, emotion theorist Paul Griffiths has likewise pointed to the existence of so-called "objectless" emotions, such as forms of depression, elation, and anxiety, as evidence in support of Tomkins's idea that certain basic or primary emotions are undetermined by beliefs.[18] So, my ability to give a reason for my feeling something must be a mistake, because what I feel is just a matter of my physiological condition. This is a core materialist claim, and Tomkins's emotion theory is therefore a materialist theory that displaces or suspends considerations of intentionality or meaning in order to produce an account of the affects as inherently corporeal in nature.

A striking aspect of Tomkins's theory in this regard is that it introduced a radical dissociation between feeling and cognition in ways that emphasize the role of contingency and error in emotional life. For him, it is *because* the affects are self-rewarding and self-punishing, *because* they inherently lack knowledge of the world and of objects, that we are so prone to error, so liable to be wrong about ourselves. He therefore argued for a "radical dichotomy between the 'real' causes of affect and the individual's own interpretations of these causes" (*AIC*, vol. 1–2, 248). In other words, the price the affective system pays for its plasticity and flexibility is "ambiguity and error" (*AIC*, vol. 1–2, 23). As he stated:

> We have stressed the ambiguity and blindness of this primary motivational system to accentuate what we take to be the necessary price which must be paid by any system which is to spend its major energies in a sea of risk, learning by making errors. The achievement of cognitive power and precision requires a motivational system no less plastic. Cognitive strides are limited by the motives which urge them. Cognitive error can be made only by one capable of commit-

18. Paul E. Griffiths, *What Emotions Really Are: The Problem of Psychological Categories* (Chicago and London, 1997), 28, 243.

ting motivational error, i.e., being wrong about his own wishes—their causes and outcomes. (*CSP*, 17–18)

Tomkins's approach to the affects has proved attractive to recent commentators in the humanities, such as Sedgwick, precisely for its stress on the play or slippage between stimulus and response, or what she has characterized as the "unexpected fault lines between regions of the calculable and the incalculable" (*TF*, 106). "Where cognitive psychology has tried to render the mind's processes transparent through and through from the point of view of cognition," she writes in a characteristically brilliant passage,

> where behaviorism has tried to do the same from the point of view of behavioral "outcome," where psychoanalysis has profited from the conceptual elegance of a single bar (repression) between a single continuous "consciousness" and a single "unconscious," Tomkins's affect theory by contrast offers a wealth of sites of productive opacity. The valorization of feedback in systems theory is also, after all, necessarily a valorization of error and blindness . . . Tomkins emphasizes that the introduction of opacity and error at the cognitive level alone would not be sufficient even for powerful cognition . . . Thus it is the inefficiency of the fit between the affect system and the cognitive system—and between either of these and the drive system—that enables learning, development, continuity, differentiation. Freedom, play, affordance, meaning itself derive from the wealth of mutually nontransparent possibilities for being wrong about an object—and, implicatively, about oneself. (*TF*, 106–8)

Tomkins's commitment to the idea that there is a gap or disjunction between emotion and cognition was accompanied by a commitment to the hypothesis that the emotional system comprises eight—or is it nine? or perhaps fifteen?—discrete emotions or "affect programs," infinitely recombinable, yet rooted in the body in distinctive and irreducible ways. Yet, although such a claim is an inextricable part of Tomkins's anti-intentionalism, even Sedgwick, Tomkins's most sophisticated advocate in the humanities, is not sure whether that aspect of his theory is credible.[19] "At some level," she acknowledges, "we have not demanded even of ourselves that we ascertain whether we believe this hypothesis to be true" (*TF*, 117).

19. For a further critique of Sedgwick's commitment to Tomkins, see Leys, *From Guilt to Shame*, chapter 4.

I suggest that, in spite of her enthusiasm for Tomkins's work, Sedgwick was right to raise the question of the validity of that hypothesis. In the rest of this chapter, I aim to discuss the trajectory of Tomkins's career and to highlight certain key features of his affect theory in order to understand his influence on the postwar sciences of emotion while raising questions about the validity of his ideas.

TOMKINS'S ANTI-INTENTIONALISM

All sources on Tomkins's life and work agree that he was more of a theorist than an experimentalist. Born in Philadelphia in 1911, he completed both his undergraduate and graduate training at the University of Pennsylvania. Early on he abandoned the study of drama first for psychology and then for philosophy, completing his PhD in philosophy in 1934. In 1936 he began postdoctoral work in philosophy at Harvard, before coming into the orbit of the "personality" theorists Henry Murray and Robert W. White at the Harvard Psychological Clinic, a lively center of psychological research. Tomkins spent the better part of eight years there as a postdoctoral fellow and research assistant, during which time he underwent a long psychoanalysis, before in 1947 accepting a position as an associate and then full professor of psychology and director of the clinical training program at Princeton University. That same year, he published *The Thematic Apperception Test*, a book that Brewster Smith has described as a brilliant application of John Stuart Mill's inductive inference to the analysis of such Thematic Apperception Test stories but as "too obsessive-compulsive in interpretive style to suit many clinicians."[20] The fit between Tomkins and Princeton was not a happy one, and he eventually moved to the City University of New York and then to Rutgers University before retiring in 1975. He died in 1991, a few days after his eightieth birthday.

As Brewster Smith's remark suggests, Tomkins had neurotic tendencies that manifested themselves in several ways, especially in his inability to bring his magnum opus, *Affect Imagery Consciousness*, to completion in a timely fashion. After publishing the first two large volumes in rapid

20. M. Brewster Smith, "Introduction," *Exploring Affect: The Selected Writings of Silvan S. Tomkins*, ed. E. Virginia Demos (Cambridge, 1995), 6. Longtime friend Irving Alexander has made a similar assessment of Tomkins's TAT book, commenting that "I doubt if anyone ever used it in order to learn how to interpret a TAT record" (cited by Sedgwick, *TF*, 99). For an interesting recent history of the American projective test movement, see Rebecca Lemov, "X-Rays of Inner Worlds: The Mid-Twentieth-Century American Projective Test Movement," *Journal of the History of the Behavioral Sciences* 47 (3) (Summer, 2011): 251–78.

succession in 1961 and 1962, he repeatedly postponed the publication of the last two volumes, with the result that by the time they appeared thirty years later—volume 3 just before his death in 1991, and volume 4 posthumously in 1992—he had lost most of his readers.[21] Tomkins's writings also are repetitive: the same passages and formulations appear almost verbatim again and again in texts spanning many years. Nor did he help his reputation by providing only the spottiest of citations; he was so careless in this regard that he often failed to draw attention, or give references, to those of his own contributions or more recent research of others relevant to the topics he was discussing.

Moreover, although Sedgwick and Frank have described his writings as difficult but "compelling," many readers have found them, and still find them, tedious and off-putting. In a somewhat backhanded compliment, Brewster Smith has observed of Tomkins that he was never comfortably at home in academia: "Tomkins was a theorist in the grand style, a style that has become passé in psychology. He had the chutzpah, the grandiosity, to try to sketch a general theory of human beings and their place in the cosmos. This was surely unwise, but it led him like Freud into paths of creative originality, seeing connections that others had missed. Much of his exposition reads like pronouncements of a seer: obiter dicta about psychological phenomena and relationships asserted without evidence. This is not the stuff of completed scientific work with its emphasis on verification."[22] He owed his influence to his turn to biology at a time when American psychology was ready for it after the dominance of behaviorism with its emphasis on stimulus-response learning, and also to his canny ability to get others—notably Paul Ekman and Carroll E. Izard—to take on the work of trying to test his claims experimentally. Indeed, it was largely because of Ekman's and Izard's research efforts that Tomkins's ideas entered the mainstream of American psychology. Later, his reputation was further bolstered by the publication of two books in 1995: a selection of his writings edited by E. Virginia Demos in a series sponsored by Paul Ekman and Klaus Scherer; and a book by Eve Kosofsky Sedgwick and Adam Frank, also containing a selection of Tomkins's

21. As his admirer Donald Nathanson has commented, "[t]he bulk of his audience had died along with the enthusiasm generated by his ideas." Nathanson even suggests that Tomkins was superstitious that the completion of his work equaled death and therefore delayed publication (*AIC,* vol. 1–2, xi).

22. Brewster Smith, "Introduction" (*Exploring Affect*, 11). Nathanson has remarked that "there is a special sort of 'nerve' or 'guts' required for a close reading of Tomkins's overwhelming masterpiece" (*AIC*, vol. 1–2, xxvi).

writings, with an introductory essay that signaled the beginning of the turn to affect that has marked literary and cultural studies ever since. A complete edition of Tomkins's *Affect Imagery Consciousness* was published in 2008.[23]

Tomkins stated that his awareness of the centrality of affect began as early as the 1940s but that it took time for his ideas to develop. This was because the field of emotions was in "deep trouble and disrepute at that time" ("Q," 308). In his view, several factors had contributed to the squelching of interest in emotion prior to his intervention. One factor was the contradictory results of more than five decades of experimental research on the recognition of facial expressions in posed photographs (a topic on which I will have more to say in the next chapter). Another was the dominance of drive theory, according to which organisms behave as they do in order to reduce drives such as hunger or the sexual appetites. Freud and Hull especially were perceived as advocating such drive-reduction views, and Tomkins was by no means alone in regarding those views as inadequate.[24] One result of these various trends was the view that the concept of emotion itself had become useless for scientific purposes.[25] Although he does not mention it here, Tomkins also considered the long ascendancy of behaviorism and its hostility to the idea of mental states as yet another impediment to the appreciation of affect. Psychologists Maria Gendron and Lisa Feldman Barrett have recently questioned that argument on the grounds that a long tradition of psychological constructionism in the study of the emotions had continued to flourish even during the zenith of behaviorism.[26] Against this, it is worth emphasizing that criticisms of behaviorism were made not just by Tomkins but by many of his contemporaries when they reflected on the obstacles they felt had had to be overcome in the 1950s in order to open up psychology to the problem of affect. As Brewster Smith has observed: "That was the heyday of neobehaviorism, of the predominance of psychoanalytic ideas in the emerging postwar field of clinical psychology, and of Dollard and Miller's

23. *Exploring Affect: The Selected Writings of Silvan S. Tomkins*, ed. E. Virginia Demos (Cambridge, 1995); *Shame and Its Sisters: A Silvan Tomkins Reader*, eds. Eve Kosofsky Sedgwick and Adam Frank (Durham, NC, and London, 1995).

24. For a useful contemporary discussion of the failings of drive theory, see Robert W. White, "Motivation Reconsidered: The Concept of Competence," *Psychological Bulletin* 66 (5) (1959): 297–333.

25. Elizabeth Duffy, "An Explanation of 'Emotional' Phenomena Without the Use of the Concept of 'Emotion,'" *Journal of General Psychology* 25 (1941): 283–93.

26. Maria Gendron and Lisa Feldman Barrett, "Reconstructing the Past: A Century of Ideas About Emotion in Psychology," *Emotion Review* 1 (4) (October 2009): 316–39.

(1950) brave attempt to marry Freudian psychoanalysis and Hullian behaviorism" (*Exploring Affect*, 1–2).[27]

In their efforts to topple behaviorism, Tomkins and other critics were of course the beneficiaries of, as well as contributors to, the postwar rise of cognitivism in psychology, a movement that challenged behaviorism by making the study of mental processes respectable after decades of neglect. Tomkins would later accuse cognitivism of overreaching as certain scientists, such as Arnold, Schachter, Singer, and Lazarus, began to propose various cognitive theories of emotion. But at the beginning of his career, the turn to cognition in psychology was helpful to him. As I have already noted, he was especially stimulated by the concomitant rise of cybernetics, which from the start exerted a fundamental influence on the cognitive movement by suggesting that mental processes could be theorized in computational-mechanical terms, even if research on artificial intelligence remained largely indifferent to the problem of affect.[28] He reported that he experienced a breakthrough when he came across Norbert Wiener's early writings on cybernetics, leading him to the question of how it might be possible to design a truly humanoid machine ("Q," 309).[29] Tomkins's fundamental cybernetic insight was that the human personality must be constituted by the co-assembly of numerous inherently autonomous systems. "One could not engage in such a project," he wrote, "without the concept of multiple assemblies of varying degrees of independence, dependence, interdependence, and control and transformation of one by another. It was this general conception which, one day in the 1940s, resulted in my first understanding of the role of the affect mechanism as a separate but amplifying co-assembly" ("Q," 309). Or, as he later observed:

27. For some speculations about the causes of the relatively sudden rise of interest in the emotions starting in the 1960s, see Plamper, *The History of Emotions*, 201–2. The author mentions among possible contributing factors internal developments in psychology; the emphasis on the emotions in the philosophy of existentialism; general social developments such as the rise of the women's movement; and the influence of psychotherapeutics, self-help groups, and New Age talk about one's feelings.

28. Before accepting too quickly the idea of a complete break between behaviorism and the cognitivist movement, it is worth noting the affinities between them in their analyses of mental processes, a point stressed by Shanker in his superb discussion of the behaviorist origins of AI. Stuart Shanker, *Wittgenstein's Remarks on the Foundations of AI* (London and New York, 1998).

29. It is possible that Tomkins's reading in the 1940s included Rosenblueth, Wiener, and Bigelow's influential paper "Behavior, Purpose, and Teleology" (1943), in which the authors proposed a mechanical model of teleological behavior based on the feedback principle. But I have found no reference to this paper in Tomkins's writings, nor have I found any references to specific publications by Wiener.

"I felt then, and now continue to believe, that an understanding of the separate, but matched, components of the human being is a necessary foundation for understanding the more complex integration which we call personality. General, experimental psychology had, I felt, abdicated its responsibility to provide the theoretical base for the study of personality. It was my intention to provide such a base" (*AIC*, vol. 3–4, xxxi).

Armed with this insight, Tomkins argued that affects and cognition are two separate components or systems that usually are co-assembled but are inherently independent of each other. For him, it is the emotions that give energy or "heat" to cognitions that, without such affective amplification, are "cold," and it is the cognitive system that enlightens or gives meaning to the affects that without cognitive input are hot but lack cognitive significance. As Tomkins put it in a statement that stands as the epigraph to this chapter: "*Cognition without affect is weak; affect without cognition is blind.*" (I take it that this is an adaptation toward very different—indeed opposed—ends of Kant's famous statement that "Thoughts without content are empty, intuitions without concepts are blind.")[30] According to Tomkins, the blindness of the affects is due to the fact that they are inherently objectless: although the emotions may acquire objects through learning and socialization, they are not innately directed at objects and hence lack knowledge of the world. As he observed in distinguishing the cause (or "activator") of affect from its aims or objects:

> Everyman and personality theorists alike have exaggerated the dependence of the activation and awareness of an affect on an object. While ever since Freud it has been assumed that one may be mistaken about what one is afraid of, and while phenomenologically the state of objectless anxiety is assumed to be what it appears to be, there is nonetheless a continuing assumption that there is a true object for every affect. If this true object is not in awareness because the affect is free-floating, or if it is not in awareness because of some error in identification of the true object, it is nonetheless assumed that there really is a true object and that this is the cause of the activation of the affect. *We distinguish sharply the cause of an affect*

30. Kant's famous statement is the starting point for John McDowell's influential analysis of the way concepts mediate the relationship between mind and world. See especially McDowell, *Mind and World* (Cambridge, MA, and London, 1994), 1–23; idem, *Having the World In View: Essays on Kant, Hegel, and Sellars* (Cambridge, MA, and London, 2009). McDowell's claim that conceptuality is coextensive with perceptual experience is directly opposed to Tomkins's position, which views affect program phenomena as nonconceptual and noncognitive.

from its object. We will argue that every affect arousal has one or more causes but that it may or may not have an object. (*AIC*, vol. 1–2, 515–16; my emphasis)[31]

Tomkins cited not only the existence of objectless emotions as proof of his position, he regarded the seeming passivity of our emotional experiences—the idea that we do not appear to initiate or intend our passions but experience them as bodily eruptions that are beyond our voluntary control—as further evidence of their non-intentional character.[32] He also appealed to a neuronal theory of activation to explain the involuntary character of our affect program responses. According to Tomkins, differences in affective experience are the result of variations in the density of neural firing over time. He therefore treated qualitative emotional alterations as the outcome of innately determined, quantitative neuronal changes (*AIC*, vol. 1–2, 135–49). Tomkins's views on the innate activators of affect were developed in dialogue with, and some distantiation from, Konrad Lorenz and Niko Tinbergen's influential, ethological "releaser" theory, a theory that will play an interesting role in Ekman's neurocultural theory (see chapters 2 and 5).[33]

There is an interesting moment in volume 1 of *Affect Imagery Conscious-*

31. Tomkins thus claimed that "I may arise feeling ashamed, defeated, discouraged but about nothing in particular" (*AIC*, vol. 1–2, 516), and he applied the same argument to fear, anger, and all the other primary affects, each of which was held by him to be capable of manifesting itself in an objectless manner, although, as he was aware, even in his own time there was room for debate on the topic; indeed, the arguments have continued to this day.

32. In connection with the passivity of the emotions, Tomkins observed: "Most of us achieve, at best, a negative kind of control in the inhibition of overt affective responses. We learn *not* to *show* our anger or grief or fear in external behavior, but we find it difficult not to *feel* angry if someone affronts us; we find it difficult not to feel afraid when we are in danger and not to feel grief upon the loss of a loved one. Still more difficult is it for most of us to achieve positive control over our own affective responses. Few of us can turn on love, fear, anger, in the way that we have achieved control of our limbs—that is 'intend' to walk from one place to another and simply walk. We cannot in the same way 'intend' to feel love or anger or fear and simply initiate these responses, or, if they have already been initiated, continue them or turn them off at will" (*AIC*, vol. 1–2, 79–80). Tomkins's description of how culturally learned procedures for managing our emotional behaviors, which are themselves defined as passively experienced, barely disciplined or regulated involuntary eruptions, is close to Ekman's picture of the relation of culture to nature in the latter's neurocultural theory, as we shall see in chapter 2.

33. For Tomkins's remarks on Lorenz and Tinbergen's releaser theory, see *AIC*, vol. 1–2, 138–49.

ness when Tomkins discusses the difficulty of investigating the emotions by way of a review of the results of a study he had conducted on the effects of electrical shock on experimental subjects. Tomkins gives no details beyond stating that the shocks were administered to eleven male subjects and some controls. Sedgwick and Frank are interested in Tomkins's experiment because in their view it undoes the "presumptive simplicity of the experience of electric shock, which had classically been considered the transparently aversive stimulus par excellence." Instead, they insist with reference to Tomkins's shock experiment that

> in the body of his work density of neural firing is virtually never a direct translation of some external event that could be discretely segregated as "stimulus." Rather, it already itself reflects complex interweaving of endogenous and exogenous, perceptual, proprioceptive, and interpretive—causes, effects, feedbacks, motives, long-term states such as moods and theories, along with distinct transitory physical and verbal events. Against the behaviorists, Tomkins consistently argues that relevant stimulus for the affect system includes internal as well as external events, concluding firmly that there is no basis—and certainly not the basis of internal versus external—for a definitional distinction between response and stimulus.[34]

As a description of Tomkins's findings, this is reasonable enough, but it risks obfuscating the crucial fact that, however much he accepted the idea of the mutual entanglement of cognition and affect, he remained committed to the idea of their autonomy as separate subsystems and hence as two distinct components in the overall response. Thus, although in his discussion of his results Tomkins noticed that the applied shock meant different things to the different experimental subjects—a fact that would seem to illustrate the cognitive dimension of affective reactions—he was not interested in the idea that cognition must therefore be considered an integral part of the emotional reaction, as a cognitivist such as Lazarus would soon claim. Rather, he treated the different meanings generated by the shock simply as "something more" than the experimenter intended, by virtue of the fact that the shock represented different things to the different subjects, in order to stress the multiplicity of the affects aroused and hence the difficulty of controlling affective induction experimentally. "One had only to listen to the spontaneous exclamations throughout an experimental series to become aware of the difficulty of evoking one and only one affect by the use of what seems an appropriate stimulus" (*AIC*, vol. 1–2,

34. Sedgwick and Frank, *Shame and Its Sisters*, 10–11.

106). In other words, in accordance with his anti-cognitivism, Tomkins treated the affects and cognition as separate contributing factors in the reaction, factors that might be difficult to disentangle but which were in principle independent components. "This is a sample of the variability of interpretation and affective responses to the simplest of experimental attempts to induce fear by the threat of electrical shock," he wrote. "It is not that fear is not ordinarily induced by such a threat, but rather that it is not all that happens nor even the most central affective response. The visibility of simple fear under these particular experimental conditions is not great. This is not to say that the experimental investigation of affects is hopelessly complex but rather that the investigator must proceed with unusual caution and imagination if he is to catch fleeting affect on the wing" (*AIC*, vol. 1–2, 110).

One way of putting Tomkins's claims about the intrinsic absence of cognition and the objectlessness of the affects is to say that, although the affects may be activated by reasons, this is only because they have been co-assembled with the separate cognitive system. In themselves they lack rationality, which is to say that their functions do not belong to what the American philosopher Wilfred Sellars (1912–1989) called the "logical space of reasons." As Tomkins stated: "Everyman has been puzzled for centuries at the irrationality of affect investment, that this one who has every reason in the world to be happy is miserable, whereas that one, whose lot is unrelieved misery, seems nonetheless to be full of zest for life. In part, such confusion is a derivative of the failure to understand the basic freedom of object of the affect system" (*AIC*, vol. 1–2, 74). Or, as Tomkins later complained about contemporary cognitivist approaches to the emotions in this regard:

> Cognitive theory is in close accord with common sense in its explanation of how affect is triggered—too close in my view. For some thousand years Everyman has been a "cognitive" theorist in explaining why we feel as we do. Everyone knows that we are happy when (and presumably because) things are going well and that we are unhappy when things do not go well. When someone who "should" be happy is unhappy or suicides, Everyman is either puzzled or thinks that perhaps there was a hidden reason, or failing that, insanity. There are today a majority of theorists who postulate an evaluating, appraising homunculus (or at the least an appraising process) that scrutinizes the world and declares it as an appropriate candidate for good or bad feelings. Once information has been so validated, it is ready to activate a specific affect. Such theorists, like Everyman, cannot imagine

> feeling without an adequate "reason." There must indeed be a cause or determinant of the affective response when it is activated, and the determinant *might* be a "reason," but it need not be. ("Q," 316)

On Tomkins's model, intention and meaning belong only to the cognitive system, whose function Tomkins described in cybernetic, feedback terms not unlike those suggested by Miller, Galanter, and Pribram. "There is a sharp distinction between affects as the primary motives and the aims or intentions of the feedback system," Tomkins stated.

> The purpose of an individual is a centrally emitted blueprint called the *Image*. This image of an end state to be achieved may be compounded of diverse sensory, affective, and memory imagery, or any combination or transformation of them . . . Despite the fact that there may be intense affect preceding and following the achievement of the Image, there may yet be a high degree of phenomenological independence between what is intended and the preceding, accompanying, and consequent affect . . . Motives or motivating experiences may or may not become organized with cognitions to form purposes.[35]

Accordingly, Tomkins accused contemporary cognitivists of ideological bias when they insisted on the cognitive-intentional character of emotional responses. He explicitly criticized appraisal theorists in this regard, at different times dismissing the work of Magda Arnold, Richard Lazarus, James R. Averill, Stanley Schachter, and Jerome Singer, who, as we shall see, on various grounds treated the emotions as intentional states.[36] "The cognition of the ideologue," Tomkins complained, "burns with a gem-like flame" ("Q," 307).

35. Carroll E. Izard and Silvan S. Tomkins, "Affect and Behavior: Anxiety as a Negative Affect," in *Anxiety and Behavior*, ed. Charles D. Spielberger (New York, 1966), 92–93. I have found no reference to Miller, Galanter, and Pribram's book *Plans and the Structure of Behavior* (1960) in Tomkins's publications. But the book was cited several times in Tomkins's edited volume *Computer Simulation of a Personality* (1962), at one point with specific reference to Tomkins's ideas (*CSP*, 67). By making use of the language of information processing to describe the cognitive functions, Tomkins implicitly embraced a form of computational psychology that, like that of Miller, Galanter, and Pribram, imagined it was possible to analyze even the higher mental functions in mechanical-cybernetic terms.

36. For Tomkins's criticisms of Arnold and Lazarus, see Silvan S. Tomkins, "Affect Theory," in *Emotions in the Human Face*, 2nd ed., ed. Paul Ekman (Cambridge, 1982), 362; for his criticisms of Schachter and Singer, and Averill, see Tomkins, "Q," 310–12, 319.

It is tempting to see in Tomkins's anti-intentionalism a reflection of the views of the philosopher Willard Van Orman Quine (1908–2000), whose courses he took when he was a postgraduate at Harvard in 1935–1937 and whose name he evoked in passing more than once, though without indicating any specific debt to his ideas. Thus, when Tomkins claimed he had abandoned the field of philosophy for psychology in part because of "an impatience with the inherent difficulty of empirical validation of the basic questions raised by philosophy" ("Q," 307), he sounded as though he had been persuaded by Quine's philosophical naturalism according to which no sharp distinction should be made between philosophy and the empirical sciences. Moreover, in *Word and Object* (1960), Quine famously questioned the value of Brentano's notion of intentionality and of intentional locutions and expressions. "One may accept the Brentano thesis either as showing the indispensability of intentional idioms and the importance of an autonomous science of intention," Quine wrote, "or as showing the baselessness of intentional idioms and the emptiness of a science of intention. My attitude, unlike Brentano's, is the second." He added: "Not that I would foreswear daily use of intentional idioms, or maintain that they are practically indispensable. But . . . if we are limning the true and ultimate structure of reality, the canonical scheme for us is the austere scheme that knows . . . no propositional attitudes but only the physical constitution and behavior of organisms."[37] Tomkins appears to have taken Quine's argument to heart when he theorized the emotions as intrinsically independent of intentionality. According to his affect theory, the words "emotion" or "affect" do not connote what people ordinarily mean when they use those terms but something else: facial and visceral responses organized by a set of putatively objective "affect programs" located subcortically in the brain.

THE PSYCHOLOGY OF KNOWLEDGE

Tomkins's anti-intentionalism informed a related aspect of his thought that has been largely neglected as irrelevant to the modern scientific study of the emotions but that from the start he regarded as central to

37. W. V. Quine, *Word and Object* (1960; Mansfield Centre, CT, 2013), 221. I have found the following references especially useful: Peter Hylton, "Willard van Orman Quine," *Stanford Encyclopedia of Philosophy* (Winter 2014 ed.), ed. Edward N. Zalta; http://plato.stanford.edu/archives/win2014/entries/quine/; Jaegwon Kim, "What Is 'Naturalized Epistemology'?," *Philosophical Perspectives* 2 (1988): 381–406; Williams, *Wittgenstein, Mind and Meaning*, chapter 8.

his enterprise. This aspect concerned the nature of ideology and belief, or what he called "the psychology of knowledge." Tomkins argued that, as a result of systematic differences in the socialization of each of the primary affects in childhood, our emotions congeal into relatively stable "ideo-affective postures," or what he later called "scripts"—long-standing emotional attitudes of body and mind that determine what we come to believe about a wide range of topics. Indeed, he claimed that, depending on how we are treated in childhood, our ideologies tend to be polarized between a left-wing set of beliefs or a right-wing set of beliefs, or what he also labeled a "humanistic" ideology or a "normative" ideology. He also argued that it was possible to predict a person's attitudes or beliefs on the basis of his or her early affective experiences. "Even when an individual is completely innocent of any ideology," he stated, "if one knows his ideo-affective posture, one can predict what his ideological posture will be if one asks him to consider an ideological question."[38] In short, Tomkins treated beliefs, including our political, literary, scientific, and religious ones, as essentially affective phenomena defined in non-intentional, non-cognitive terms. As Bartulis, a rare commentator on this feature of Tomkins's thought, has recently observed: "For beliefs, no less than emotions, are triggered, so goes the positive dimension of Tomkins's account, by genetically causal mechanisms, which are *themselves* triggered by the involuntary perceptual intake and 'imaging' of a merely affective subject's physical environment."[39]

Of course, there is nothing wrong in thinking that our feelings can influence our beliefs. The mistake only arises when psychological causes are substituted for normative judgments in order to treat beliefs as non-epistemic, affective states. This is the mistake Tomkins made when he proposed a psychobiographical explanation of people's ideas in terms of their affective backgrounds—as if an understanding of the putative emotional experiences that lead people to hold the beliefs they do can stand in for an independent assessment of the truth or falsity of those same beliefs. I have found only one place in all his writings where Tomkins

38. See Tomkins, "The Right and the Left: A Basic Dimension of Ideology and Personality," in *The Study of Lives*, ed. R. W. White (New York, 1963), 391. See also idem, "Affect and the Psychology of Knowledge," in *Affect, Cognition, and Personality*, eds. Carroll E. Izard and Silvan S. Tomkins (New York, 1965), 72–97, hereafter abbreviated as "APK"; idem, "Script Theory," in *The Emergence of Personality*, eds. J. Arnoff, A. I. Rabin, and R. A. Sucker (New York, 1987), 147–217; and "Affect and Ideology," *AIC*, vol. 3–4, 763–77.

39. Jason Bartulis, "The (Super)Naturalistic Turn in Contemporary Theory," *nonsite.org*, no. 8 (2012–2013).

comes close to acknowledging the difference between an analysis of the causal antecedents of belief, and the justification of our claims to knowledge. This is when in 1975 he appeared to concede that it is one thing to identify the ostensible psychological sources of people's theories and another to determine the correctness of those theories. As he remarked at the end of a discussion of the influence of subjective factors on the production of various accounts of personality theory itself: "All research in this area of human personality, the present paper included, is associated with a background of personal and subjective influences. Therefore the speculations developed above should not be construed as attempts to explain anyone's concepts away. A theory of personality is more than just a subjective product, and its value as an explanatory system is something which must be assessed independently of its creator."[40] The implication would appear to be that the epistemological soundness of any set of beliefs or theory, including Tomkins's own affect theory, needs to be weighed independently of any claims about its ostensible psychological causes or origins.

But Tomkins did not follow his own good advice. Instead, in the same article and elsewhere, he proceeded as if it ought to be possible to establish a unifying framework of ideas in the field of personality theory simply by recognizing the subjective forces determining various theorists' beliefs—as if a sufficiently self-aware understanding of the psychobiographical origins of any theorist's commitments ought to be enough to establish their validity or invalidity. In other words, in conformity with Quine's anti-intentionalism, Tomkins continued to suggest that the best way to proceed was to treat ideology or beliefs in terms of subpersonal, hardwired corporeal-causal processes viewed, in Bartulis's words, as "both explanatory and predictive of world-historical developments." As the latter has rightly pointed out, this meant that, according to Tomkins, my beliefs about politics or religion or any other topic are not made intelligible by giving my reasons for holding them, but by describing my views as "idea-affective postures," "theories," or scripts" that are themselves defined as emotional states resulting from various triggering situations.[41] In

40. George E. Atwood and Silvan S. Tomkins, "On the Subjectivity of Personality Theory," *Journal of the History of the Behavioral* Sciences 12 (1976): 177.

41. Bartulis, "The (Super)Naturalistic Turn in Contemporary Theory." The contrast I am drawing between reasons and causes does not preclude the possibility that reasons may themselves be understood as a certain kind of cause. The philosopher Donald Davidson made this case in his influential paper "Actions, Reasons, and Causes," *Journal of Philosophy* 60 (23) (1963): 685–700. For an interesting discussion and critique of Davidson's arguments, see Fred Stoutland, "Interpreting Davidson on Intentional

a chapter on "Affect and Cognition: 'Reasons' as Coincidental Causes of Affect Evocation" in *Affect Imagery Consciousness,* Tomkins put this point in the following way:

> There must indeed be a cause or determinant of the affective response whenever it is activated. The critical question is whether the apparent "reason" is ever or always, that cause. I will argue that the apparent reason is *never* the cause in any case for the simple reason that the affect mechanism is a general one that can be "used" by *any* other mechanism, motor, perceptual, memorial, or even by another affect. More importantly, however, I will argue that even when it *does* truly evoke an affect that it does so *coincidentally* via its abstract profile of neural firing rather than through its apparent content. In other words, it is neither the content of a joke which makes us laugh, nor its unexpectedness, but the sharp drop in rate of neural firing following the sharp increase in neural firing, which is correlated first with one expectancy and then by its violation. (*AIC*, vol. 3–4, 651; his emphasis)

In this statement, Tomkins suggests that reasons are explanatorily impotent or mere epiphenomena because the real causal work in our affective reactions is performed by quantitative changes in the density of neural firing. According to him, even when cognitions enter the picture, their content role is merely concurrent and not causative: "It is my belief," he wrote, "that cognitive processes *can* and *do* activate anger, but primarily through their level of neural firing rather than through their meaning or content. Content, therefore, is *coincidental* to the level of neural firing" (*AIC*, vol. 3–4, 696; his emphasis).[42] In sum, since Tomkins defined ideology as "any organized set of ideas about which human beings are at once most articulate and most passionate, and for which there is no evidence and about which they are least certain," it appears he thought our ideas and beliefs cannot be defended by good arguments because they are not rational but nonrational, non-intentional affective phenomena ("APK," 73).[43]

Action," in *Dialogues with Davidson on Acting, Interpreting, Understanding* (Cambridge, MA, 2011), 297–324.

42. For a critique of the doctrine of content epiphenomenalism, which views intentional mental content as irrelevant to a causal explanation, see Jason Bridges, "Teleofunctionalism and Psychological Explanation," *Pacific Philosophical Quarterly* 87 (December 2006): 403–21.

43. As a consequence of his position, Tomkins even reproached psychotherapists for being seduced by "an exclusive concern with the apparent 'meaning'" of their pa-

In a recent article, Adam Frank, Sedgwick's coauthor in her writings on Tomkins, advocates the superiority of Tomkins's biologically based theory of the affects to Freud's psychoanalytic theory for understanding the relations between emotion and ideology. On the basis of Tomkins's contestable assertions, Frank explains the ways in which, according to Tomkins, people's affects influence their beliefs and ideologies.[44] But of course, as I have already noted, no one has ever doubted that our feelings can influence our beliefs. The problem is that by adopting Tomkins's separation of the affects from our cognitions and by treating the affects as nonintentional states, Frank, like others among today's new affect theorists in the humanities and social sciences, implicitly deflates or eliminates disagreements over ideas in favor of an emphasis on what we feel or who we are, a position that allows concern with affect and identity to trump debates over the rightness or wrongness of what we believe. In chapter 7, I provide a further elaboration and critique of those views.

THE FACE OF AFFECT

In the course of his many writings on the psychology of knowledge, Tomkins offered what he considered robust empirical evidence in support of the postulated relationship between personality and ideology. As he reported in a paper in 1965, to test his theory he constructed a Polarity Scale designed to differentiate "humanistic" or left-leaning subjects from "normative" or right-leaning subjects on the basis of their responses to fifty-nine items. The Polarity Scale included items such as "humans are basically good" versus "humans are basically evil" and so on, through a wide range of such beliefs. Tomkins's claim was that humans tend to idealize or to denigrate themselves and that if one knows which one of these items is selected, then the experimenter will know a great many other ideological options that will also be selected because these options are correlated with the individual's "ideo-affective posture." According to him, a detailed analysis of results with approximately 500 subjects confirmed the existence of such positive correlations between items on the scale: thus, a subject who agreed that "human beings are basically good" also agreed with eighty percent of all the other items on the scale keyed

tients' affects in a quest for interpretive insight, instead of sensitizing their clients to the abstract quantitative features of their emotions (*AIC*, vol. 3–4, 704).

44. Adam Frank, "Some Affective Bases for Guilt: Tomkins, Freud, and Object Relations," *ESC* (*English Studies in Canada*) 32 (1) (2006): 11–25.

"humanistically," while a subject who agreed that "human beings are basically evil" likewise agreed with eighty percent of all the other items keyed "normatively."

Building on those findings, Tomkins took photographs of the facial expressions of models who were asked to simulate the primary affects according to his theoretical requirements. The best set of photos of all the affects posed by one of the models, as judged by untrained judges, was selected for presentation in the form of slides in a stereoscope. In each trial, the experimental subject was presented simultaneously with one slide of an affect photo in the right eye and another slide of a different affect photo in the left eye, and so on through a series of six posed affects, with each affect expression being pitted against every other one. After the presentation of each pair of slides, the subject was asked to describe the posed affect he or she had just seen. Then one of the two slides was shown separately to him and he was asked whether it resembled the face he had just seen. Next the other slide was shown to him and he was again asked whether it was like what he had seen before. The idea behind the experiment was that when the brain or perceptual system was confronted with two incompatible faces, the response would be either a suppression of one face, a fusion of the two, or a rivalry and alternation between them. According to Tomkins, the results confirmed that left-leaning subjects, as defined by their performance on his Polarity Scale, unconsciously selected a dominance of the smiling face over all the other affects, whereas the right-leaning subjects unconsciously produced a dominance of the contemptuous face ("APK," 88–97).[45]

The interest for us in Tomkins's undoubtedly simplistic and reductive ideas about the psychology of knowledge and in the experiments he performed on this topic lies chiefly in what they reveal about the importance he attached to the face as the site of the primary emotions, for it is through his ideas and experiments on this specific topic that his influence on psychology has been chiefly felt. A critical discovery for him occurred early on, when in 1955 he observed his newborn son on a daily basis. As he reported, he was struck by the massiveness of his infant's crying response,

45. In a dissertation directed by Tomkins, Vasquez claimed she had predicted and confirmed differential facial affective responses in left-leaning and right-leaning subjects. She videotaped subjects previously selected on the basis of their performance on Tomkins's Polarity Scale and found evidence held to support Tomkins's conception of "humanistic" versus "normative" ideologies reflecting differences in affect orientations. J. F. Vasquez, *The Face and Ideology*, unpublished doctoral dissertation, Rutgers University, 1975.

which seemed to center on the face. He rejected Freud's suggestion that the birth cry was the prototype of anxiety, concluding instead that the crying face was a sign of distress, in his view one of the primary affects. Tomkins also observed intense excitement on his child's face and a smiling response, which seemed to him to be expressions of two distinct affects, interest and joy ("Q," 309).

Tomkins's emerging views on the importance of the face as the site of the expression of discrete emotions were reinforced by his reading of Darwin's *The Expression of the Emotions in Man and Animals* (1872). In this book Darwin provided the first evolutionary treatment of facial displays, emphasizing the continuities and resemblances between humans and nonhuman animals. On the basis of his many observations of human and nonhuman animal displays, as well as the results of an informal, anecdotal questionnaire designed to investigate commonalities in judgment of facial expressions made by observers around the world, Darwin concluded that people in different cultures made the same expressive movements in similar circumstances, thereby helping confirm his theory of the universality of emotional expressions based on shared phylogeny. Even though his list of the primary affects did not completely agree with that of Tomkins, Darwin seemed to Tomkins to lend support to his own intuitions concerning the existence of discrete, universal emotions, to the point that Tomkins called not only for a return to Charles Darwin's classic text (*AIC*, vol. 1–2, 114) but also to a faculty psychology: "Commonplace and nineteenth-century-ish as the distinctions are between the 'faculties,'" he observed. "[It] is our belief that they must be reintroduced (in the form of sub-systems) and taken seriously in personality theory if we are to understand the complex structure of personality" (*AIC*, vol. 1–2, 488). (The same year Tomkins published his theory of the universality of the primary affects, the psychologist Robert Plutchik brought out a small book advancing a similar theory of basic or primary emotions on similar evolutionary grounds—although his influence on future researchers appears to have been negligible.)[46]

46. Robert Plutchik, *The Emotions: Facts, Theories, and a New Model* (New York, 1962). "Plutchik (1962) proposed another theory of the basic emotions at the same time [as Tomkins]," Ekman later recalled, "but I do not believe his work influenced researchers who produced the findings I have cited." Paul Ekman, "Are There Basic Emotions?," *Psychological Review* 99 (3): 552, n. 5. In 1991 Brown cited the existence of universal facial expressions as one of six crucial cases for bringing back the concept of a universal human nature. See D. E. Brown, *Human Universals* (New York, 1990), as noted in Alan J. Fridlund, *Human Facial Expression: An Evolutionary View* (San Diego and New York, 1994), 188, 192–93.

Darwin's book led Tomkins to Duchenne de Boulogne's important late nineteenth-century attempt, *The Mechanism of Human Facial Expression* (1862) and related texts, to make a detailed map of the facial muscles based on their selective responses to direct electrical stimulation. In volume 1 of *Affect Imagery Consciousness,* Tomkins reprinted several extracts from Duchenne's work that concerned the latter's efforts to identify the particular muscles involved in specific emotions; he also included reproductions of two of the latter's figures showing the distribution of the muscles and the motor nerves of the face (*AIC*, vol. 1–2, 125–33). Tomkins was the first person in the post–World War II period to take a serious interest in Duchenne's work and may be credited for calling Ekman's attention to its value for a Basic Emotion Theory.

As is well known, many of Duchenne's experiments on facial expression were conducted on a man who suffered from an anesthetic condition of the face, with the result that it was possible for Duchenne to stimulate individual groups of facial muscles without causing other muscles to respond and without causing the subject too much pain. (Some of Duchenne's experiments were also conducted on fresh cadavers, the muscles of which were still responsive to the electrical probe.) Duchenne's ostensible aim was to experimentally isolate the expressive effects of the different facial muscles by treating the muscle contractions as, in modern terminology, the controlled variables, and the expressions that followed as the dependent variables. He demonstrated that in some cases only one muscle in complete isolation from all the others was necessary to produce an expressive effect. Duchenne's discovery of "muscle minimalism" was congruent with another discovery, that of the "optical illusion," according to which, when a single muscle was all that was necessary to produce a particular expression, it nevertheless seemed to the observer, because of the illusion of apparent movement, that the whole face had been modified. In cases where a few muscles were involved in the production of an expression, the same illusion prevailed.

As Dupouy has shown in a superb discussion, at first glance Duchenne's texts can be conceived as works of analysis or methodological "decomposition" focused on the muscles, so that the resulting expressions appeared to be a direct consequence of electrical stimulation.[47] But an attentive reading of Duchenne's narrative reveals that the desired expressive effects often shaped the management of the experiments themselves. The outcome

47. Stéphanie Dupouy, *Le visage au scalpel: L'expression faciale dans l'oeil des savants (1750–1889)*, PhD thesis, Université Paris 1 Panthéon-Sorbonne, 2007; hereafter abbreviated as *LV*.

was that, during the course of the work, Duchenne's analytic project was progressively transformed into another enterprise: that of "simulation," by which Dupouy means that in carrying out his experiments Duchenne "composed" the face of his subject in order to obtain certain effects. In other words, in order to achieve a required expression, Duchenne adjusted the lighting, décor, timing, and other aspects of the electrical stimulation process as well as his photographic and staging procedures in such a way as to impose a certain legibility on the results—a legibility that conformed to his preexisting physiognomic views about the existence of the universal passions. It was thus through the method of electrical simulation itself that Duchenne obtained portraits that he presented as exact imitations of nature and hence as expressions of the natural language of the emotions.

Moreover, as Dupouy observes, precisely because the subject of Duchenne's electrical experiments did not himself mime or experience the passions manifested on his face but was the passive, somewhat anesthetic recipient of the mechanical-electrical stimulation, the judgment of the success of the expression, as fixed in the photographic plate, rested on the observer who was invited to recognize the resulting facial display as a convincing representation of the emotion. This meant that the observer alone became the arbiter of the identity of the expressed passion and even, paradoxically, of its sincerity—of whether the artificially produced smile resembled an authentic smile or the false grimace of the hypocrite. Thus, the observer himself or herself became the object of a new kind of scientific observation and scrutiny, as he or she was liable to experience "illusions" of seeing and had to be instructed and trained in how to view the photographic images. One device in this regard was to provide labels designed to help the viewer classify the emotions represented in the photographic plate. The implication was that observers were not necessarily reliable witnesses to the truth of the depicted passions and had to be helped to see the truth.

An interesting aspect of Duchenne's project concerns the question of the pose. On the one hand, Duchenne presented his electrical stimulation method for producing expressions as enjoying an advantage over photography itself because those methods did not involve the posing of the subject for the camera, as did photography, where slow exposure times inevitably involved the subject's awareness of being observed and hence introduced an element of self-consciousness and artificiality into the portrait. Duchenne held that the expressions he produced were less forced and studied than those photographers at the time were able to create, because in his experiments facial expressions were produced by mechanical stimulation of the individual's facial muscles independently of any attempt at voluntary posing or miming by the experimental subject and, it was claimed, in part without the latter's awareness. The whole tendency of Duchenne's project was therefore to suggest that there was an exact match between natural emotional expressions and the expressions he artificially produced through electrical stimulation. Paradoxically, Duchenne held that his manipulations guaranteed the truth of the resulting facial movements by creating the illusion of profound emotion on the part of the experimental subject in the absence of any actual subjective feelings.

As Dupouy also notes, a long debate had ensued over the nature of acting ever since the great eighteenth-century *philosophe* Denis Diderot had framed the topic in his *Paradox on the Actor* (*Paradoxe sur le comédien*, published posthumously in 1830). The issue raised by Diderot was whether the actor should proceed by adopting such a total identification with, or immersion in, his or her role that he or she effectively becomes the character to be portrayed on the stage, or whether instead he or she should aim to produce the appropriate signs of feeling or emotion so as to elicit the appropriate response from the audience, but without becoming caught up in the depicted passions. Diderot strongly recommended the second method, arguing that the soundest and most consistently effective practice for the actor was to be able to seek to create persuasive representations of the emotions while remaining affectively detached.

But Duchenne explicitly rejected Diderot's arguments, maintaining that the actor could not succeed in successfully miming the passions unless he actually experienced them for himself. His argument was that there is a difference between genuine emotional expressions and those that are merely simulated or posed. Indeed, a crucial element in Duchenne's argument about the emotions was that they are difficult or impossible to feign. He made the smile a particular focus of attention, arguing that there was a fundamental difference between the natural smile and one that had been artificially contrived: "In creating certain situations," he wrote, "actors can, in virtue of a special talent, call on these artificial emotions. Nevertheless, it would be easy for me to demonstrate that it is not given to man to simulate or paint a certain emotion on his face, and that the attentive observer can always, for example, discover and confound the lying smile" (*LV*, 331).[48]

48. I have translated from Duchenne's French text, as cited by Dupouy: "'En se créant une situation imaginaire, [les acteurs] peuvent, en vertu d'une aptitude spéciale, faire appel à ces emotions artificielles. Cependant, il me sera facile de démontrer qu'il n'est pas donné à l'homme de simuler ou peindre un certain emotion sur son visage, et que l'observateur attentive peut toujours, par exemple, découvrir et confondre un sourire menteur'" (*LV*, 331).

Emphasizing the difference between the naturalness of the involuntary smile and the forced character of the voluntary or posed one, Duchenne observed that "the muscle of the true smile does not obey the will: it is only put into play by a true feeling, by an agreeable emotion of the soul. Its inertia, in the smile, unmasks a false friend" (*LV*, 331).[49] He therefore suggested that the genuine smile was always the result of the concurrent contraction of two muscles, the *zygomatic major* muscle that lifts the corners of the mouth, and the sphincter or *orbicularis oculi* muscle governing the wincing or protective reflex of the eyes. According to Duchenne, the wincing movement is not under the control of the will, which is why the voluntary or posed smile never exactly matches a spontaneous or natural one.

As will emerge in subsequent chapters, the question of whether there are objectively discernible differences between honest or authentic facial signals versus dishonest or simulated ones haunts the entire postwar science of the emotions. Ekman later wrote a paper on Duchenne's views precisely because they matched his own opinions concerning the difference between authentic and false smiles (for further discussion, see chapter 2).[50] Here I simply want to emphasize the agreement between certain of Duchenne's ideas and those of Tomkins. In particular, we shall see that Tomkins shared Duchenne's assumption that the muscle configurations of the face mapped by Duchenne reflected true distinctions in nature. Tomkins also shared Duchenne's claim that there is a detectible difference between expressions that are triggered involuntarily and those that are produced voluntarily through the influence of learning and the demands of culture.[51] On the one hand, Tomkins repeatedly stressed the ways in which the effects of socialization make it difficult to separate innate or unlearned affective responses from learned reactions. He used the concept of "backed-up" affect for the suppressive effects of culture, effects that according to him cover over the genuine, innate emotional responses.

49. Duchenne's original text, as cited by Dupouy, reads as follows: "[le muscle du vrai sourire] n'obéit pas à la volonté: il n'est mis en jeu que par une affection vraie, par une émotion agréable de l'âme. Son inertie, dans le sourire, démasque un faux ami" (*LV*, 331).

50. Paul Ekman, "Duchenne and Facial Expression of Emotion," in G.-B. Duchenne de Boulogne, *The Mechanism of Human Facial Expression*, edited and transl. R. Andrew Cuthbertson (Cambridge, 1990), 270–84.

51. In the course of his discussion of Duchenne's work in the first volume of *Affect Image Consciousness*, Tomkins observed that Darwin had credited Duchenne not only with the careful study of the contractions of the muscles in expression but also with having demonstrated "'which muscles are least under separate control of the will'" (*AIC*, vol. 1–2, 125).

In the case of anger, he commented, "[t]he empirical determination of the nature of the anger response has been complicated by the universal confusion of the experience of backed-up affect with that of biologically and psychologically authentic innate affect" (*AIC*, vol. 3–4, 688–89).

This statement preserves a distinction between backed-up and biological affect while suggesting that in adult life the appearance of the innate response is a rare phenomenon. As Tomkins also remarked in this connection: "[T]he unmodulated ungraded innate response is in fact relatively rare in acculturated human beings. Normally, what we see on the face of the adult even when an innate affect is triggered, is some transformation of the innate response which is superimposed on the innate response. Thus much innate affect is neither experienced nor communicated as such. It is often suppressed in the form of what I have called backed-up affect."[52] On this formulation, backed-up affect is emotion that has been suppressed by the individual in conformity with the demands of socialization (but not repressed, a Freudian concept that Tomkins rejected). Backed-up affect is thus the result of what Ekman would subsequently call cultural "display rules," that is, the consequence of the rules or conventions that particular societies explicitly and implicitly impose on individuals in order to moderate and control their emotional expressions in public. As Tomkins emphasized, backed-up affect complicates the picture of the emotions by making it difficult to sort out the contributions of culture from the innate response.

On the other hand, in specific cases and situations Tomkins believed it was possible to discern the difference between innate and backed-up affect, or between involuntary and voluntary emotional behaviors. With reference to the smile, for instance, he observed in terms close to those of Duchenne:

> It is not an uncommon phenomenon within the nervous system that the same organ is capable of multiple innervation and multiple inhibition. In general this enables more precise and graded control of each organ and also sensitivity to numerous other sub-systems . . . Not the least important of these alternative types of innervation are those subserving voluntary and involuntary control of the same organ. It is almost always possible for the voluntary control to modulate the involuntary control mechanism and even to successfully imitate it . . . However, it is also the case that the voluntary inner-

52. Silvan S. Tomkins, "The Phantasy Behind the Face," *Journal of Personality Assessment* 39 (1975): 558.

vation of organs may not precisely duplicate the innate involuntary patterning of response. This seems particularly so in the case of affective facial expressions. It will eventually be possible to discriminate between a voluntary and an involuntary smile by means of the high speed camera as it was by Landis and Hunt in the case of the voluntary and involuntary startle. (*AIC*, vol. 1–2, 175)

Tomkins's reference was to Landis and Hunt's study *The Startle Pattern* (1939), in which the authors had examined the responses of subjects startled by a pistol shot and had used a high-speed camera to detect the differences between involuntary and voluntary startle reactions.[53] Tomkins regarded Landis and Hunt's study as a model for the empirical study of affects (*AIC*, vol. 1–2, 114) and devoted several pages of the first volume of *Affect Imagery Consciousness* to a description of their results (*AIC*, vol. 1–2, 278–85). What interested him especially about Landis and Hunt's findings was precisely the idea that the voluntary duplication of the stereotypical startle response was not always a correct imitation but contained "gross exaggerations and inaccuracies of movement," though as he reported, even with the high-speed camera Landis and Hunt did not always find it possible to separate these two responses because they sometimes merged into each other (*AIC*, vol. 1–2, 175–76). In other words, what impressed Tomkins was the evidence Landis and Hunt provided for the difference between the involuntary and voluntary affect reaction. (Tomkins regarded startle-surprise as one of the primary affects, although not everyone then agreed or agrees today about the status of the startle reflex or its relation to surprise. Fridlund, for one, in his critique of Ekman's Emotion Theory to be discussed in later chapters, has questioned the relationship between startle reflex and the surprise response, suggesting that the differences between the reactions cast doubt on the idea of a startle-surprise continuum as posited by Tomkins.)[54]

Tomkins acknowledged that the commingling of innate and backed-up affect made testing the respective contributions of the two components difficult—but not unmanageable. As he stated: "With very few exceptions, no single neural message is an island unto itself. It characteristically encounters or recruits allies which may summate with it and competitors which may attenuate or inhibit it . . . This radically complicates the test of any theory of affect activation, but it does not make it impossible" (*AIC*,

53. Carney Landis and William A. Hunt, *The Startle Pattern* (New York, 1939).

54. Fridlund, *Human Facial Expression: An Evolutionary View* (San Diego, CA, 1994), 118–19; hereafter abbreviated as *HFE*.

vol. 3–4, 711). He suggested that the difference between involuntary and voluntary affective responses could be seen in various situations or conditions, such as certain pathological states (*AIC*, vol. 1–2, 176). The behavior of very young infants especially interested Tomkins, precisely because newborns appeared to exhibit the primary affects in a pure form. Thus, the cry of distress in neonates seemed to him strong evidence of the unlearned character of the primary affects and hence to present a strong argument against the cognitivists, such as Arnold and Lazarus.[55] As early as 1969, in one of his earliest papers acknowledging Tomkins's influence, Ekman suggested that further research on the development of affect displays in blind and sighted children would help clarify the origins of the primary affects.[56] Starting in the 1980s, Carroll E. Izard made infant emotional reactions a special focus of research and to this day maintains that the findings support his version of Tomkins's affect theory. In particular he claims that infants express a set of basic emotions that are represented in a set of corresponding facial expressions according to his own Differential Emotions Theory.[57] But, as in so many areas of research in the emotion field today, infant emotional expressions remain a topic of considerable dispute.[58]

55. Tomkins, "Affect Theory," 361–62.

56. Paul Ekman and Wallace V. Friesen, "The Repertoire of Nonverbal Behavior: Categories, Origins, Usage, and Coding," *Semiotica* 1 (1969): 73.

57. In a long list of publications, see C. E. Izard, R. R. Huebner, D. Risser, G. C. McGinnes, and L. M. Dougherty, "The Young Infant's Ability to Produce Discrete Emotion Expressions," *Developmental Psychology* 16 (2) (1980): 132–40; C. E. Izard and C. Malatesta, "Perspectives on Emotional Development. I. Differential Emotions Theory of Early Emotional Development," in *Handbook of Infant Development*, 2nd ed., ed. J. Osofsky (New York, 1987), 494–554; C. E. Izard, C. A. Fantauzzo, J. M. Castle, O. M. Haynes, M. F. Rayias, and P. H. Putnam, "The Ontogeny and Significance of Infants' Facial Expressions in the First 9 Months of Life," *Developmental Psychology* 31 (6) (1995): 997–1013; Carroll E. Izard, Elizabeth M. Woodburn, and Kristy J. Finlon, "Extending Emotion Science to the Study Discrete Emotions in Infants," *Emotion Review* 2 (2) (April 2010): 134–36. Similarly, E. Virginia Demos has adopted Tomkins's affect theory in her account of infant emotional responses. Among numerous publications, see E. Virginia Demos, "Facial Expressions of Infants and Toddlers," in *Emotion and Early Interaction*, eds. T. Field and A. Fogel (Hillsdale, NJ, 1982), 127–60; idem, "Crying in Early Infancy: An Illustration of the Motivational Function of Affect," in *Affective Development in Early Infancy*, eds. T. B. Brazelton and M. Yogman (Norwood, NJ, 1986), 39–73; idem, "Affect and the Development of the Self: A New Frontier," in *Frontiers of Self Psychology: Progress in Self Psychology*, ed. A. Goldberg (Hillsdale, NJ, 1988), 27–53; idem, "Silvan Tomkins's Theory of Emotion," in *Reinterpreting the Legacy of William James*, ed. M. E. Donnelly (Washington, DC, 1992), 211–19.

58. For a recent debate about emotional expressions in infancy, see Linda A. Camras and Jennifer M. Shutter, "Facial Expressions in Infancy," *Emotion Review* 2 (2) (April

In an attempt to provide his own test of the role of the face in the emotions, Tomkins did not turn to the study of neonate reactions but made use of an experimental method involving the judgment of posed facial expressions, a method that, as I noted earlier, had produced contradictory results in both the pre- and early post–World War II period. But Tomkins's study turned out to have a much better lease on life than might have been predicted from the previous checkered history of attempts of that kind, largely owing to the influence of his experiment on Carroll E. Izard's and especially Paul Ekman's campaigns of research. It is to Tomkins's experiment on facial expression that I now turn.

"THE UNIVERSAL GRAMMAR OF EMOTION"

In 1964 Tomkins published, with coauthor Robert McCarter, the results of an experiment on the judgment of facial expressions. In this experiment, the authors presented to a group of twenty-four firemen sixty-nine photographs of facial expressions posed by models to simulate either a neutral feeling or one of the eight emotions according to Tomkins's views concerning the nature of the primary affects. The authors stated that they used child and adult "models" of both sexes. (Tomkins did not say whether the "models" posing the expressions were professional actors or amateurs.) The photographs were the same ones Tomkins used in the experiment, published in 1965 and discussed above, which was designed to test the validity of his theories about ideology. The photos were mounted on index cards and the firemen were asked, while sitting at a table in small groups of four to six at a time, to classify the stimuli into eight primary affect categories plus one neutral category, using a short list of alternative words for each category of affect provided for that purpose. According to Tomkins and McCarter, the firemen demonstrated a remarkable degree of consensus in their judgments of the affect expressions: the average cor-

---

2010): 120–29; Matthew J. Hernstein, "Comment: Cautions in the Study of Infant Emotional Displays," ibid., 130–31; Stefanie Hoehl and Tricia Striano, "Comment: Discrete Emotions in Infancy: Perception Without Production?," ibid., 132–33; Carroll E. Izard, Elizabeth M. Woodburn, and Kristy J. Finlon, "Comment: Extending Emotion Science to the Study of Discrete Emotion in Infants," ibid., 134–36; Jennifer M. Shutter and Linda A. Camras, "Author Reply: Complexities in the Study of Infant Emotional Development," ibid., 137–38. For the persistence of competing views on this topic, see also Sherri C. Widen, "Children's Interpretation of Facial Expressions: The Long Path from Valence-Based to Specific Discrete Categories," *Emotion Review* 5 (1) (January 2013): 72–77; and Pamela M. Cole and Ginger A. Moore, "About Face! Infant Facial Expressions of Emotion," *Emotion Review* 7 (2) (April 2015): 116–20.

relation between the intended affect expressions as posed by the models and the judgments of the untrained observers was .86.[59]

I will have more to say about the use of photographs of posed expressions for the scientific study of emotions in the next chapter, when they take on a central role in Paul Ekman's cross-cultural judgment studies—studies undertaken at the suggestion of Tomkins, using some of Tomkins's photographs, and on the basis of the latter's theoretical assumptions. As we shall see, it was Ekman rather than Tomkins who would attempt to address the methodological and interpretive issues involved in the employment of such photographs and who would also reanalyze the quite substantial earlier experimental literature on this topic. This is a literature that Tomkins in 1964 simply ignored. It is impossible to tell from his meager references to that earlier literature how much of it he had read; it is hard to believe he had not consulted it before embarking on his judgment studies. Had he paid more attention to it, he would have found some thoughtful discussions about the challenges of conducting emotion judgment experiments of this kind, discussions that might have cast doubt on the validity of the procedures he decided to adopt. It may be that it was precisely for this reason that he disregarded the earlier work. The only study that appears to have interested him and which he cited regularly was Landis and Hunt's 1939 study of the startle response.

Ekman would subsequently remark that Tomkins and McCarter's 1964 paper had little direct impact on the field of psychology, then or later, but that "it had enormous impact on me. It is amazingly rich in theoretical ideas. And it is the first empirical study I had encountered which showed that observers *could* reach very high levels of agreement in judging which of eight emotions was shown in an expression" (his emphasis).[60] I note the ambiguity in Ekman's comment, for Tomkins's judgment study did

59. Silvan S. Tomkins and Robert McCarter, "What and Where Are the Primary Affects? Some Evidence for a Theory," *Perceptual and Motor Skills* 18 (1964): 119–58; hereafter abbreviated as "WWPA."

60. Paul Ekman, "Silvan Tomkins and Facial Expression," *EA*, 209. Ekman likewise observed that "Tomkins . . . provided a theoretical rationale for studying the face as a means of learning about personality and emotion. He also showed that observers could obtain very high agreement in judging emotion if the facial expressions were carefully selected to show what he believes are the innate facial affects . . . Tomkins greatly influenced myself and [Carroll] Izard, helping each of us plan our initial cross-cultural studies of facial expression. The resulting evidence of universality in facial expression rekindled interest in this topic in psychology and anthropology." Paul Ekman, "The Argument and Evidence About Universals in Facial Expressions of Emotion," in *Handbook of Social Psychopathology*, eds. H. Wagner and A. Manstead (London, 1989), 145.

not and could not address the question of the relationship between the posed expressions and the "emotions," if by that term he meant internal states of feeling, since the emotions experienced by the models posing the expressions for photographic purposes were not assessed. As the psychologist James Russell in a brilliant critique would later argue, judgment studies of this kind are really emotion "attribution" studies rather than "recognition" studies. As he pointed out, the commonly used but misleading term "recognition" presupposes that emotions are in the face to be recognized, whereas in judgment studies of this kind observers are only asked to "cognize" or interpret posed facial displays by attributing a label to them, without regard to any hypothesized causal link between the facial displays and internal affective states and without regard also to the question of the universality of a set of primary emotions.[61] (The term "accuracy" is also ambiguous in a somewhat similar way, since it can mean that observers are accurate in judging the emotion actually being expressed in the face, or it can merely refer to the extent to which observers accurately label an emotion according to the intended expression but without regard to what the poser really felt.) In short, the meaning of the Tomkins–McCarter findings was ambiguous, and that ambiguity would find its way into Ekman's lifelong research enterprise.

A crucial innovation in the Tomkins–McCarter experiment was Tomkins's frankly admitted decision to preselect photographs of posed expressions that best illustrated his theories about the nature of the primary affects and to eliminate those that did not satisfy his criteria. (From now on, when describing this paper I shall attribute the experiment and its findings only to Tomkins, who was the instigator and principal author of the study.) In other words, Tomkins chose those photographs from an array of posed images that represented what he considered on the basis of his theoretical assumptions each of the expressions of the primary affects ought to look like, and used those preselected photographs as "stimuli" for his judgment study. In an important sense, Tomkins's procedures predetermined the results, because those results were the outcome of judgments of photographic images of expressions that were already thought to be good examples of the pure form of the eight primary affect categories. As Russell has observed of the preselection of images and other methodological biases in judgment studies of this kind, when the preselection of

61. James A. Russell, "Is There Universal Recognition of Emotion from Facial Expression? A Review of the Cross-Cultural Studies," *Psychological Bulletin* 115 (1994): 102–41; and James A. Russell, "Facial Expressions of Emotion: What Lies Beyond Minimal Universality?," *Psychological Bulletin* 118 (1995): 379–91.

the stimulus material is not so extreme, recognition scores tend not to be so high.[62] But far from regarding his preselection procedures as a weakness of his investigation, Tomkins treated them as a strength. Indeed, he would subsequently criticize Izard's cross-cultural judgment studies, undertaken at Tomkins's suggestion, precisely because Izard's photograph selections were guided by empirical criteria rather than the kinds of theoretical considerations that shaped his own experiments. By contrast, he was proud of Ekman's success in replicating his results when the latter used Tomkins's theoretically based selection methods in his own cross-cultural judgment studies.[63]

In discussing his findings, Tomkins stressed the high degree of consensus among the untrained firefighter-observers in their judgments of the photographs of the posed expressions. As he reported, the consensus was especially striking when only the three best photographs for each affect category were used as a subset. He acknowledged that the "guidance" in the "answer booklets," which provided four alternative words for deciding which affect category a particular photograph belonged to, might have heightened the degree of consensus among the observers. He also noted that because each affect was described with more than one word, different subjects might have selected the same affect for different reasons. Tomkins stated in this connection that preliminary studies on other subjects who were asked to respond freely to the photographs of posed expressions, without the benefit of the suggested labels, achieved much less consensus, the exact degree of consensus depending on how gross or refined was the coding scheme the investigators employed when grouping the free responses. Russell would also later criticize the use of "forced choice" labeling methods in experiments of this kind precisely on the grounds that such methods alert subjects to the experimenter's expectation that a facial expression is to be interpreted in terms of emotion and even suggest which emotion is to be selected (*HFE*, 220–24). But the problem of forced-choice labeling methods does not appear to have unduly concerned Tomkins. What interested him was that by his methods he had obtained confirmation of his hypothesis that subjects are able to

62. James A. Russell in Fridlund, *HFE*, 215.

63. The Tomkins and McCarter 1964 paper was accepted and then rejected by the editor of the journal to which it was first submitted for publication, on the grounds that the study did not meet minimal scientific standards. Tomkins took this incident as a sign of the resistance of the field at that time to treating affect as a central phenomenon ("Q," 310).

judge the primary affects from posed expressions with a high degree of consensus, and he quickly dropped the labeling problem.

Tomkins argued that such a high degree of consensus was possible because of the great familiarity of humans with the human face. He suggested that owing to the importance of the information communicated by the human face and because of the affective responsiveness of human beings to the faces of others, individuals become so skillful and efficient at organizing facial information that they are capable of interpreting an extraordinary amount of information from momentary facial reactions. According to Tomkins, our interpretive skills are aided by an isomorphism between the face of the other and the "interoceptive face of the self" ("WWPA," 125), by which he meant that through facial feedback we learn the "rules of translation" between what someone else's face looks like and what our own face feels like, with the result that we are eventually capable of "imitating either what the face looks like or what it feels like" ("WWPA," 125). In short, in terms not unlike those that would be put forward by mirror neuron theorists thirty years later, Tomkins proposed the existence of a mechanism of facial matching according to which the self is able to mimetically "read" or react to the face of another and accordingly feel the other's affective expressions.[64]

But Tomkins's experiment also proved that the judgment of facial expressions was not unanimous, and that finding seems to have interested him just as much as the fact of consensus itself. As he reported, even in the case of expressions of enjoyment—the posed affect that obtained the highest degree of consensus among the observers—the consensus was not perfect, because joy was sometimes confused with interest or some other emotion category. In attempting to explain these results, Tomkins never doubted his assumptions concerning the existence of eight primary affects, each of which manifested itself in characteristic facial expressions. Those assumptions were a given: they were the "theoretical foundations" ("WWPA," 136) on the basis of which he interpreted his findings. The entire burden of his analysis therefore fell on the various factors that prevented perfect unanimity. We might put it that in analyzing the results of his

64. For a discussion and critique of an experiment on disgust and empathy that makes use of mirror neuron theory to explain the results, see Leys, "'Both of Us Disgusted in *My* Insula': Mirror-Neuron Theory and Emotional Empathy," in *Science and Emotion After 1945: A Transatlantic Perspective*, eds. Frank Biess and Daniel M. Gross (Chicago and London, 2014), 67–95. A longer version of this paper can be found at *nonsite.org*, no. 5 (2012).

experiment, Tomkins took for granted the truth of his claims concerning the existence of eight innate primary affects and their associated facial expressions and, when the results of his judgment study fell somewhat short of expectations, shifted the "blame" on to certain inadequacies in the stimulus (the posed expressions) and the observers (the fireman judges).

According to Tomkins, then, the first factor preventing unanimous consensus was the insufficiency of the photo "stimulus" itself. In particular, he reported that the stimulus failed in various ways to represent the affective response as he had described it to the model posing the expression. For one thing, the skill of the model was rarely good enough to perform exactly what he or she was asked to do: "He may smile with his mouth but fail to produce the 'smiling eyes' which characteristically accompany the smile of the mouth in the innately activated smile" ("WPPA,"125). This statement echoes Duchenne's claim that the authentic or natural smile involves the contraction of both the mouth muscles and the eye sphincter muscles associated with wincing or the blink reflex, whereas the fake smile does not involve the latter. As we shall see, the dichotomy between the genuine or felt "Duchenne" smile, versus the phony or unfelt "non-Duchenne" smile, which is a dichotomy between what is natural and what is owed to culture, will be an important feature of Ekman's neurocultural version of Tomkins's affect theory. Whether it is a true dichotomy is another matter.

Another reason Tomkins gave for the lack of unanimous consensus in the judgment of expressions was that each model tended to be inhibited in miming certain facial displays. For example, he reported that the six- or seven-year-old children in his experiment denied that they had ever felt angry and found it "all but impossible to mimic our angry expression" (his wording here suggests that he himself posed the expression of anger for the children to copy). Adults experienced the same inhibitions: "Some adults cannot sneer [in contempt, presumably]. Others cannot look ashamed" ("WWPA," 125). Moreover, according to Tomkins, each face has a principal expression which "shines through and contaminates the attempt to pose an affect which is underdeveloped relative to the major affect in each model. Thus a predominantly hostile model does not have a completely warm smile. A predominantly friendly model does not have a completely pure sneer of contempt. His mouth's readiness to smile contaminates the pose of disgust" ("WWPA," 125).

Here I want to emphasize the obvious fact that the expressions Tomkins deployed in his experiment were not "spontaneous" facial displays, in the sense of occurring in natural contexts, but voluntary simulations posed by "models" under the direction of Tomkins himself. As posed expressions, they were held to be more or less accurate approximations of what Tomkins thought the expressions of the primary affects should look like, based on his theory of the existence of eight different innate affects, a theory he briefly outlined at the start of the paper. Since, as we have seen, Tomkins believed that there was a difference between the innate, involuntary expressions of affect and voluntary simulations, from his theoretical perspective it would seem that the deliberately posed facial displays used in his experiment were never likely to be completely accurate imitations of the basic or innate affect response. But even though Tomkins acknowledged in his paper that the use of posed expressions in his experiment was "not ideal" ("WWPA," 121), he nevertheless proceeded as if he thought it should be possible for his models to simulate or pose "natural" expressions on command, and as if he thought such posed expressions could provide important evidence about the innate primary affects. He had nothing to say about the problems associated with the practice of using posed expressions in judgment experiments of this kind, such as the tendency of such expressions to look artificial or unnatural, even though the topic had attracted considerable attention in previous research. On the contrary, he appeared to think that, with a few adjustments, accurate and persuasive expressions ought to be achievable. As he wrote: "It should be possible in the future to obtain more convincing poses if we restrict ourselves to that single affect which is the predominant affect for each person, which is held in a state of readiness on the face at all times, and ask each model to pose only this his major affect" ("WWPA," 125).[65]

Tomkins also suggested that another factor contributing to the failure

65. Tomkins was not coherent on the topic of the significance of the voluntary posing of emotion. He argued that the facial displays triggered by the affect programs in turn determined emotion experience by way of facial feedback innervation. In his reply to a failed attempt by Tourangeau and Ellsworth to prove the facial feedback hypothesis using voluntarily posed expressions, Tomkins condemned such voluntary poses as artificial and hence rejected their use as an inadequate test of his theory. On the other hand, when Ekman and his colleagues appeared to demonstrate that voluntarily posed expressions and facial displays produced emotion-specific autonomic nervous system responses, Tomkins welcomed their findings as further proof of the validity of his affect theory. The contradiction was not lost on Ekman who, close as he was to Tomkins, remained unconvinced about the facial feedback hypothesis. See R. Tourangeau and P. Ellsworth, "The Role of Facial Response in the Experience of Emotion," *Journal of Personality and Social Psychology* 37 (1979): 1519–31; Silvan S. Tomkins, "The Role of Facial Response in the Experience of Emotion: A Reply to Tourangeau and Ellsworth," *Journal of Personality and Social Psychology* 40 (2) (1981): 355–57; Tomkins, "The Quest for Primary Motives," 315–16; Ekman, "Silvan Tomkins and Facial Expression," *EA*, 212–13.

of observers to achieve consensus in the judgment of posed expressions was the photograph itself. There are many complex issues at stake in the use of photographic stills in the study of facial expressions, some of which I will address in the next chapter. Here I simply note that Tomkins restricted his comments to the fact that facial expressions consist of a set of movements that occur over time, whereas the photograph congeals those movements into a static image that, as he put it, is "never entirely adequate as a recognizable affective response" ("WWPA," 126). If Tomkins had in mind the supposed tendency of still photographs to represent facial expressions in a somewhat odd or exaggerated manner because of the subject's consciousness of being photographed, or because of technical and other factors, he did not say so. He was not the first to note the inadequacies of the photographic image as a stimulus for emotion judgment studies. In 1954, in an influential paper, Bruner and Tagiuri had objected to the use of such images precisely on the grounds that a face frozen in time in this way was not representative of the kind of faces we judge in ordinary circumstances, suggesting that such photographs lacked what would now be called ecological validity. They had observed that "From the point of view of the adaptiveness of social behavior, it is rare to the vanishing point that judgment ever takes place on the basis of a face caught in a state similar to that provided by a photograph snapped at 20 milliseconds." "Historically speaking, we may have been done a disservice by the pioneering efforts of those who, like Darwin, took the human face to be in a state of arrested animation as an adequate stimulus situation for studying how well we recognize human emotion. If research on this topic is to be revivified, it is plain that a more catholic view will have to be taken about the nature of the cues we use in judging whether a man is sad, in pain, grieved, or in love."[66] Tomkins briefly noted the problem of the still, though without reference to Bruner and Tagiuri's paper, and then let the matter go. His qualifications about the use of the photographic still did not prevent him from employing such photos in his own research or from advocating their use when Ekman subsequently set out to conduct cross-cultural judgment studies informed by Tomkins's affect theory.

Tomkins next turned to the analysis of problems causing a lack of consensus in judgment studies owing to difficulties on the observer side of such experiments. His suggestion was that even if an individual were to be confronted with an accurate emotion stimulus—for example, with a moving picture or "a person responding with real 'live' affect"—there would

66. J. S. Bruner and R. Tagiuri, "The Perception of People," in *Handbook of Social Psychology*, vol. 2, ed. G. Lindzey (Cambridge, 1954), 638.

still remain several impediments to unanimity.[67] The first impediment was the problem of "semantic confusion." "The relationship between the primary affects and the words for these is problematic," Tomkins stated. "The relationship between an affect and its name may be quite variable and somewhat ambiguous, depending on the circumstances surrounding the learning of the name." By this he meant the erroneous ways in which children are often brought up. As he explained, if a child is angry and shouts rebelliously, only an unusual parent will take this as an opportunity to teach the child to recognize his feelings correctly by saying, "I know you are angry because I'm busy and can't play with you." More frequently the child is taught instead that his anger is something bad, a sign of moral delinquency, with the result that many years later he will misrecognize his own anger and that of others for what it really is. As Tomkins put it: "If these words have been taught and more specific affects have not been taught, it is altogether possible for a human being never to attain a working knowledge of the correct names for the specific affects with which he and others respond" ("WWPA," 126). According to Tomkins, the isomorphism between the face of the other and our own face on which our skill in judging faces normally depends may be grossly impaired if parents unduly punish the facial expressions of affect in their children.

Further reducing the visibility and identifiability of the affects, Tomkins argued, was not only the absence of "standardized instruction for such learning," but the nature of affective responses themselves. For one thing, the affects are variable, both innately—a person may be angry very briefly or for a few minutes or constantly, or he may be mildly irritable or very angry—and because of changes in expression due to learning. Indeed, according to Tomkins, the innate affective response "is usually so transformed that by the time the human being attains adulthood there are many different learned ways of responding facially with fear, anger,

67. Tomkins observed that even moving pictures did not solve the problem of providing an accurate stimulus for judgment studies of posed expressions. As he observed, "The research of Landis and Hunt (1939), employing a moving picture camera with speeds up to 3000 frames per second, has made it clear that the speed of response of facial muscles is such that some responses, such as partial eyelid closures, are too fast to be seen by the naked eye, and that the patterning of both facial and gross bodily movements is so complex that one must resort to repeated exposure of the same moving pictures if one is to extract the information which is emitted by human beings responding with affect in changes of facial and bodily movement" ("WWPA," 126). He later reported that he had carried out many studies using a high-speed camera but confessed that, in spite of a large expenditure of time and effort along these lines, his findings had fallen short ("Q," 314).

or distress" ("WWPA," 127). In other words, the effects of socialization blur the distinction between the innate primary affect response and the learned reaction.

Moreover, Tomkins added, combinations of affects may also reduce the visibility of affects in general and of the primary affects in particular, by producing blended expressions that correspond to no simple affect and for which "there may or may not be a name in the common language." Here Tomkins explained (or explained away) the failure of consensus among observers by positing the existence of "blends" or combinations of discrete expressions in faces, owing to the effects of local cultural conditions. This is an argument he would take up when in his own research he rejected the use of any posed expressions that were deemed insufficiently "pure" owing to the presence of blends of two or more primary emotions. The theory of blends has subsequently enjoyed a considerable success as a way of accounting for emotions that do not closely fit the affect program model. Thus, according to the theory of blends, affect program phenomena are held to be common to all individuals, and cultural and individual differences are treated as the result of blends of the six or seven or eight basic emotions.[68] Tomkins also suggested that the formal languages of communication—that is, speech—might block or decelerate the skill in learning the language of affect, because those messages may be sufficiently complex and urgent as to reduce the visibility of the face, as when a motorist asks directions from a stranger and listens to the words but barely notices the stranger's facial expressions. Finally, as a further impediment to accurate judgment of facial affect, Tomkins observed in a revealing statement that it is a "somewhat culture-bound skill. The individual who moves from one class to another or one society to another is faced with the challenge of learning new 'dialects' of facial language to supplement his knowledge of the more *universal grammar of emotion*" ("WWPA," 127; my emphasis).[69]

68. See for example Griffiths's discussion and critique of the theory of blends in *WERA*, 101–2.

69. For an important recent discussion of the in-group advantage that obtains when observers judge expressions made by people of the same culture as themselves, see Pamela J. Naab and James A. Russell, "Judgments of Emotion from Spontaneous Facial Expressions of New Guineans," *Emotion* 7 (4) (2007): 736–44. In this paper, the authors report the results of a recognition study using the twenty spontaneous facial expressions from Papua New Guinea originally photographed, coded, and labeled by Ekman (1980). Their aim was to reexamine Ekman's claims for a high recognition agreement in the case of such spontaneous expressions. They state that the observers achieved rather low levels of agreement when compared to those obtained with posed expressions.

Altogether this is a remarkable way of handling the question of why observers fail to make accurate emotional judgments of posed facial expressions. Tomkins seemed to think not only that there are eight or nine discrete primary affects in nature but that there are correct names for each of them, names he knows but that many people fail to learn properly because of inadequate education or because of certain confusions in the affective display itself. It is as if he imagined there ought to be a "standardized instruction" for learning the "language of nature" or what he called the "universal grammar of emotion," but that in the absence of such a standard method of education, and owing also to the innate and learned variability of affect expression, semantic confusions and mistakes in judgment will inevitably occur. How he himself learned the correct names for the affects, he does not say. His entire enterprise depended on the claim that people are pretty good at judging the innate primary affects in facial displays but that, when they fail, this is because they have not learned the correct names for the affects or because of related "common confusions." All of Tomkins's considerable efforts, in the second part of his paper, to show that such common confusions take a systematic form, owing to certain theoretical considerations or to individual biases, cannot hide the fact that his experimental exercise was designed to yield positive results and, when the results fell somewhat short, to explain those failures away on the grounds of various inadequacies in the stimulus or the observer. The fundamental presupposition of the existence of eight innate affects in nature remained intact.

Nevertheless, it is worth remarking that many years later, in spite of the success of his 1964 experiment in providing Ekman with a blueprint for the further investigation of the emotions, and in spite also of the influential support his own claims about the primary affects had by then received from the research of Ekman and others, Tomkins cautioned that scientists *still* did not know the exact difference between the innate primary affect response and the effects of culture. Writing in the third volume of *Affect Imagery Consciousness*, published just before his death but probably mostly written several years earlier, he observed of the negative affect of anger that the empirical determination of the innate response of anger was complicated by the "universal confusion of the experience of backed-up affect with that of biologically and psychologically authentic innate affect" (*AIC*, vol. 3–4, 688–89). By this he meant that since no society allowed individuals to express rage whenever they wanted to, anger was suppressed or "backed-up" in various ways that ultimately confused not only the individual himself but the scientific investigator. "The more serious theoretical problem," Tomkins wrote, "is that we do not know the exact differences between innate affects and backed-up affects." Thus, in the case of anger, pseudo- or backed-up affect interrupts breathing and vocalization in ways that transform or distort the innate anger response and the experience of anger to the point that the individual may experience some ambiguity about what anger really feels like. "The psychological consequences of such suppression would depend upon the severity of the suppression," he wrote:

> Even the least severe suppression of the vocalization of anger must result in some distortion and bleaching of the experience of anger and some ambiguity about what anger really feels like. Such confusion also occurs in the look of the anger of the other. Since the closed mouth, with tightly pressed lips and clenched jaws, is one universal form of backed-up anger, we tend to share a worldwide illusion that this *is* the appearance of innate anger and that the other is really angry when in fact he is experiencing backed-up anger. To be "boiling mad" is *not* to be angry, either in the appearance to the other or in the experience of the self. It is in some as yet indeterminate way more than anger, and in some way less than anger, and, most critically, in some way different from anger . . . The only way to determine what is innate in the anger response would be to examine it in earliest infancy, before it suffers the imposed control which backs it up. (*AIC*, vol. 3–4, 689; his emphasis)

But this is a rather amazing way to solve the problem of the meaning of facial expressions. If the only authentic expression of the innate affects is to be found in pre-socialized, pre-linguistic newborn infants, what paradigm of the emotions does that provide? Tomkins's affect theory seems to depend on an impossible ideal of authenticity. According to him, even quite young children are not good enough models for determining the innate affects because they have already been influenced by their parents and the demands of society to inhibit and transform their natural expressions of feeling. Only the neonate manifests the innate affects in their pure form. Such a claim puts Tomkins's theory of the primary affects under great strain: Are we to imagine that the newborn feels shame when it defecates? How could that be determined?[70] How are we to decide

70. Indeed, I note that as long ago as 1963, in his comments on Tomkins's paper on the simulation of personality with which I began this chapter, Donald W. MacKinnon stated that he found it difficult to believe that shame or humiliation were given innately rather than in large part "learned and acquired as a result of interpersonal experience" (*CSP*, 62).

the difference between the newborn's cry of distress and, say, its cry of anger? It is hard enough for adults to determine the difference between these emotions; how much harder it is in the case of tiny infants. And yet, this is the model Tomkins bestowed upon his heirs and successors. It is true that he also recommended the use of brain stimulation experiments, combined with film recordings of the face, as another way of studying the primary affects, a suggestion that reveals the extent to which for Tomkins the basic emotions are fundamentally brain processes.[71] It is also the case that throughout his career Tomkins encouraged Ekman to continue the study of the emotions by the use of posed expressions, an approach that Ekman made the signature method of his career.

AN "AFFECT REVOLUTION"?

A final note. In 1981 Tomkins published a paper entitled "The Quest for Primary Motives: Biography and Autobiography of an Idea." It was his attempt to give an account of the social and personal influences in American psychology that had shaped not only postwar attempts to explain the nature of motivation, but his own development and growth as a scientist. It is the most revealing statement Tomkins ever made about his personal and research trajectory, amounting to a retrospective summing up of his life's work. (He retired in 1975.)

The "idea" whose biography and autobiography Tomkins wanted to tell was the idea of affect. What bears remarking on, though, is that he presented his article as a contribution to what he had long called the "psychology of knowledge"—that is, as an investigation into "the personal as well as social influences on the ebb and flow of affect investment in ideas and ideology, in methods and styles of investigation, and in what are considered acceptable criteria of evidence" ("Q," 306). And he defined ideology in the same terms he had deployed in many different publications over the years, namely, as "any organized set of ideas about which human beings are at once both articulate and passionate and about which they are least certain" ("Q," 306). Moreover, Tomkins also suggested that there is no escaping ideology, because, as he put it, "beneath the surface of any domain of knowledge one finds ideology" ("Q," 306). Even more striking, he asserted that ideology is fundamentally a form of religious belief or

71. But for a critical summary of the past sixty years of the effects of electrophysiological research on emotion, see S. A. Guillory and K. A. Bujarski, "Exploring Emotions Using Invasive Methods: Review of 60 Years of Human Intracranial Electrophysiology," *Social Cognitive Affective Neuroscience*, Epub: March 20, 2014.

faith: "But if one goes deeper and higher, below and above ideology is always theology" ("Q," 306), The implication of Tomkins's introductory remarks thus seemed to be that, since "at the growing frontier of science there necessarily is a maximum of uncertainty" ("Q," 306), even his own passionately held beliefs about the nature of affect had to be considered more a matter of faith than empirical truth. Nor did he disdain the mantle of theologian. On the contrary, he described his essay as a contribution "not only in history and science and in ideology," but "above all . . . a sermon, a cautionary tale, lamenting and denouncing both the godless and those who worship false idols, warning of the trials and perils of the new faith, and presenting a vision of the promised land for the chosen people, in the classic prophetic mode" ("Q," 306).

This was an unusual way for a scientist to proceed when summarizing the origin and development of his or her achievements. Tomkins seemed to think that what was demanded in 1981 was a sermon designed to condemn the nonbelievers and apostates in the emotion field and to convert listeners to a new credo. But since in this same essay he also declared that "science will never be free of ideology, *although yesterday's ideology is today's fact or fiction*" ("Q," 306, my emphasis), perhaps he thought that, while he needed to deliver a religious message in order to convert others to the cause, his own doctrine would nevertheless turn out to be closer to empirical fact than to untruth or invention.

The timing of Tomkins's sermon or paper is also interesting. His article appeared one year after the publication of Robert Zajonc's influential article "Feeling and Thinking: Preferences Need No Inferences," the opening shot in the important battle that would ensue between Zajonc and Richard Lazarus over the role of cognition in affect.[72] Tomkins's claim in his own essay was that, although the cognitive revolution had played a necessary role in emancipating the study of cognition from its cooption and distortion by behaviorism and psychoanalytic theory, it had overplayed its hand. He therefore argued that an "affect revolution" was required to emancipate psychology from an "overly imperialistic cognitive theory" ("Q," 306), a revolution that would be accomplished if researchers now converted to his own views. On this scenario, for Tomkins the "godless" were the behaviorists who had trivialized emotion, and the worshipers of "false idols" were the scientists who now paid attention to emotion but erroneously believed that cognition played a far more central role in its production than was warranted. From his perspective, Zajonc was on the

72. Robert B. Zajonc, "Feeling and Thinking: Preferences Need No Inferences," *American Psychologist* 35 (1980): 151–75.

right track when he argued against Lazarus that feeling and thinking (or cognition) are two independent mechanisms, a position that, as Tomkins noted, he himself had been urging for more than twenty years. For Tomkins, Zajonc was thus one of the few scientists who had seen the promised land ("Q," 316–17). He also commended Izard and Ekman for taking up and promoting his work, praising Ekman for his "brilliant experiment" on the facial expressions of Japanese and American students videotaped and designed to show how Tomkins's theory of the universality of the affect programs or basic emotions could be reconciled with the findings of cultural differences in the expression of affect.

But Tomkins did not mention Lazarus. As a prominent cognitivist or "appraisal" theorist, Lazarus rightly understood that his views were a main target of Zajonc's article and he responded accordingly (see chapter 3). But Tomkins paid no attention to Lazarus's considerable body of work. Rather, in his eyes the chief culprit in the cognitivist camp—the reason, as he saw it, for the neglect or resistance to his own ideas about affect—was the theory of emotion proposed by two scientists nearly twenty years earlier. This was the theory proposed by Schachter and Singer in 1962—the very same year Tomkins published the first volume of his big book on the affects. In a famous experiment, Schachter and Singer had injected adrenaline into unsuspecting experimental subjects in order to induce physiological arousal. The scientists had then manipulated the same unsuspecting subjects and their environments in various ways in order to alter the subjects' emotional experiences. On the basis of their results, Schachter and Singer had proposed a "two-factor" or "cognition-arousal" theory according to which an emotional state is the product of two independent components: diffuse, undifferentiated physiological arousal, and cognition about the arousing situations. On this model, and in disagreement with Tomkins's affect program theory, there are no discrete emotions, only general arousal states that are labeled differently depending on the individual's cognition of the perceived context or situation.[73] Schachter and Singer's experiment had come to be considered a classic in social psychology and generated research not only in the field of emotion but in other areas as well.

But how accurate was Tomkins's picture of the emotion scene in 1981 with its focus on the cognitive Schachter–Singer model of affect?? My sense is that he may have been somewhat out of touch. It is characteristic of Tomkins that he would hark back twenty years, to the moment in 1962

73. S. Schachter and J. E. Singer, "Cognitive, Social, and Physiological Determinants of Emotional State," *Psychological Review* 69 (1962): 379–99.

when he and Schachter–Singer published their competing paradigms of affect. It's as if he was still living in 1962 and still begrudged the success of his rivals, a success he attributed to an ideological overestimation of the importance of cognition in psychology. In his eyes, the Schachter–Singer was a cognitive theory that had "needed to be believed" ("Q," 310). Yet it is worth noting that by 1981 the flaws in the Schachter–Singer experiment were beginning to be quite well known.[74] Moreover, Tomkins surely underestimated the success of his own affect "idea" by that date. As a grand theorist who since the mid-1960s had stood largely apart from ongoing laboratory research, he was perhaps unaware of the strong influence his disciple Ekman (and to a lesser extent Izard) had already exerted on the emotion field.

A rather different picture of the state of emotion research at the moment of Tomkins's "sermon" is given by laboratory researchers James Russell and José-Miguel Fernández-Dols in their instructive book on facial expression, *The Psychology of Facial Expression* (1997). There they trace the modern era of psychology's study of facial expression back to the 1962 publication of books by Tomkins and Plutchik and go on to observe that "Stimulated by these books, the pace of research on facial expression accelerated through the 1970s. By 1980, research on the face was dominated by the Facial Expression Program, centered on a list of specific 'basic' emotions as the cause and the signal received from facial expressions."[75] Thus, according to Russell and Fernández-Dols, the emergence of the Tomkins-inspired Basic Emotion Theory as the dominant research paradigm occurred in 1980, at the very moment Tomkins was complaining about the unwarranted influence of the cognitivists.

No doubt the situation and circumstances were confused, with both sides competing for control of the field. What can be said with some confidence is that, as Russell and Fernández-Dols observe, from the moment

74. For an important discussion of the flaws in the Schachter–Singer theory and a critical assessment of the research literature, see Rainer Reisenzein, "The Schachter Theory of Emotion: Two Decades Later," *Psychological Bulletin* 94 (2) (1983): 239–64; and idem, "Schachter–Singer Theory," in *The Oxford Companion to Emotion and the Affective Sciences,* eds. David Sander and Laus R. Scherer (Oxford, 2009), 352–53. Tomkins cited the criticisms of the Schachter–Singer position by C. Maslach, "The Emotional Consequences of Arousal without Reason," in *Emotions in Personality and Psychopathology*, ed. C. E. Izard (New York, 1978), 563–96 ("Q," 310). Richard Lazarus also critiqued the Schachter–Singer theory in his *Emotion and Adaptation* (New York and Oxford, 1991), 18–19.

75. James A. Russell and José-Miguel Fernández-Dols, eds., *The Psychology of Facial Expression* (Cambridge, 1997), 9–10.

of its ascendancy in the 1980s, Tomkins's affect program theory generated "more research than any other in the psychology of emotion."[76] As these authors also recognize, the triumph of the Facial Expression Program or basic emotion program had been largely due to Ekman's research efforts. Far from it being the case that the cognitivists were in the ascendant in 1981, as Tomkins suggested, Russell and Fernández-Dols treat Tomkins's affect program theory, and its successor, the Basic Emotion Theory as elaborated by Ekman, as the principal research program in the field—and for them, the program to challenge. It is to Ekman's work that I now turn.

76. Russell and Fernández-Dols, *The Psychology of Human Expression*, 10.

[CHAPTER TWO]

PAUL EKMAN'S NEUROCULTURAL THEORY OF THE EMOTIONS

In 1998 Paul Ekman, the foremost theorist of the concept of "basic emotions" in the United States, published a new edition of Charles Darwin's *The Expressions of the Emotions in Man and Animals*. Ekman used the occasion to offer a highly personal account of the dispute between the "culturalists" and the "universalists" over the nature of emotional expression, a dispute that had dominated post–World War II research on the affects in the United States. The drama, as Ekman portrayed it, had involved the clash of powerful personalities and the eruption of strong feelings about Darwin's legacy regarding the expression of the emotions. The principal actors on one side of the debate had been Margaret Mead, Gregory Bateson, and Ray Birdwhistell, influential anthropologists who, suspicious of biology in the wake of Nazism and of biological determinism generally, during the immediate post–World War II years believed in the malleability and cultural specificity of social behavior and accordingly viewed human facial expressions as learned forms of communication, not innate expressions of discrete emotions. Their chief antagonist had been the "philosopher and psychologist" Silvan Tomkins who, almost alone among emotion researchers at the time, in his unwieldy volumes *Affect Imagery Consciousness*, had incorporated Darwin's ideas by arguing that facial expressions are innate and universal. Ekman presented himself as "the last actor" in the drama, entering the fray as a young, unknown scientist when he began his research on facial expression in 1965, but also as a "protagonist" who had taken up the cause of universalism once he had become convinced by the scientific evidence.[1]

1. P. Ekman, "Afterword," in Charles Darwin, *The Expression of the Emotions in Man and Animals*, 3rd ed., introduction, afterword, and commentaries by Paul Ekman (Oxford, 1998), 370; hereafter abbreviated as "A." (For ease of reference, when citing Darwin's *Expression* I shall use Ekman's third edition throughout my book.) Margaret Mead had been the editor of the previous, now out of print, 1955 edition of Darwin's *Expression*, and in his afterword, Ekman berated her editorial comments, including her failure in her introduction to say anything about Darwin's claims for the universality of facial expressions and, with reference to Ray Birdwhistell's studies of nonverbal behavior, substituting the term "communication" for Darwin's term "expression." But

Ekman reported that before starting his research on facial expression he had sought out the anthropologists, only to be met with incomprehension and hostility when he told them of his plans to reopen the question of the universality of the emotions. Tomkins, on the other hand, was one of the heroes of the drama, as a researcher who not only offered a difficult but compelling theory about the universality of the primary affects but also possessed such a "magical" ability to read facial expressions accurately without any contextual clues that, on demonstrating his skill, he singlehandedly convinced the self-described "atheoretical" Ekman that the latter was capable of undertaking cross-cultural research in order to test Tomkins's ideas. In the rest of his afterword, Ekman recounted the success of that research and his related scientific studies in proving the validity of Darwin's universality thesis, and asserted the value of his "neurocultural" theory of the emotions in reconciling the warring parties in the dispute.[2]

But the story Ekman had to tell was not one of complete triumph. Starting in the late 1980s, objections to his neurocultural theory had begun to

---

for those readers unconvinced by Ekman's views, his own intrusive editorializing throughout his own 1998 edition of Darwin's text based on his Basic Emotion Theory can be criticized on the same grounds. Political motives also animated the debate, in which Ekman's research was dismissed as "fascist" by some anthropologists sensitive to postwar worries about racism and Ekman defended his position as antiracist because his universalist position implied unity and fraternity. On these points see Ekman, "A Life's Pursuit," 18–19, and Jan Plamper, *The History of Emotions: An Introduction* (Oxford, 2015), 166–67.

2. Ekman reported that he and Wallace V. Friesen had been loaned a large corpus of motion picture film taken by Carleton Gajdusek and Richard Sorenson showing the behavior of members of two different preliterate or "stone age" cultures in Papua New Guinea and that he and Friesen spent about six months inspecting the facial behavior shown in this film, using slowed- and stop-motion procedures. "We were struck by two 'findings,'" he noted, "which suggested that *both* the universal and relative views on facial expression might be correct." Near the end of this time, Tomkins visited his laboratory, and was shown short samples of the facial behavior from each of the two cultures but without any contextual information. Ekman and Friesen were very impressed when, on the basis of facial expressions alone, Tomkins was able to infer many aspects of the differences in behavior in the two cultures as regards childrearing, marital practices, and adult interactions, differences that he and Friesen knew to be correct from information provided by Gajdusek and Sorenson. "These experiences convinced us that there must be both universal and culture-specific facial expressions. We set about developing a theoretical framework which could explain the occurrence of both, thus reconciling the differences in the past controversy over this issue" (Paul Ekman and Wallace V. Friesen, "Universal and Cultural Differences in Facial Expressions of Emotion," in *Nebraska Symposium on Motivation, 1971*, ed. James. K. Cole [Lincoln, NE, 1972], 210–11).

be raised, and in his afterword Ekman attempted briefly to answer his critics. His text thus offers a revealing self-portrait of a man with a mission whose life's work had begun to be threatened by criticisms that he tried to defeat or deflect but that in fact he would not find easy to answer. The purpose of this chapter is to analyze the theoretical assumptions and methods informing Ekman's approach to the emotions in an attempt not only to understand his enormous influence in the United States but also to begin raising certain questions about the validity of his ideas and experimental practices.

Among those who have shared many of Ekman's presuppositions is the celebrated neuroscientist Antonio Damasio. In the course of this chapter, I shall be discussing an important case, investigated by Damasio and colleagues, of a young woman who, it was claimed, was unable to experience fear because she had an abnormal brain. More precisely, the patient suffered from a genetic disorder that had caused bilateral calcification of the amygdala, a subcortical group of neurons that has been widely held to be implicated in rapid emotional responses, especially fear. Crucial to Damasio's interpretation of the case was his use of the methods developed by Ekman to test a person's ability to judge emotional expressions. In particular, Damasio employed a standardized set of pictures of people intentionally presenting, that is, posing such expressions drawn from Ekman's portfolio of such items, in order to evaluate the deficits in the amygdala-damaged patient's skill at judging threatening faces.

I find Ekman's pictures at once interesting and puzzling (see Figures 1 and 2).

One of my aims in this chapter is to make such images historically intelligible while also bringing out what seems to me their sheer strangeness as scientific documents.

EMOTIONS AS NON-INTENTIONAL STATES: TOMKINS'S AFFECT PROGRAM THEORY

To begin: When in the 1960s Ekman began studying nonverbal behavior, including facial expressions, the emotions after years of neglect were just beginning to become a topic of renewed concern among scientists. Ekman (b. 1934) has attributed his own lifelong fascination with emotion to the death of his mother from mental illness when he was fourteen years old. He went to the University of Chicago one year later and, having decided to study psychology and psychoanalysis, transferred to New York University to complete his undergraduate degree. By the time he finished his doctoral work in clinical psychology at Adelphi University, he had become

Figures 1 and 2. From Paul Ekman and Wallace V. Friesen, *Pictures of Facial Affect* (Palo Alto, CA, 1976), reproduced by permission.

interested in nonverbal behavior, a rather unfashionable topic at the time, completing a year's clinical internship in 1957–1958 at the Langley Porter Institute at UCSF and the University of California–San Francisco (UCSF). He has worked in San Francisco ever since, first as a research assistant at UCSF and then as a research professor. He has now retired from his academic position but maintains a highly visible scientific presence through the activities of the Paul Ekman Group and other endeavors.[3]

Two experiences proved crucial in Ekman's early years as he tells it: almost by accident he received a grant that permitted him to undertake cross-cultural studies of nonverbal behavior; and in 1964 or 1965, just before he embarked on those studies, he met Tomkins. He was so impressed with Tomkins, whom he has described as the major intellectual influence in his life, that the two men continued to meet once a year and talked on the phone at least once a month for many years.[4]

3. The Paul Ekman Group offers online training in emotion regulation, microexpression detection, and related emotion skills and research tools.

4. For details concerning Ekman's biography, see P. Ekman, "A Life's Pursuit," in *The Semiotic Web, 1986*, eds. Thomas Sebeok and Jean Umicker-Sebeok (Berlin, 1987),

Ekman initiated his research on emotions at a time when, as we have seen in the previous chapter, Tomkins was proposing a new way of thinking about the emotions. Influenced by several important trends in the human sciences, especially a resurgent interest in Darwinian evolution and the rise of cybernetics, he had turned his back on the then-reigning orthodoxies of psychoanalysis to advocate instead a cybernetic- and Darwinian-inspired theory of the emotions. To summarize Tomkins's views: he maintained that there exists a small number of primary "affect program phenomena" or "basic emotions" defined as universal or pancultural adaptive responses of the organism. (Ekman used Tomkins's concept of "facial affect programs" from the start of his work on facial expression. As far as I have been able to determine, he used the term "basic emotions," already employed by Magda Arnold in her 1960 book, as early as 1972.)[5]

3–45; and Paul Ekman and Robert W. Levenson, "Inside the Psychologist's Studio," *Perspectives on Psychological Science*, 1 (3) (2006): 270–76. In the earlier text, Ekman dates his first encounter with Tomkins to 1965; the later text gives the date of 1964. In the later text he states that in the last ten years of Tomkins's life the two men had a serious falling out, a rupture Ekman attributes to Tomkins's jealousy of Ekman's rapid publication success at a time when Tomkins was publishing nothing. Already in 1976–1977, in a series of letters Tomkins accused both Ekman and Izard of failing to adequately acknowledge his priority, explaining that he felt humiliated when audiences sometimes accused him of having plagiarized from them the very theories he had originated. Ekman—described by Tomkins in his letters as his chosen intellectual son—managed to mollify Tomkins more successfully than Izard, to the point that at the end of the correspondence Tomkins decided to suspend further collaboration with the latter. It appears that Tomkins's own failure to continue publishing in the years leading up to the 1976–1977 exchange contributed to his feelings of isolation and neglect. The correspondence is in the Silvan S. Tomkins Papers on deposit at the Drs. Nicholas and Dorothy Cummings Center for the History of Psychology, University of Akron, Ohio.

5. Magda Arnold titled chapter 10, "Basic Emotions," in her book, *Emotion and Personality*, vol. 1, *Psychological Aspects* (New York, 1960), 193–215. For Ekman's first use of the "basic emotions" concept, see P. Ekman, W. V. Friesen, and P. Ellsworth, eds., *Emotion in the Human Face: Guidelines for Research and an Integration of Findings (Cambridge, 1972)*, 57; hereafter abbreviated as *EHF*. Ekman later especially embraced the term when it came into common usage in the 1990s during the debate instigated by Ortony and Turner over the existence of such "basic emotions." See A. Ortony and T. J. Turner, "What's So Basic about Basic Emotions?," *Psychological Review* 97 (1992): 315–31; Paul Ekman, "Are There Basic Emotions?," *Psychological Review* 99 (1992): 550–53; T. J. Turner and A. Ortony, "Basic Emotions: Can Conflicting Criteria Converge?," *Psychological Review* 99 (3): 566–71; P. Ekman, "An Argument for Basic Emotions," *Cognition and Emotion* 6 (3–4) (1993): 169–200; P. Ekman, "All Emotions are Basic," in *The Nature of Emotion: Fundamental Questions*, eds. P. Ekman and R. Davidson (New York, 1994), 146–49; and P. Ekman, "Basic Emotions," *Handbook of Cognition and Emotion,* eds. T. Dalgleish and M. J. Power (London, 1999), 45–60.

Tomkins proposed that there were eight or nine such primary emotions, namely, fear, anger, distress, disgust, interest, shame, joy, and surprise (he later added contempt). He argued further that each evolved, discrete emotion was controlled by an "affect program" in the brain that when activated triggered distinct physiological and behavioral responses, and especially characteristic facial expressions.[6] Central to Tomkins's approach was the idea that although the emotions can and do combine with the cognitive and other information-processing systems in the brain, they are essentially separate from these. In other words, he posited a disjunction between our *emotions* and our *knowledge* of what causes and maintains them by treating feeling and cognition as two separate systems. He thus claimed that there is a "radical dichotomy between the 'real' causes of affects and the individual's own interpretations of these causes."[7] In short, for Tomkins the affects have no inherent knowledge of, or relation to, the objects that trigger them—which is why, he thought, we are so liable to be wrong about our feelings and ourselves.

By separating the affects from cognition, Tomkins broke with the paradigm that had previously governed much of the work on the emotions. According to Tomkins's predecessors, emotions are intentional states: they are directed toward objects and depend on our desires and beliefs about the world. In Sigmund Freud's case of "Little Hans," to take one famous example, the five-year-old Hans became terrified of horses and refused to go anywhere near them. The immediate, precipitating cause of his fear was his experience of seeing a horse fall down, as if dead. But what interested Freud was not the contingent role of the event of the falling horse, but the *meaning* the horse had for Hans, especially the meaning it had for him as a substitute object for his conflictual desires, wishes, and beliefs. In the course of Freud's narrative of the case, it emerges that the little boy not only feared horses but also unconsciously identified with them, to the point of wanting to *be* the very object of his terror. According to Freud, the reason the horse had become so salient or important for the

6. The emotion field has benefited by invoking Darwin's writings on facial expression, but it has frequently misinterpreted his ideas. Darwin is typically thought to have viewed expressions as evolved, selected, or adapted for the purposes of communicating emotions, in the sense of involuntarily expressing internal feelings that are manifested in external displays that can be understood or "read" by others. But as Fridlund has pointed out, Darwin's conclusion was the opposite: he believed that most facial displays were not evolutionary adaptations, but vestiges or accidents. *HFE*, 15, n. 2.

7. S. S. Tomkins, *Affect Imagery Consciousness*, *The Complete Edition*, 4 vols. (New York, 2008), vol. 1–2, 137; hereafter abbreviated as *AIC*, vol. 1–2, or *AIC*, vol. 3–4.

child was that it had once been an object of interest and desire for him: "what is today the object of phobia must at one time in the past have been the source of a high degree of pleasure." The child had liked horses and enjoyed playing at being a horse before the onset of his phobia. By identifying with or becoming the horse in play, he could "bite" the father whom he (loved and) dreaded. As Freud would later argue, the child's unconscious yielding to, or identification with, the powerful, aggressive father solved the problem of the child's conflictual wishes and intentions toward him, albeit at the price of the child's experiencing the incorporated paternal aggression in the form of a guilty, anxious conscience. Hence Freud joined together fear, the phobic object, identification, guilt, anxiety, and the subject in a single explanatory complex. And of course for Freud the wishes and intentions that underlay the phobic reaction could only be brought into the subject's awareness through the transferential-narrative process of the "talking cure."[8]

I am not concerned here with the validity of Freud's analysis of the case of Hans, which has been widely dismissed in scientific circles ever since Wolpe and Rachman in 1960 critically reviewed it and reinterpreted phobia as a conditioned response.[9] My purpose is simply to indicate that psychoanalysis—and more generally "cognitivist," "propositional," "appraisal," or "phenomenological" approaches to the emotions—operates within an intentionalist paradigm that makes questions of meaning and belief of fundamental importance. The intentionalist approach contrasts strikingly with that of Tomkins, for whom, as we saw in the previous chapter, the affects are non-intentional states. His account of fear goes something like this: If I run from a snake I do not do so because I believe there is a dangerous object in front of me and desire or intend not to be bitten by it. I run because I am terrified of snakes. The threat of the snake lies out there in the object: snakes make me frightened because they were once terrifying to our ancestors in evolutionary history. The snake on this

8. Sigmund Freud, "The Analysis of a Phobia in a Five-Year-Old Boy" (1909), in *The Standard Edition of the Complete Psychological Works of Sigmund Freud*, trans. and ed. James Strachey (London, 1953–74), 3, 149–289. For Freud, it is because Hans fears the powerful father's retaliation against his own unmasterable hostility that he (the child) identifies with him to the point of incorporating the father's aggression. That incorporated aggression fuels the child's superego, so that the aggressivity of the scenario is played out in the mode of the subject's experience of a relentless self-reproach. Aggressivity, fear, and guilt on this interpretation thus have a deep connection, one that is lost in modern interpretations of fear, such as those by Tomkins and Ekman.

9. J. Wolpe and S. Rachman, "Psychoanalytic 'Evidence': A Critique Based on Freud's Case of Little Hans," *Journal of Mental and Nervous Diseases* 131 (1960): 135-48.

model does not function as an object of my beliefs or desires but as a "trigger" or trip wire for an involuntary, hardwired response that is rapidly discharged without the affect system's knowledge of the object that triggered it. Tomkins's affect theory thus suspends or displaces considerations of intentionality and meaning in order to produce an account of the emotions as discharges of facial behaviors and related autonomic responses that are fundamentally corporeal in nature.[10] On this model, the nervous system is understood to be "wired directly to the onset of the danger."[11]

For Ekman in the 1960s, the most intriguing aspect of Tomkins's affect theory was his claim that emotions are universal categories or "natural kinds" that are accompanied by distinct facial expressions. Tomkins's views were at odds with the prevailing consensus among psychologists and particularly anthropologists, who doubted that facial behaviors had the same meaning in all cultures. The contradictory results of more than five decades of experimental research on the recognition of facial expression reinforced their skepticism on this point.[12] But Tomkins was determined to prove them wrong. Especially telling for Ekman in this regard was Tomkins and McCarter's 1964 experiment, discussed in the previous chapter, demonstrating that observers could obtain a high degree of consensus in the judgment of emotions if posed facial expressions were carefully selected for this purpose to show what Tomkins believed were the innate facial affects.[13] Decisively influenced by that experiment, Ekman in these early years developed a "neurocultural" version of Tomkins's theory of affect. According to Ekman, socialization might influence the range of elicitors that can trigger the innate "facial affect programs" in the brain

10. Because the affect programs are said to be located in primitive, subcortical parts of the brain which function independently of knowledge, they are held to bypass the slower cognitive pathways of the higher cortical centers and respond instead in a quick and mindless way to the innate or learned triggers that elicit them. Much recent work on the emotions has been devoted to validating such an approach to the emotions, which has also found its way into the popular media. For example, in an article in *Newsweek* about how people respond to terrorist threats and other dangerous situations, we find this statement: "The strongest human emotions are fear and anxiety. Crucial to survival, they are programmed into the brain's most primitive regions, allowing them to trump rationality but not for rationality to override them." "When It's Head versus Heart, the Heart Wins," *Newsweek*, February 11, 2008, 35.

11. B. Massumi, "Fear (The Spectrum Said)," *Positions* 13 (1) (2005): 37.

12. For an important assessment from the 1950s, see J. S. Bruner and R. Tagiuri, "The Perception of People," in *Handbook of Social Psychology*, vol. 2, ed. G. Lindzey (Reading, MA, 1954), 634–56.

13. S. S. Tomkins and R. McCarter, "What and Where Are the Primary Affects? Some Evidence for a Theory," *Perceptual and Motor Skills* 18, (1) (1964): 119–58.

and can moderate facial movements according to the cultural norms or "display rules," but under certain conditions the innate emotions might nevertheless "leak out." The aim of his neurocultural theory was therefore to reconcile the war between the universalists and the culturalists by suggesting how both universal and culture-specific facial expressions could occur.[14] He was convinced on this basis that it ought to be possible to find ways of distinguishing the involuntary, biologically determined facial expressions from those that are influenced by learning and culture. With Tomkins's encouragement and help, Ekman in the late 1960s embarked on a series of highly influential cross-cultural judgment studies aimed at determining whether the basic emotions, as manifested in photographs of posed facial expressions, could be universally recognized by literate and preliterate people alike.[15]

14. For accounts of Ekman's neurocultural model of the emotions, see for example P. Ekman and W. V. Friesen, "Nonverbal Leakage and Clues to Deception," *Psychiatry* 32 (1) (1969): 88–106; P. Ekman, "Universals and Cultural Differences in Facial Expressions of Emotion" (1972); and idem, "Biological and Cultural Contributions to Body and Facial Movement in the Expression of Emotions," in *Explaining Emotions*, ed. A. O. Rorty (Berkeley, 1980), 73–99, 100–101. Ekman adopted Tomkins's notion of "affect programs" early on, defining them as "innate subcortical programs linking certain evokers to distinguishable, universal facial displays for each of the primary affects" (P. Ekman, R. Sorenson, and W. V. Friesen, "Pan-Cultural Elements in Facial Displays of Emotion," Science 164 [1969], 87). As he also stated: "The term *affect program* refers to a mechanism that stores the patterns for these complex organized responses, and which when set off directs their occurrence . . . The organization of response systems dictated by the affect program has a genetic basis but is influenced also by experience. The skeletal, facial, vocal, autonomic, and central nervous system changes that occur initially and quickly for one or another emotion, we presume to be in largest part given, not acquired" (P. Ekman, "Biological and Cultural Contributions to Body and Facial Movement in the Expression of Emotions," 82). In 2003 he referred to affect programs as "inherited central mechanisms," observing that "stored in these central mechanisms there must be sets of instructions guiding what we do, instructions that reflect what has been adaptive in our evolutionary past" (P. Ekman, *Emotions Revealed: Understanding Faces and Feelings* [London, 2003], 65).

15. Carroll I. Izard developed a similar theory to that of Ekman around the same time and based on similar cross-cultural matching research. His theory was focused on early growth in childhood, but was not as developed structurally as Ekman's neurocultural views, so it did not inspire experiments or the kind of critical discussion stimulated by Ekman's work. For this reason, I have chosen to focus on Ekman's contributions. Tomkins did not inform Izard and Tomkins that he had encouraged both of them to carry out similar cross-cultural studies of facial expressions at the same time, with the result that there was a natural rivalry between the two men. For Tomkins's views on his relations with Izard, see the former's "The Quest for Primary

PHOTOGRAPHING EMOTIONS

The use of photographs for the scientific study of expression goes back to the groundbreaking work of the French scientist Duchenne de Boulogne, and of course to Charles Darwin's *The Expression of Emotions in Man and Animals* (1872). Early on in *Expression,* Darwin—among the first to use photographs for scientific purposes—described the challenges confronting anyone wishing to study the face in an objective way. In a brief but telling passage, he observed:

> The study of expression is difficult, owing to the movements being often extremely slight, and of a fleeting nature. A difference may be clearly perceived, and yet it may be impossible, at least I have found it so, to state in what the difference consists. When we witness any deep emotion, our sympathy is so strongly excited, that close observation is forgotten or rendered almost impossible.[16]

One implication that was drawn from this state of affairs was the need for "instantaneous photography" because, it was thought, only high-speed photography would be capable of capturing objectively—that is, of arresting and "freezing" in time—the rapid play of facial muscles in a single image.[17] It might be objected that Darwin's statement has rather different implications. For instance, one might want to argue on the basis of his remarks that expressions are all about movement, especially the subtlety and volatility of the movements of the muscles of the face, and that these movements—perhaps *actions* would be a better word—are what we ordinarily mean by emotion. From this point of view, any technology for objectifying the movements of the face in a still photograph is likely to be useless, because by freezing an expression the camera isolates a single, static moment in an overall flow of events that alone gives that moment its meaning. Another implication of Darwin's remarks would seem to be that there is an ineluctable sympathetic-identificatory dimension to our perceptions of other people, a dimension that will inevitably be ignored or suppressed by techniques of observation based on the assumption of an

---

Motives: Biography and Autobiography of an Idea," *Journal of Personality and Social Psychology* 41 (1981): 312–13.

16. Charles Darwin, *The Expression of Emotion in Man and Animals*, 19.

17. Of course, "instantaneous" is a relative term in photography, as Phillip Prodger makes clear in *Time Stands Still: Muybridge and the Instantaneous Photography Movement* (Oxford, 2003).

absolute separation between subject and other, between the viewer and the object of his or her gaze.

For the photographers to whom Darwin turned for help to illustrate the expressions of emotion, however, the solution to the difficulty was to be found in the use of "instantaneous" photographic techniques capable of capturing and fixing once and for all the subtle, transient movements of the face invisible to the naked eye. But, as Phillip Prodger has observed, although advances in photochemical technology had reduced exposure times, thereby making it possible for the camera to capture relatively small movements in time, by contemporary standards they were still long.[18] This meant that the subject was required to remain completely still for several seconds while his or her picture was taken. One consequence of this, of course, was that the subject was aware of being observed. In its fidelity to the scene, the photograph inevitably registered this fact, which was often perceived in terms of an aspect of artificiality, conventionality, exaggeration, or "theatricality" in the sitter's expression as he or she self-consciously held the desired pose.[19] Darwin accepted the need for posing and indeed for photographic manipulation and editing in order to produce convincing portraits of the emotions. Figure 3, for example, shows a picture of the Swedish photographer Oscar Rejlander, well known for his mastery of photographic manipulation, himself posing the emotion of surprise in one of the thirty photographs Darwin included in his book. Figure 4 shows Darwin's picture of fear or terror. The latter is not a photograph; rather, it is a slightly altered engraving after a photograph. The original was taken by the French scientist Duchenne de Boulogne (Figure 5). In it, one can see that the muscles of the man's face have been artificially stimulated by electrical currents to produce the facial movements allegedly expressive of terror; in the engraving he commissioned for his book, Darwin had the electrodes removed in order to give the face a more natural appearance (*IS*, 169–70).

Prodger has remarked that Darwin "considered the fleeting nature of expression to be the principal difficulty in observing facial movements," as if for Darwin the problems he encountered were merely technological

18. Phillip Prodger, "Illustration as Strategy in Charles Darwin's 'The Expression of the Emotions in Man and Animals,'" in *Inscribing Science: Scientific Texts and the Materiality of Communication*, ed. Timothy Lenoir (Stanford, 1998), 156; hereafter abbreviated as *IS*.

19. "With exposure times in the tens of seconds, a 'true' rendering of emotional expression would have required the subjects to achieve naturally a representative emotional expression, then hold it unadulterated for the photographer to record on film. This could not be done reliably or with precision" (*IS*, 177).

Figure 3 Oscar Rejlander posing the emotion of surprise, from Charles Darwin, *The Expression of Emotions in Man and Animals*, 3rd ed. (1998), 287.

Figure 4 "Terror," from Charles Darwin, *The Expression of the Emotions in Man and Animals*, 3rd ed. (1998), 301.

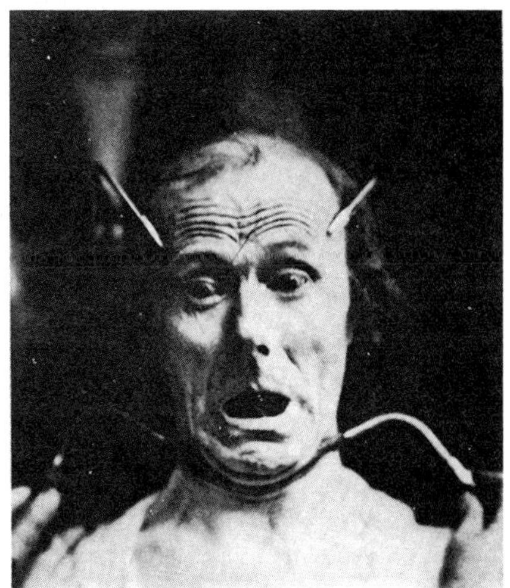

Figure 5 "Horror and Fear," from G.-B. Duchenne de Boulogne, *The Mechanism of Human Facial Expression*, ed. and trans. R. Andrew Cuthbertson (Cambridge, 1990), plate 61.

(*IS*, 177).[20] This seems to be Prodger's view as well, since he observes in this connection that "the limitations of photographic technology precluded objective recording of facial expressions" (*IS*, 177). There is an important sense, of course, in which the limitations Darwin confronted were not merely or only technological, but had a deeper, more philosophical or, say, ontological cause. A passage in a text by Johann Caspar Lavater, a famous eighteenth-century predecessor in the physiognomic tradition to which Darwin, Tomkins, and Ekman can all be seen to belong, bears directly on this point. "I want to discover in the mild-mannered person the traits of this manner, in the humble person the signs of humility," Lavater observed in 1772, exactly a hundred years before Darwin. He goes on:

> But my observations must be exact; they must be repeated and tested often. How can that be possible if I have to make these observations on the sly? Isn't it presumptuous to analyze faces? And if a humble person notices that she is being observed, won't she turn away and hide her face? Indeed, it is here that I encountered one of the greatest obstacles to my studies; anyone who notices that he or she is being observed either puts up resistance or dissimulates. How can I get around this problem? Perhaps in part in the following way.
>
> I retire into solitude; I place before me a medallion or a piece of antique sculpture, the sketches of a Raphael, the apostles as depicted by Van Dyck, the portraits of Houbraken. These I can observe at will, I can turn them and view them from all sides.[21]

On the one hand, in this passage Lavater recognizes that the spectatorial gaze of the observer is *itself* the problem, since it inevitably makes the subject aware of being seen. From this it follows that the risk of a certain posing or conventionalizing of expression inheres in the very situation of

20. However, Prodger continues this passage in a remark suggesting that Darwin was aware of the risk of theatricality or conventionalizing inherent in the situation of being beheld: "Darwin was concerned that an observer might taint a subject's actions through his or her own reaction to the emotion displayed, and that an observer might read into his or her observations because of the circumstances of the occasion" (*IS*, 177).

21. Johann Caspar Lavater, *Von Der Physiognomik* (1772), cited by R. T. Gray, *About Face: German Physiognomic Thought from Lavater to Auschwitz* (Detroit, 2004), 49–50. In his book, Gray makes many illuminating observations about the ideals of immobility, stasis, and repose in Lavater's physiognomic-observational strategies, ideals that likewise inform the Tomkins–Ekman paradigm—as if only the human form stripped of vital, intentional actions and movements can reveal the authentic, primordial essence of the natural human being.

being beheld, which is to say that the problem of posing can't be solved by technological means, such as the invention of the high-speed camera. On the other hand, and this is the point I especially want to stress, Lavater's belief that the truth shows itself on someone's face when he or she believes he or she is alone is what might be called a theoretical or ideological claim. It does *not* follow from the fact that the face is a privileged site of expression for us that if only it could be seen in solitude it would tell us the deep truth about ourselves. To believe this is to succumb to the view that a distinction can be strictly maintained between authentic and artificial signs, based solely on the state of awareness of the subject in question. More precisely, it is to surrender to a physiognomic ideology that claims we can draw an absolute differentiation between the masks we allegedly put on for others and the genuine faces we have when we are alone and no one is watching. (One limit case here, I suppose, would be to photograph persons when they are asleep. The expressions of blind people have also been imagined as free from all social taint, which is why they have figured in discussions of the Tomkins–Ekman Basic Emotion Theory.)[22] It's as if we were to imagine that we cannot find out the truth about people in the course of interacting with them in daily life, or by discovering how they act in an emergency, or on the basis of intimate conversations with them, but only when they are entirely removed from

22. Thus Ekman cited J. S. Fulcher, "'Voluntary' Facial Expression in Blind and Seeing Children," *Archives of Psychology* 38 (1942): 272; F. I. Goodenough, "Expression of the Emotions in a Blind-Deaf Child," *Journal of Abnormal and Social Psychology* 27 (1932): 328–33; J. Thompson, "Development of Facial Expression of Emotion in Blind and Seeing Children," *Archives of Psychology* 37 (1942): 264. See P. Ekman and W. V. Friesen, "Constants Across Cultures in the Face and Emotions," *Journal of Personality and Social Psychology* 17 (2) (1971): 128, where the authors state in reference to these sources that "Evidence of constants in facial behavior and emotion across cultures is also consistent with early studies which showed many similarities between the facial behavior of blind and sighted children." In a later publication, on methodological grounds Ekman and Oster offer a more cautious evaluation of both the earlier and the more recent literature on the topic, concluding that "it is still not known when the distinctive, universal facial expressions corresponding to certain emotions first appear, nor how they develop" (P. Ekman and H. Oster, "Facial Expressions of Emotion," *Annual Review of Psychology* 30 [1979]: 533). See also I. Eibl-Eibesfeldt, "The Expressive Behavior of the Deaf-and-Blind Born," in *Social Communication and Movement*, eds. M. von Cranach and I. Vine (San Diego, 1973), 163–94. For an evaluation of research on congenitally blind children by a critic of the Basic Emotion Theory, see Fridlund, *HFE*, 112–14. Fridlund comments that the only definitive test of the innateness of certain infantile facial displays would be provided by studies of children born blind and reared by blind caretakers who would have to restrict their interactions with children so drastically to control for learning that the prospects for such studies are likely to be zero.

human intercourse. But if someone is untrustworthy, or inconsistent, or indeed honorable, is it plausible that the truth about him will emerge simply because he is alone?[23]

And of course if you think this way, if you believe that the answer to this question is yes, then you will have to worry about whether the expressions you have recorded by photographic means for scientific observation really are natural and unposed. You will find yourself continually haunted by the suspicion or charge that the images you are studying are not truly spontaneous because you have not adequately isolated them from the contamination of others.[24] As we have seen, Lavater in the pre-photograph era thought one solution to the problem of the subject's consciousness of being beheld was to study artistic illustrations of expression in the privacy of his study. His idea seems to have been that, since the subjects of such artists' representations were not physically present to him, the problem of posing could not arise. Lavater's solution could hardly satisfy Ekman, and it is not surprising that in the latter's discussion of methods for recording emotional expressions he ruled out of consideration the results of *any* experiments that had made use of artists' drawings and illustrations, precisely on the grounds that such pictures were likely to be too stereotyped or subjective for scientific purposes.[25]

But Ekman *was* committed to the project of distinguishing between nature and culture, between the "natural signs" of emotional expression that might be so small or rapid as to be easily overlooked and those culturally coded displays or "artificial signs" that he thought tended to

23. The classic statement of the traditional view is by Susan Sontag: "There is something on people's faces when they don't know they are being observed that never appears when they do" (*On Photography* [New Haven, CT, 1977], 37). This claim is complicated and indeed refuted more than once in Michael Fried, *Why Photography Matters as Art as Never Before* (New Haven and London, 2008).

24. Darwin justified his study of expression in nonhuman animals by stating that "In observing animals, we are not likely to be biased by our imagination; and we may feel safe that their expressions are not conventional" (Darwin, *The Expression of Emotion in Man and Animals*, 24). I take Darwin to mean, not that animals are unaware of being seen, but that they are not aware of our specifically human interests and motives in observing them, a point caught by Prodger when he observes of Darwin's statement that "Animals are unaware of the scientist's purpose, and are unlikely to modify their behavior in reaction to experimental observation" (*IS*, 149).

25. "Until a systematic means of coding facial behaviors had been widely accepted and drawings as well as large numbers of actual photographs and films have been scored, there is no way of knowing to what extent they represented fantasy or reality" (*EHF*, 50–51).

mask or disguise them.[26] The whole point of his Tomkins-inspired "neurocultural" theory of the emotions was to reconcile the warring sides in the dispute between the universalists and the culturalists by developing a framework capable of explaining the occurrence of *both* universal and culturally specific facial expressions.[27] According to Ekman, the display rules of any given culture are the social norms that govern the way a person intentionally and consciously manages and controls the expression of his or her emotions. Thus, when a person is fearful, he might exaggerate, constrain, or neutralize his feelings and expressions in conformity with the requirements of the social context (Ekman had been reading Erving Goffman on the performance of social roles).[28] But Ekman believed that it should be possible to identify the pan-cultural or universal, involuntary signs of emotion behind the culturally determined display rules and even outright dissimulations ("lies") that mask the truth.[29] For his pur-

26. For Ekman's denial that his approach bears any relation to the older physiognomic tradition, see his "Duchenne and Facial Expression of Emotion," in *Duchenne de Boulogne: The Mechanism of Human Facial Expression*, ed. and trans. R. Andrew Cuthbertson (Paris, 1991), 271. But Ekman's fundamental assumption that we can draw a distinction between natural and cultural facial expressions *is* the physiognomic assumption.

27. It is worth noting that Ekman tended to exaggerate the conflict between nature and nurture in pre-Tomkins thinking about faces. He painted his predecessors as extreme cultural relativists, but that portrait was somewhat of a distortion that served his ends.

28. Ekman has reported that he first met Goffman in 1971 and that they subsequently became close friends, in spite of substantial differences between them. He has observed in this regard that "I found Erving's views challenging, provocative, and almost always on a different plane from my own. When we examined a social incident, our analyses and interpretations were not contradictory, but unrelated" (Ekman, "A Life's Pursuit," 31). Although Ekman's concept of "display rules" has been traced back to the work of Otto Klineberg, it seems possible that he also owed a debt to Goffman's discussion of the regulation and management of social interactions in the latter's *Behavior in Public Places: Notes on the Social Organization of Gatherings* (1959), a book Ekman cited in P. Ekman and W. V. Friesen, "Nonverbal Leakage and Clues to Deception," 1969, 89–90. Ekman also cites the same Goffman book in Ekman and Friesen, "Universals and Cultural Differences in Facial Expressions of Emotion," 228. Ross Buck associates Ekman's concept of display rules and deception with Goffman's text in Buck, *The Communication of Emotion* (New York and London, 1984), 231–32, 244, 254.

29. As Ekman and Friesen wrote in a preliminary statement of the neurocultural theory: "We believe that some of the impressions of cultural differences in affect display have arisen from a failure to distinguish adequately the pan-cultural elements from the circumstances governing the display of affects which are markedly influenced by social learning and vary within and between cultures. We believe that, while

poses, photographs of persons posing expressions were an indispensable research tool because, by fixing expressions in still images, they made possible the testing of people's ability across the world to recognize and identify particular emotions. Still pictures were also crucial if he was to realize his project of identifying and coding the component muscle movements responsible for the facial expressions supposedly linked to each of the hypothesized basic emotions—not an easy project, as he recognized, since facial movements seemed to lack natural divisions and the selection of variables was unclear.[30]

---

the facial muscles which move when a particular affect is aroused are the same across cultures, the evoking stimuli, the linked affects, the display rules and the behavioral consequences all can vary enormously from one culture to another." P. Ekman and W. V. Friesen, "The Repertoire of Nonverbal Behavior: Categories, Origins, Usage, and Coding," *Semiotica* 1 (1969): 73. In this paper, following Tomkins's lead, the authors offer a preliminary statement of the neurocultural theory, which included the following elements: the existence of seven (later six) universally recognizable, pan-cultural "primary affect states" (happiness, surprise, fear, sadness, anger, disgust, and interest, but not shame) governed by involuntarily triggered "affect programs" in the brain; "display rules" defined as culturally specific social procedures or norms for the management of affect displays in various social settings; a theory of "leakage" based on a distinction between intentional or deliberate facial signals that are consciously meant to communicate a message and that may be used deceptively; and involuntary facial behaviors that, because *they express* emotions rather than communicate information, escape efforts to deceive owing to their non-intentional, involuntary character, thereby providing cues to deception. In a note, Ekman and Friesen observe in this regard: "Intentionality is a concept traditionally avoided by psychologists interested in human communication on the grounds that it is not possible to operationalize intentions, or that the investigator will become lost in questions of levels of intention or unconscious intention. We believe that there are some nonverbal behaviors which the sender usually consciously intends as communicative signals to convey messages . . . and that through naturalistic or experimental methods it is possible to differentiate these acts from other forms of nonverbal behavior" (53–54, n. 2).

30. See Ekman's early discussion of this issue in "A Methodological Discussion of Nonverbal Behavior," *Journal of Psychology* (1957) 43: 141–49. Ekman recognized that still photographs "freeze" expressions by slicing them out of what is naturally a sequence of movements in time. Jerome Bruner and R. Tagiuri had criticized the use of such photographs because, as Ekman quoted them as stating, "'judgment based on a frozen millisecond of exposure' is not representative of the type of judgment made in naturally occurring conditions." But Ekman et al. argued that if the purpose of the study did not require a judgment of sequential behavior, then still photographs "may be useful for some research questions" (*EHF*, 49). The authors especially defended the use of posed still photographs for their savings in cost and ease of use, remarking that in the case of posed facial behavior, where the poser holds the pose for the camera, "a still will provide the same information as five seconds of film or videotape of the frozen position. Certainly, there is considerable evidence that the frozen few millisec-

EKMAN AND THE PROBLEM OF THE POSE

Starting in the late 1960s, then, Ekman set out to test the idea that the emotions could be correctly judged by everyone, literate and preliterate people alike. On the basis of studies purporting to demonstrate that American, Brazilian, Japanese, and other observers, including observers from among the last of the isolated, preliterate cultures of New Guinea, could all successfully distinguish the primary affects as represented in photographs of people posing expressions, he concluded that such facial expressions must contain universal or pan-cultural elements linked to the hypothesized primary or "basic" emotions. In order to uncover the pan-cultural elements, he needed photographs of facial movements that, as he put it, were free of cultural differences because of learned evokers, display rules, and consequences. He therefore attempted to select such photographs, in order to prove that observers from different cultures recognized the same emotion from the same photograph. In other words, to test the validity of the universality thesis, Ekman preselected photographs of posed expressions that he theorized were *already free* of cultural influence. The Facial Action Scoring Technique, or FAST, which he and Tomkins developed at this time as a system for measuring facial expressions held to be characteristic of the hypothesized affect categories, played a role in this preselection process. For, partly on the basis of FAST (and partly on the basis of personal intuition), the two men examined over three thousand pictures of posed expressions employed by themselves or others in previous experiments, and culled only those that had obtained the highest agreement about the emotion portrayed and that seemed to them to exhibit the emotional expressions in their pure ("unmixed") and unacculturated form.[31] Since posed facial expressions were critical to es-

---

onds of a still photograph *can* provide quite a bit of information" (*EHF*, 49)—which was of course precisely the point at issue.

31. Tomkins's theory of "blends" played an important role here. He and Ekman assumed the existence of a core set of affects that might combine in various ways with the result that much of the time the face expresses mixtures of two or more emotions. The theory of blends had the advantage of explaining away any disagreements in previous judgment studies as "confusions" on the part of observers owing to the existence in posed photographs of facial expressions of blends resulting from local, cultural conditions. For the argument about blends and "confusions," see Tomkins and McCarter, "What Are the Primary Affects?" (1964); Ekman and Friesen, "The Repertoire of Nonverbal Behavior: Categories, Origins, Usage, and Coding," 75–76; Ekman et al., *EHF*, 25, 62–64; and Ekman, "Universals and Cultural Differences in Facial Expressions of Emotion," 222–24.

tablishing FAST in the first place, posing was institutionalized in Ekman's first system of measurement from the start. His procedures were therefore highly recursive, to say the least.

Thus, if we follow Ekman's writings during these important early years when he was attempting to lay the foundations of a science of emotion, we find him arguing that the facial expressions in the photographs he employed in his experiments must have been free of cultural taint *because* they were universally recognized. At the same time, he suggested that those facial expressions were universally recognized *because* they were free of cultural taint. Similarly, Ekman presented FAST as if it were the result of the finding that photographs of posed expressions could be distinguished correctly by literate and preliterate observers. But there is an important sense in which observers from literate and preliterate cultures were capable of distinguishing these emotions correctly *because* FAST (plus subjective intuition) had already predetermined the "correct" or representative posed facial expressions to be used in his judgment tests.

Ekman, Friesen, and Tomkins decided from the outset to develop FAST in terms of emotion categories (happiness, sadness, surprise, anger, disgust, and fear) rather than emotion dimensions (pleasantness-unpleasantness, active-passive, etc.). To produce FAST, Ekman, Friesen, and Tomkins divided the face into three areas (brows-forehead area; eyes-lids-bridge of nose; lower face, consisting of cheek-nose-mouth-chin-jaw). Lists of facial components within each of the three facial areas for each of the six emotions were then compiled. The authors based these lists on the past literature as well as on their own observations and intuitions. To specify the appearance of the relevant components, they obtained "visual-photographic" definitions of each item by requesting several actors to pose the facial movements within each facial area (not the emotions) after they were told and shown what to do with the face. FAST thus consisted of three sets of photographs, one set for each facial area.

The faces to be scored by FAST were selected on theoretical grounds to show only one of the six hypothesized discrete emotions from photographs of posed expressions used by investigators in the past (including most of those used by Ekman and Friesen in their cross-cultural studies), each of which had obtained at least seventy percent agreement among observers about the presence of a single emotion. These selected photographs—the aim was to obtain ten pictures of each of the six emotions, although this proved impossible for disgust and fear—were projected life-size onto a screen, which was masked to show only one facial area at a time, and trained scorers assessed them by selecting the FAST item that best matched the area in question. The results for reliability and validity were found to be "very encouraging" for predicting the recognition of posed faces, although FAST had relatively poor success with fear faces as compared to the other emotion category faces, owing to the problem of blends. FAST's applicability to spontaneous expressions was justified on the grounds that, since it had been shown by Ekman and Friesen that posed expressions could be judged as showing the same emotions across different cultures, posed expressions must have some intrinsic relation to spontaneous ones. "In light of this evidence," the authors concluded, "it seems reasonable to assume that posed behavior differs little from spontaneous facial behavior in form."[32]

I can sum up Ekman's procedures by saying that photographs of posed emotional expressions were central to his research enterprise, even as he was also committed to maintaining that such photographs were unconventionalized, unacculturated representations of the universal truth of the primary emotions. It was especially urgent for him to argue this point, since the status of still, posed expressions had been a topic of much contention during the previous decades of emotion research. Ekman's concern with this theme is especially evident in his book *Emotion in the Human Face: Guidelines for Research and an Integration of Findings* (1972). Here, he and his coauthors reanalyzed many of the experiments on the judgment of facial expression performed between 1914 and 1970 and presented their own findings in order to argue that, contrary to the opinion of the majority of his contemporaries, the preponderance of the data yielded consistently positive evidence in support of the universality thesis. But the problem of posing never really disappears from his work.[33]

Although Ekman moved with seeming confidence in this terrain, his arguments are marked by considerable tension and even, at times, incoherence. In effect, he adopted two different, apparently competing responses to the problem of posing. His *first response* to the problem of posing was to argue on the basis of the empirical evidence that, since people all over the

32. P. Ekman, W. V. Friesen, and S. S. Tomkins, "Facial Affect Scoring Technique: A First Validity Study," *Semiotica* 3 (1971): 37–53. For further discussions of FAST, see also Ekman et al., *EHF*, 114–19; Ekman, "Universals and Cultural Differences in Facial Expressions of Emotion," 248–53; and Ekman and Oster, "Facial Expressions of Emotion," 527–54.

33. For a defense of the use of posed expressions, see for example M. O'Sullivan, "Measuring the Ability to Recognize Facial Expressions of Emotion," in *Emotion in the Human Face*, 2nd ed. (Cambridge, 1982), ed. P. Ekman, 298–99. See also P. Ekman and E. Rosenberg, eds., *What the Face Reveals: Basic and Applied Studies of Spontaneous Expression Using the Facial Action Coding System*, 2nd ed. (Oxford, 2005), where the issue of posing comes up frequently.

world, regardless of their cultural origins and background, consistently labeled photographs preselected to depict the basic emotions the same way, facial expressions must be universal. "Observers in both literate and preliterate cultures chose the predicted emotion for photographs of the face, although agreement was higher in the literate samples," Ekman wrote. "These findings suggest that the pan-cultural element in facial displays of emotion is the association between facial muscular movements and discrete primary emotions, although cultures may still differ in what evokes an emotion, in rules for controlling the display of emotion, and in behavioral consequences."[34] In other words, the fact that, as his cross-cultural experiments seemed to demonstrate, literate and preliterate people alike could agree when looking at photographs that certain facial expressions connoted the same emotional categories, argued against the idea that posed expressions were determined by conventions, as the cultural relativists claimed. Rather, posed expressions must contain pan-cultural elements and must be in some way "based on the repertoire of spontaneous facial behaviors associated with emotion" (*EHF*, 106). "We will argue," he stated, "that posed facial behavior, while special, is not unique and does have generality to spontaneous facial behavior" (*EHF*, 21).

Ekman granted that posed and spontaneous expressions were not completely identical. Rather, as he explained, posed expressions were similar in appearance to "that spontaneous behavior which is of extreme intensity and unmodulated, although it may differ in onset, duration, and decay time."[35] According to Ekman, if posed expressions do look a bit strange, or odd, or extreme, this is just because they are what spontaneous expressions would look like if they were not fixed in time, or not blends, or not altered or modified ("modulated" or "moderated") by culturally inflected display rules. His idea seems to have been that social constraints suppress or damp down or deflect our facial expressions from conveying the inward "truth" of our feelings, and that the emphaticness and single-mindedness of posed expressions (that is, their exaggerated, caricatural, even comic-book quality) are the outcome of the successful avoidance or

34. P. Ekman, E, R. Sorenson, and W. V. Friesen, "Pan-Cultural Elements in Facial Displays of Emotion," *Science* 164 (1969): 86.

35. P. Ekman and W. V. Friesen, "Constants Across Cultures in the Face and Emotion," *Journal of Personality and Social Psychology* 17 (1971): 128, n. 6. Or, as Ekman et al. also observed: "[P]osed behavior differs little from spontaneous facial behavior in form . . . Posed behavior differs from spontaneous facial behavior in duration, in the lack of attempt to control or otherwise moderate the behavior, and in the frequency of single emotion faces as compared to blend faces" (Ekman, Friesen, and Tomkins, "Facial Affect Scoring Technique: A First Validity Study," 53).

outflanking or overcoming of that suppressive effect, though it is not clear how that feat is managed. For him, the tendency of such expressions to look exaggerated is not the result of convention but exactly the opposite: the absence of those conventions that in everyday life constrain or moderate or mask behavior. In effect, Ekman weirdly claimed that *posed expressions in their very caricatural intensity are among the best examples we have of what we would look like if we were entirely alone.* "Our view is that posed facial behavior is similar to, if perhaps an exaggeration of, those spontaneous facial behaviors which are shown when the display rules to deintensify or mask emotion are not applied . . . Posed behavior is thus an approximation of the facial behavior which spontaneously occurs when people are making little attempt to manage the facial appearance associated with intense emotion" (*EHF*, 167). It followed for Ekman that when an investigator asks someone to pose an emotion, all he is doing is implicitly requesting the subject to show that emotion without attempting to "deintensify, mask, or neutralize" his facial appearance as the display rules would normally require.[36] The subject simply interprets the instruction as an "occasion to display an uncontrolled version of the emotion" (*EHF*, 167). But of course how this is supposed to work is a mystery: it's as if Ekman imagined that the poser ordinarily follows a set of explicit rules and conventions about how he or she is meant to act but is able to suspend these at will when asked to do so—or indeed when asked to pose!

But Ekman also dealt with the problem of posing in a different way. According to what I am calling his *second response*, and in an apparently contradictory gesture, he admitted that the subject's consciousness of being beheld when posing for the camera inevitably introduced what was, according to the logic of his project, a problematic component of conventionality. Ekman not only knew and accepted this, he even used the issue of the subject's awareness of being observed in certain experiments as a reason for rejecting some of his predecessors' findings in the field. He also recognized that so-called "candid" photos, widely understood as solving the problem of posing because the subject was assumed to be oblivious of the camera, might not be so candid after all. In other words, he conceded what he elsewhere denied, that posed facial expressions are conventionalized behaviors.

We can track Ekman's arguments on this point by noting that in 1940 a psychologist named Norman L. Munn had objected to some experiments by Carney Landis on the grounds that what the latter had taken to be spontaneous emotional expressions of his subjects responding to various

36. Ibid., 106.

situations created in the laboratory, such as witnessing a rat having its head cut off, might not have been so spontaneous after all because the subjects had been filmed and therefore might have been "self-conscious, and hence not so spontaneously expressive as under more natural circumstances."[37] This is an objection with which Ekman agreed, and he did so precisely on the grounds that Landis's subjects knew they were being photographed. As Ekman wrote:

> A number of aspects of the experimental setting indicate the operation of display rules either to neutralize the facial responses or to mask them with a positive affect. All of Landis' subjects knew him; most were psychologists who had had other laboratory experiences. Not only did they know they were being photographed, but, because Landis had marked their faces with burnt cork in order to measure the components of facial behavior in his other use of these records, they knew Landis was interested in their facial behavior. (*EHF*, 81)

In short, Ekman here endorsed the idea that in posed expressions the subject's awareness of being photographed was likely to induce the operation of masking conventions or "display rules"—the very idea that according to his first response to the problem of posing he appeared to reject.

In order to meet the criterion of "spontaneity without conventionalization," Munn had decided that "candid" photographs might do the job.[38] Ideal material of this kind, he felt, would comprise candid shots of the same persons in a variety of emotional situations, together with verbal reports of the emotions they had experienced in each case. Munn remarked, though, that to obtain many different expressions of the same individual under such conditions would be practically impossible because "the subject, aware of the photographer's purpose, would be no more naturally expressive than are subjects in the laboratory."[39] He therefore resorted to the use of "candid-camera" images published in *Life* and other journals of the day. He characterized such candid shots as "obviously unposed emotional expressions" and reported significant agreement among experimental subjects concerning the emotions those candid pictures were

37. Norman L. Munn, "The Effect of Knowledge of the Situation Upon Judgment of Emotion from Facial Expressions," *Journal of Abnormal and Social Psychology* 35 (1940): 325.

38. The phrase "spontaneity without conventionalization" is William A. Hunt's in his discussion of Munn's work in "Recent Developments in the Field of Emotion," *Psychological Bulletin* 38 (1941): 261.

39. Munn, "The Effect of Knowledge of the Situation Upon Judgment of Emotion from Facial Expressions," 326.

thought to express.[40] But Munn's solution to the problem of posing was soon held to be no solution at all when in 1941 Hunt complained that most of Munn's "candid" photographs involved social situations where conventionalization could be expected. "Even the man holding the drowned person's hand," he commented, "seems aware of the camera."[41] He therefore dismissed findings based on the use of posed expressions that appeared to support the thesis of the universal recognition of emotions. In Hunt's view, emotions were not defined by distinct, universal facial expressions or even distinct physiological responses but were situational reactions involving a complex, learned relationship between the individual and his environment.

To Ekman, Hunt epitomized everything that was wrong with the culturalists' approach to the emotions and facial expression. Nevertheless—and this is what I want to emphasize—he did not question the validity of Hunt's criticisms of the candid shot. On the contrary, Ekman acknowledged as a methodological limitation in Munn's experiments the fact that, as he observed, "the behavior studied (candid photographs taken from magazines) may not all actually have been spontaneous. The person shown in the photographs may have been aware of the photographer, or, even worse, might have completely re-enacted or staged the behavior for the press" (*EHF*, 94). With this admission one might have expected Ekman to reconsider his use of posed expressions in his research. But he did not. Instead, he sidestepped the difficulty, essentially by changing the topic. That is, he proceeded as if after all the real problem was not that of posing *per se*, but the merely methodological-technical one of the representativeness of findings based on the use of posed expressions.

The result was a tendency on Ekman's part to divert attention from the problem of posing, and to justify the continued use of posed expressions in experiments on emotion as long as the latter were carried out with the appropriate methodological safeguards. For example, Ekman insisted that it was important to make sure that adequate sampling procedures had been followed in selecting photos for use in experiments of this kind, and that the investigator did not rely on an atypical group of persons to pose expressions, such as those who, because they were "highly extroverted" or specially trained, were particularly good at producing what he continued to call "spontaneous" facial behavior—as if the problem of posing could be solved by avoiding the use of professional actors in experiments of this kind (*EHF*, 95). These suggestions for good experimental

40. Ibid.
41. Hunt, "Recent Developments in the Field of Emotion," 261.

practice notwithstanding, he continued to use trained actors in his own subsequent research.

Above all, Ekman returned to the (by-now-familiar) argument that posed behavior was not a product of social learning because he had demonstrated that posed facial expressions could be accurately judged across cultures (*EHF*, 105–6). In other words, Ekman's conclusion once again was that posed expressions are simply slightly exaggerated because they are unmodulated or uncontrolled versions of expressions of the kind that occur naturally when subjects are alone.

There would come a moment when Ekman had to respond to criticism on precisely this point. In 1975, in a scathing review of his work, the anthropologist Margaret Mead—no lover of the universalist thesis—protested that all that Ekman appeared to have demonstrated with his cross-cultural studies was that simulated expressions of emotion—by which she meant pantomimed, or highly theatricalized, facial movements—could be recognized across cultures.[42] This was not a trivial rebuke in light of Ekman's commitment to distinguishing between natural and artificial expressions and hence to isolating authentic faces hidden behind the codes of convention. Nor did Ekman dismiss her objection.[43] On the contrary, he ac-

42. "Margaret Mead Calls 'Discipline-Centric' Approach to Research an 'Example of the Appalling State of the Human Sciences,'" review of *Darwin and Facial Expression: A Century of Research in Review*, ed. P. Ekman (1973), *Journal of Communication* 25 (10) (1975): 212. This left open the question of why, if Ekman's experimental findings could be trusted, everybody, regardless of cultural origin, could be persuaded to recognize such "simulated" or theatricalized expressions. Mead was prepared to entertain the idea that there must be some universal, presumably innate, element involved, perhaps of the kind Ekman had postulated between the nervous system and specific facial muscles. Ekman for his part expressed surprise that Mead was willing to suggest that "what is innate would be apparent only in simulation and not in actual emotional experience" and suggested that the answer must come from data on spontaneous expressions of the kind he had first reported in his hidden-camera study of American and Japanese students (below). P. Ekman, "Biological and Cultural Contributions to Body and Facial Movement," in *The Anthropology of the Body*, ed. J. Blacking (London, 1977), 69.

43. For Ekman's account of his difficult relations with Mead, see also his afterword, where he characterized Mead's review of his 1973 book as a "denunciation" ("A," 364). Mead herself adopted a communicative analysis of facial expression, with a debt to cybernetics and "kinesics" (the study of body movements as forms of nonverbal communication), an analysis that now appears closer to the more recent "paralanguage" theories of facial communication advocated by Fridlund than to Ekman's basic emotions approach (see chapter 5). Mead's position on this topic was closely aligned with the work of the anthropologist Ray Birdwhistell (1918–1994), who denied the existence of universals in facial and other bodily gestures and movements. Birdwhistell empha-

knowledged as a drawback of his cross-cultural studies precisely that the expressions he had studied were not authentic, or the real thing.[44] But Ekman had a solution to the difficulty. The physiognomist Lavater in the passage I quoted earlier had remarked: "But my observations must be exact, they must be repeated and tested often. How can that be possible if I have to make these observations on the sly? . . . [A]nyone who notices that he or she is being observed either puts up resistance or dissimulates." In response to the problem of posing raised by Mead, Ekman now pointed to the results of an experiment he had already carried out using a device for recording facial expressions "on the sly" unavailable to Lavater: *the device of the hidden camera*. By secretly recording the emotional expressions of solitary individuals as they watched stressful films, Ekman felt he had discovered the key to detecting authentic, spontaneous expressions. As he replied to Mead's objection to his use of posed expressions: "The next type of research design answered this criticism."[45] Or, as he also stated: "We avoided the display rule pitfall by videotaping when the subjects thought they were completely alone and unobserved."[46] It is to Ekman's "next type of research design" involving the use of a hidden camera that I now turn.

SPONTANEOUS EXPRESSIONS

As we shall see in chapter 3, starting in the mid-1960s, in an interesting and important series of research studies, psychologist Richard S. Lazarus

---

sized the need to study body movement as a "language" of communication between people, using film and videotape as methods of recording in naturalistic settings and without experimental intervention or manipulation.

44. "A limitation of these cross-cultural experiments is that the facial expressions presented were not genuine but posed by subjects instructed to show a particular emotion or to move particular facial muscles. One interpreter of this literature [referring to Mead] suggested that universality in judgments of facial expression might be limited to just such stereotyped, posed expressions" (Ekman and Oster, "Facial Expressions of Emotion," 530). Again in 1987 Ekman acknowledged as one of three limitations of his cross-cultural studies the fact that "the facial expressions were posed, and Mead (1975) argued that establishing that posed expressions are universal need not imply that spontaneous facial expressions of emotion are universal" (P. Ekman et al., "Universals and Cultural Differences in the Judgments of Facial Expressions of Emotion," *Journal of Personality and Social Psychology* 53 [4] [1987]: 713).

45. Ekman et al., "Universals and Cultural Processes in the Judgments of Facial Expressions of Emotion," 713.

46. Ekman, "Biological and Cultural Contributions to Body and Facial Movement in the Expression of the Emotions," 69. In 1980 Ekman republished his paper in *Explaining Emotions*, ed. A. O. Rorty, 73–101, but placed his criticisms of Mead at 100, n. 7.

began using films in a systematic way in order to evaluate and measure the response of various subjects to stress. One film used extensively by him for this purpose was called *Subincision*. It was a short, silent, black-and-white anthropological film (with a running length of 17 minutes) that showed naked Aborigines in the Australian bush undergoing puberty initiation rites involving scenes of extensive cutting of the penis with sharp stones, bleeding wounds, and the adolescents wincing and writhing in pain. The boys appeared to volunteer for the initiation procedure, which was carried out by older men.[47] Lazarus demonstrated that the film induced subjective and autonomic signs of stress in a variety of student and other viewers. But by varying the soundtrack of the film in ways designed to minimize, or alternatively, to enhance the threatening content, Lazarus was also able to show that the film became less or more disturbing to watch. He argued on the basis of these results and related considerations that the film's threat could not be considered simply "out there as an attribute of the stimulus" but depended on the viewer's appraisal process, which is to say, on the "person's appraisal of the meaning of the stimulus for the thwarting of motives of importance to him." Thus, the same stimulus could be threatening or not, depending on the "interpretation the person makes concerning its future personal significance."[48] In terms of the opposition between intentionalist and non-intentionalist accounts of the emotions with which I began this chapter, Lazarus was on the intentionalist side.[49]

In spite of their apparently different theoretical orientations, Lazarus lent Ekman considerable research help when the latter decided to make use of *Subincision* and other stress films to test his ideas about the uni-

47. A few years earlier, *Subincision* had been used to measure the level of castration fear and anxiety it aroused in viewers; in Lazarus's work, the emphasis on castration largely disappeared and the film was treated instead as one among others (such as workshop accidents) depicting "bodily mutilation" that could be used to measure stress; in Ekman's work, the emphasis shifted yet again from the study of stress to the study of emotions. It is a sign of another cultural shift that when I recently attempted to locate a copy of this film through the Internet, I discovered that the term "subincision," which had formerly referred to a somewhat esoteric Aborigine adolescent initiation ritual held to be deeply upsetting if not traumatic for Western viewers, now chiefly refers to a popular form of penile tattooing.

48. Richard Lazarus, "A Laboratory Approach to the Dynamics of Psychological Stress," *American Psychologist* 19 (1966): 200.

49. For a discussion of the use of this film by Lazarus and Mardi J. Horowitz to study stress and of the question of the image, violence, and appraisal in the development of the diagnosis of Post-Traumatic Stress Disorder, see Leys, *From Guilt to Shame: Auschwitz and After* (Princeton, 2007), 93–122.

versality of emotions. In a variant of Lazarus's experiments, Ekman and Friesen used a hidden camera to secretly videotape the facial expressions of American and Japanese students as they watched brief sections of *Subincision* and three other stress-inducing or "neutral" films when each was alone in the viewing room. Japan was selected for comparison with the United States, Ekman explained, because of the common belief that Japanese facial behavior was sufficiently different from that of Americans to mitigate against finding universals (*EHF*, 163). Ekman reported on the basis of his findings that the responses of American and Japanese students to the stress films were very similar. But according to him, when an "authority figure" from the student's own culture (actually a graduate assistant dressed in a white coat) was introduced into the room and interviewed the student about his feelings while the latter was viewing additional stress material, the facial behavior of the Americans and Japanese diverged. Ekman stated that the Japanese students masked their negative feelings about the stress films more than the Americans when in the presence of the authority figure. Slow-motion videotape analysis allowing for direct scoring of the facial movements according to his Facial Action Scoring Technique, or FAST, Ekman declared, demonstrated at a micro level the occurrence of the Japanese students' spontaneous negative emotional expressions before these were covered over by the culturally determined display rules controlling for false, polite smiles (surprisingly, no video frames showing this transition have ever been published). Ekman thus claimed to have proved that the universal, biologically based feelings remained intact behind the culturally determined behavior, and hence to have demonstrated the validity of his neurocultural theory of the emotions.[50]

Ekman's study of apparently spontaneous expressions in American and Japanese students and their inhibition by display rules, presented by him as an answer to Mead's worries about the problem of posing, became canonical in the field. Tomkins hailed Ekman's study as a "brilliant experiment" demonstrating the complex relationships between pancultural universalities and "relativistic transformations," and the experiment soon

50. Over the years Ekman has given several descriptions of this study. The basic sources are Ekman, "Universals and Cultural Differences in Facial Expressions of Emotion"; Ekman, *Darwin and Facial Expression: A Century of Research in Review*, 215–18; and W. V. Friesen, *Cultural Differences in Facial Expressions in a Social Situation: An Experimental Test of the Concept of Display Rules*, unpublished doctoral dissertation, University of California–San Francisco, University of California Microfilm Archives (1972). See also chapter 5 for a further discussion of this iconic experiment.

began to be routinely cited in works on emotion as decisive evidence in favor of the Tomkins–Ekman–Izard approach to the emotions.[51] As reported by James Russell and José-Miguel Fernández-Dols, by 1980 research on the face was dominated by the assumptions of Ekman's neurocultural theory. Russell and Fernández-Dols added in this regard that the influence of the Basic Emotion Theory and research program has been "very great":

> Its assumptions appear in important theories of emotion (Damasio, 1994; Oatley, 1992). Its language is implicit in psychologists' discourse. Facial expressions are named by the specific "basic" emotion allegedly expressed (a "surprise" face or the "facial expression of anger"). When experimental participants select the predicted name, they are said to have "recognized" the facial expression; they are "accurate" or "correct"; those who select a qualitatively different term are said to have made an "error." In studies of the important question of autonomic differentiation of the emotions, Levenson (1992) used the directed facial action task in which discrete emotions were claimed to be induced by the creation of the corresponding facial expression . . . Although alternative conceptualizations of the link between emotion and faces exist, by the 1980s, it was the work and conclusions of the Facial Expressions Program [the authors' name for what now tends to be called the Basic Emotion Theory] that were presented to undergraduates in their textbooks. Advocates found that psychologists had accepted their conclusions . . . Even critics . . . stated: "We do not (and did not) dispute the fact that there are universal facial expressions associated with certain emotions."[52]

A key event in Ekman's rise to dominance occurred in the 1980s when he replaced the anthropologist Ray Birdwhistell (1918–1994) in the decisive role of arbiter of the National Institute of Mental Health grants for research on nonverbal communication. Birdwhistell had denied the existence of universals in facial and other bodily gestures. Instead, he had emphasized the need to study body movement as a "language" of communication between people, using film and videotape as methods of re-

51. S. S. Tomkins, "The Quest for Primary Motives: Biography and Autobiography of an Idea," *Journal of Personality and Social Psychology* 41 (2) (1981): 313. As I noted in my introduction, the influential emotion theorist Paul E. Griffiths also endorsed Ekman's experiment in *What Emotions Really Are: The Problem of Psychological Categories* (Chicago, 1997), 53–54.

52. James A. Russell and José-Miguel Fernández-Dols, "What Does a Facial Expression Mean?," in *The Psychology of Facial Expression*, eds. J. A. Russell and J. M. Fernández-Dols (Cambridge, 1997), 10.

cording in naturalistic settings and without experimental intervention or manipulation. His concern with methodology, especially the sensitivity of subjects to being observed, had led him to rule out having the cameraman present when participants were being filmed. But, as reported by Martha Davis, a member of Birdwhistell's research circle, by the end of the 1970s Ekman's attempts to apply traditional experimental methods to the study of nonverbal behavior had gained traction, not least because the micro studies of behavior of the kind Birdwhistell advocated took enormous lengths of time compared to Ekman's quicker, more economical procedures. Ekman's triumph over Birdwhistell—a triumph at once methodological, intellectual, and institutional—helped seal the destiny of emotion research in the United States for the next several decades.[53]

Nevertheless, as I have already indicated, in the late 1980s Ekman's success in setting the agenda for research on the emotions in psychology had already begun to be called into question.[54] In particular, as we shall see in chapter 5, Alan Fridlund—Ekman's former student—in a series of papers and in a book published in 1994 argued that the account given by Ekman and Friesen over the years of their Japanese-American experiment was inaccurate, and that their interpretation of the results in terms of the opposition between genuine emotional expressions versus culturally coded display rules was unsupportable.

Central to Fridlund's damning assessment was his rejection of Ekman's fundamental suggestion that the faces people make when they are alone are authentic signs of the truth of their inner emotional states. He cited Ekman's follower, Ross Buck, as putting Ekman's position this way: "When a sender is alone . . . he or she should feel little pressure to present a proper image to others, and any emotion expression under such circumstances should be more likely to reflect an actual motivational/emotional state."[55] Ekman made this same assumption in his crucial experiment on the differences between Japanese and American facial displays, stating of the results of that investigation: "In this single experiment we had shown how facial expressions are both universal and cultural. In private, when no display rules to mask expression were operative, we saw the biologi-

53. Martha Davis, "Film Projectors as Microscopes: Ray L. Birdwhistell and Microanalysis of Interaction (1955–1975)," *Visual Anthropology Review* 17 (2) (Fall–Winter, 2001–2002): 39–49.

54. In 1990, Ortony and Turner predicted that the basic emotions approach would not lead to significant progress in the field. See A. Ortony and T. J. Turner, "What's So Basic About Basic Emotions?"

55. R. Buck, *The Communication of Emotion* (New York, 1984), 20; cited in Fridlund, *HFE,* 158.

cally based, evolved, universal facial expressions of emotion. In a social situation, we had shown how different rules about the management of expression led to culturally different facial expressions."[56] Or, as Ekman likewise declared, "facial expressions of emotion do occur when people are alone . . . and contradict the theoretical proposals of those who view expressions solely as social signals."[57]

Ekman's presupposition that the truth shows itself on the face when someone believes he or she is alone is the same assumption that, as I earlier remarked, motivated Lavater's physiognomic project. It is a postulation that depends on the claim that a distinction can be strictly maintained between authentic and artificial signs, between nature and culture. But Fridlund rejected that claim, offering instead a "Behavioral Ecology View" of facial displays that stressed the implicit sociality of even so-called solitary facial movements. He treated the differences observed between the facial displays of the Japanese and American students in Ekman's well-known experiment as cultural differences in the management of facial behavior (*HFE*, 285–93).

Also in 1994, in a masterly assessment of the cross-cultural facial judgment or recognition experiments reported by Ekman and his colleagues, James A. Russell demonstrated that the results were artifactual, depending on forced-choice response formats and other problematic methods, such as within-subject experimental design, lack of variability in the order or stimulus presentation, and the use of posed expressions, which begged the questions to be proved in ways that fundamentally undermined Ekman's claims for the universal nature of the emotions.[58]

56. P. Ekman, "Expression and the Nature of Emotion," in *Approaches to Emotion*, eds. P. Ekman and L. Scherer (Hillsdale, NJ, 1984), 321.

57. P. Ekman, R. J. Davidson, and W. V. Friesen, "The Duchenne Smile: Emotional Expression and Brain Physiology II," *Journal of Personality and Social Psychology* 58 (1990): 351.

58. Alan J. Fridlund, *Human Facial Expression: An Evolutionary View* (San Diego, CA, 1994); hereafter abbreviated as *HFE*. J. A. Russell, "Is There Universal Expression of Emotion from Facial Expression? A Review of the Cross-Cultural Studies," *Psychological Bulletin* 115 (1994): 102–41 (reprinted as chapter 9, "Cross-Cultural Studies of Facial Expressions of Emotion," in Fridlund, *HFE*, 188–268). Ekman replied to Russell's criticisms in "Strong Evidence for Universals in Facial Expression: Reply to Russell's Mistaken Critique," *Psychological Bulletin* 115 (1994): 268–87; and Russell replied in turn in "Facial Expressions of Emotion: What Lies Beyond Minimal Universality?," *Psychological Bulletin* 118 (1995): 378–81. See also Russell and Fernández-Dols, *The Psychology of Facial Expression*, for a useful overview of the issues at stake in this debate; and Pamela J. Naab and James A. Russell, "Judgments of Emotion from Spontaneous Facial Expressions of New Guineans," *Emotion* 7 (2007): 736–44, in which the authors examined the recognition of

The net result of Fridlund's and Russell's analyses was to dramatically challenge the empirical and theoretical validity of the Tomkins–Ekman research program. Building on the work of Fridlund, Russell, and others, starting in the late 1990s Lisa Feldman Barrett went on to publish an impressive series of reviews of the growing empirical evidence that was inconsistent with the idea that there are six or seven or eight basic emotions in nature. She has come to the conclusion that the emotional categories posited by Ekman do not have an ontological status that can support induction and scientific generalization or allow for the accumulation of knowledge.[59] The consensus among this group of well-informed critics is that a new scientific paradigm for research on the emotions is needed.

A constant motif in critiques of Ekman's work has been the question of posing. It is sometimes said in this connection that, because the expressions on the faces of people who pose for the camera are artificial or exaggerated and do not appear in real life, the results of experiments based on such pictures lack "ecological relevance" or "ecological validity." The argument is that such posed expressions are caricatures that, unlike "prototypical" expressions that are close to the average set of features for a naturally occurring emotion, are exaggerated expressions designed to maximally distinguish each hypothesized emotion category from the other. Evidence indicates that such caricaturized expressions are easier to categorize than prototypical ones when the categories in question are otherwise hard to distinguish.[60]

The ways in which these posed expressions have been produced have also come under scrutiny. Fridlund has offered a dramaturgical analysis of experiments involving the use of posed expressions, suggesting that

---

emotion in twenty "spontaneous" expressions from Papua New Guinea, previously photographed, coded, and labeled by Ekman in 1980. The results showed that spontaneous expressions do not achieve the level of recognition obtained by posed expressions.

59. I have found the following review articles especially useful: Lisa Feldman Barrett, "Are Emotions Natural Kinds?," *Perspectives on Psychological Science* 1 (2006): 28–58; Barrett, "Solving the Emotion Paradox: Categorization and the Experience of Emotion," *Personality and Social Psychology Review* 10 (1) (2006): 20–46; Barrett, K. A. Lindquist, E. Bliss-Moreau, S. Duncan, M. Gendron, J. Mize, and L. Brennan, "Of Mice and Men: Natural Kinds of Emotions in the Mammalian Brain? A Response to Panksepp and Izard," *Perspectives on Psychological Science* 2 (2007): 297–312; and Barrett, K. N. Oscher, and J. J. Gross, "On the Automaticity of Emotion," in J. Bargh, ed., *Social Psychology and the Unconscious: The Automaticity of Higher Mental Processes* (New York, 2007), 173–217. For further comments on Barrett's "psychological constructionism" and theory of "core affect" as an alternative to the natural kinds approach to the affects proposed by Ekman, see my chapter 6 and epilogue.

60. Barrett, "Are Emotions Natural Kinds?," 38.

those experiments are social scenarios in which the investigator implicitly functions as a director and the subject as a Stanislavskian actor (or "method actor") who "'slips into role" (*HFE*, 179). In a somewhat similar fashion, Russell has observed that Ekman's universality thesis is not, or at least not directly, about posed faces, and states: "Posed faces do not express the emotion of the poser, but what the posed chooses to pretend and in a manner most likely to be understood by the observer."[61] Russell here puts an emphasis on deliberate feigning that may be somewhat misplaced, for he implies that actors or posers are never caught up in the emotional feelings they are trying to represent on their faces. The philosopher Ian Hacking may have come nearer to the truth about this matter when he suggests that what is often involved in posing is a subtle form of unconscious compliance. In a review of Ekman's 1998 edition of Darwin's *The Expression of Emotions in Man and Animals*, he has remarked of Darwin's and many of Ekman's choices of illustrative faces that they are

> quite extraordinary social documents. I am not sure that I have seen anyone in real life looking like any of these people. One can wonder if all three groups, the American students, the wild men of New Guinea and the investigators, were not collaborating to generate the phenomena. These experiments may serve as instances of another human need, namely, wanting to please, wanting to get along. This is not simulation: if you are an experimental subject, you do not behave as you do because the scientist is boss, but because you are in a situation where you feel good about accommodating to his wishes, just as he feels good about accommodating to yours.[62]

Hacking here points to the role played by unconscious demands and expectations in the production of facial expressions in terms that are compatible with Fridlund's dramaturgical interpretation of facial displays as determined by (conscious and unconscious) social motives. Seen in this light, Ekman's forced-choice labeling procedures, which oblige experimental subjects to identify emotions in photographs of posed expressions by selecting one label from a limited set of emotion terms, can be understood as methods designed to suppress the social-relational dimension involved in their making by "objectivizing" the emotions in the image and in the "correct" judgment of the objective truth of those representa-

61. Russell, "Is There Universal Expression of Emotion from Facial Expression? A Review of the Cross-Cultural Studies," 114.

62. Ian Hacking, "By What Links Are the Organs Excited?," *Times Literary Supplement*, July 17, 1998, 11.

tions. Russell has shown that when forced-choice response formats and the other methods advocated by Ekman are abandoned, the results are far less supportive, in fact are non-supportive, of Ekman's position. Indeed, Russell has demonstrated that if subjects are encouraged to give a description of such pictures of expression in their own words, they don't use categorical emotion terms but tend to produce a narrative account in terms of the situation they think or imagine is being enacted in the image.[63]

Yet, in spite of these criticisms—others could be cited—to this day Ekman's approach to the emotions continues to thrive. Why this should be so raises important questions about scientific influence, authority, and power. In the next section, I will briefly discuss some of Ekman's experiments and publications that have helped buttress his claims, before turning, in the final section, to the case described by Damasio and Adolphs of the patient with the damaged amygdala who cannot experience fear. In the course of my discussion, I hope to throw a critical light on how the Tomkins–Ekman paradigm has been and continues to be used to underwrite the scientific investigations of the emotions.

THE INNER AND OUTER MEANINGS OF FACIAL EXPRESSIONS

In 1976, in *Pictures of Affect*, Ekman and Friesen made available in slide form a set of black-and-white photographs of posed facial expressions. Confident that emotions are universally recognizable, they presented these pictures as prototypical expressions of six discrete affect categories. Two years later they published a new coding system, called the Facial Action Coding System (FACS), for measuring and analyzing facial movements. This coding system, which replaced the earlier FAST coding system and was based on years of arduous research, was designed to provide an atheoretical, anatomically based, standardized scoring system of the movements of the face that researchers could use to test their hypotheses about the relationship between emotion and facial expression. FACS was modeled on an earlier coding system made by the anatomist C. H. Hjortso. It is a "sterile," non-interpretive system that focuses on the visible changes (or "action units") caused by the underlying facial muscles.[64] The "theoretical" part of Ekman's coding system was and remains EMFACS (Emotion FACS), which makes predictions (or pronouncements) about which emotions are ex-

63. J. A. Russell, "Facial Expressions of Emotion: What Lies Beyond Minimal Universality?," 378–81.

64. P. Ekman and W. V. Friesen, *Pictures of Facial Affect* (Palo Alto, CA, 1976); P. Ekman and W. V. Friesen, *The Facial Action Coding System* (Palo Alto, CA, 1978).

pressed by which concatenation of action units. The risk has always been that the predictions in EMFACS will tend to get folded into FACS training, since Ekman believes that the truth of EMFACS has been definitively established. It is worth noting in this regard that the iterative process used by Ekman to select representative portrayals of emotions has not in fact yielded a signal for each emotion. Instead, for each emotion a range of signals achieves varying degrees of agreement. For example, in Ekman and Friesen's *Pictures of Facial Affect*, sixty-five different facial signals for anger are specified, yet as Russell and colleagues have observed, "[n]o theoretical rationale for this variety has been offered."[65]

Ekman's pictures of posed expressions and FACS have since been used in hundreds, if not thousands, of research studies, one reason for the continued success of his emotions paradigm.[66] It is impossible to discuss all of the many studies carried out by Ekman himself and his collaborators on the basis of posed expressions and FACS, but some of the general lines of his investigations can be briefly summarized. Thus, over the years he and his colleagues published several further studies of facial and other responses to stress films along the lines of his 1972 Japanese–American experiment on the spontaneous expression of emotion. For example, in 1980 in a partial replication of that 1972 study but without the cross-cultural comparison, Ekman, Friesen, and Ancoli used the new Facial Action Coding System, FACS, to code the secretly videotaped facial behavior of experimental subjects as they watched positive and stress-inducing film clips. In an improvement on the original experiment, the authors also obtained ratings from the subjects about their subjective experiences after watching the films. They claimed that facial expressions were closely correlated with the reported emotional feelings and thus provided accurate

65. James A. Russell, Jo-Anne Bachorowski, and José-Miguel Fernández-Dols, "Facial and Vocal Expressions of Emotion," *Annual Review of Psychology* 54 (2003): 333.

66. Since the publication of Ekman's *Pictures of Facial Affect*, various new sets of standardized posed expressions have been developed, using video clips and incorporating more diverse ethnic groups in order to provide greater ecological validity by offering more dynamic images and a greater variety of faces. On the one hand, these new methods are an advance over still photos because in ordinary life we don't react to still images. On the other hand, if the situations in which the video clips were made are not included in the data, it is all too easy for such video clips to be inserted into different experimental settings without regard to their contextual meanings. A mother might give a "flashbulb" smile if her child does something cute, and another person might give a slowly emerging warm smile to a partner. The differences between these smiles are lost when they are stripped of their contexts and used by basic emotion theorists as iconic "happiness facial expressions" regardless of the different situations that give them significance.

information about the person's specific subjective-affective state.[67] In a later variation of this experiment, in 1994 Rosenberg and Ekman improved their procedures with a method designed to obtain moment-to-moment reports of subjective responses to film stimuli and again claimed that the results supported the idea of a close correlation between facial expressions and specific emotions.[68] However, subsequently Russell, Fernández-Dols, Fridlund, and other researchers criticized these publications for various deficiencies in the experimental design and protocols, reporting of the findings, and the interpretation of the data proposed by Ekman and his colleagues.[69]

Ekman has also pursued the question of the relationship between the hypothesized basic emotions and central and autonomic nervous system activity, since the discovery, if true, that each primary affect is accompanied by specific patterns of physiological response would tremendously boost the foundational biological claims of the basic emotion theory. The topic has had a long history, originating in the controversy earlier in the century between William James, who had argued for specificity, and Walter Cannon, who had argued against. The issue was still unresolved when Ekman turned his attention to it, with the majority of psychologists in the 1960s favoring the idea of undifferentiated autonomic activity in emotions.[70] Using various manipulations to induce emotional responses and various measures of bodily changes, such as changes in heart rate, blood pressure, skin conductance, and micro movements of the face, Ekman and his colleagues carried out several studies of situation-specific variations in autonomic activity associated with specific emotions.

In one study by Ekman, Levenson, and Friesen, published in 1983, some scientists and trained actors serving as experimental subjects in a "directed facial action task" were asked to deliberately perform contractions of facial muscles thought to be involved in specific facial expressions. They were not asked to pose "emotional" expressions but instead were

67. P. Ekman, W. V. Friesen, and S. Ancoli, "Facial Signs of Emotion," *Journal of Personality and Social Psychology* 39 (6) (1980): 1125–34.

68. E. Rosenberg and P. Ekman, "Coherence Between Expressive and Experiential Systems in Emotion," *Cognition and Emotion* 8 (1994): 201–29.

69. See Fernández-Dols and Ruiz-Belda, "Spontaneous Facial Behavior During Intense Emotional Episodes: Artistic Truth and Optical Truth," 260–64; and Fridlund, *HFE*, 276–79.

70. W. James, "What Is an Emotion?," *Mind* 9 (1888): 188–205; W. B. Cannon, "The James-Lange Theory of Emotions: A Critical Examination and an Alternative Theory," *American Journal of Psychology* 39 (1927): 106–24; Ekman, "Expression and the Nature of Emotion," 324. Cf. Plamper, *The History of Emotions*, 191–92.

instructed to contract specific facial muscles in order to produce facial configurations that, according to theory, were held to be the universal signals of each of six target emotions. In order to pose the hypothesized prototypical facial expressions in the correct way, the subjects were provided with a mirror and a coach (Ekman himself). In the "relived emotion task," the same subjects were asked to re-experience each of the six basic emotions by reliving a past affective experience for thirty seconds. During both tasks, facial behavior was videotaped and changes in heart rate, skin temperature, skin resistance, and muscle tension were recorded using attached electrodes and other measurement techniques. After the "relived emotion" trial, subjects rated the intensity of any felt emotion on a scale from 0 to 8.

According to the authors, the hypothesis that there are autonomic differences among the six emotions was supported by the results. On both the directed facial action task and the relived emotion task, physiological activity distinguished not only between positive and negative emotions but also among four negative emotions. Ekman et al. concluded that "it was contracting the facial muscles into the universal emotion signals which brought forth the emotion-specific autonomic activity."[71] They took their findings to refute cognitivists, such as Schachter and Singer who, as we saw in chapter 1, presumed the existence of undifferentiated autonomic arousal in emotional states. Similar experiments, using the same "directed facial action" and "relived emotions task," were subsequently carried out by Ekman, Levenson, and others, with comparable results.[72]

71. P. Ekman, R. W. Levenson, and W. V. Friesen, "Autonomic Nervous System Activity Distinguishes Among Emotions," *Science* 221 (1983): 1208–10.

72. R. W. Levenson, P. Ekman, and W. V. Friesen, "Voluntary Facial Action Generates Emotion-Specific Autonomic Nervous System Activity," *Psychophysiology* 27 (1990): 363–84; Levenson, "Autonomic Nervous System Differences Among Emotions," *Psychological Science* 3 (1992): 23–27; Levenson, P. Ekman, K. Heider, and W. V. Friesen, "Emotion and Autonomic Nervous System Activity in the Minangkabau of West Sumatra," *Journal of Personality and Social Psychology* 62 (1992): 972–88; Levenson, "The Search for Autonomic Specificity," in *The Nature of Emotion. Fundamental Questions,* eds. P. Ekman and R. Davidson (New York, 1994), 252–57; Levenson, "Autonomic Specificity and Emotion," in *Handbook of Affective Sciences*, eds. R. J. Davidson, K. R. Scherer, and H. H. Goldsmith (New York, 2003), 212–24. For more recent claims about emotion-specific physiology, see Dacher Keltner and Daniel T. Cordaro, "Understanding Multimodal Emotional Expressions: Recent Advances in Basic Emotion Theory," *Emotion Researcher*, ed. Andrea Scarantino, http://emotionresearcher.com/understanding-multimodal-emotional-expressions-recent-advances-in-basic-emotion-theory/, accessed December 29, 2015, also discussed in my epilogue.

On the basis of such studies, in the 1990s Ekman included "distinctive physiology" among the defining characteristics of the basic emotions.[73]

It is not difficult to appreciate, however, that in spite of the widespread acceptance of the validity of experiments of this kind, they involved several confounds that call the findings into question.[74] For one thing, as Fridlund would later object, in their 1983 experiment Ekman, Levenson, and Friesen failed to control for physiological changes due to the general arousal resulting from the facial exertions required to hold the poses (*HFE*, 170)—a point that Ekman himself conceded. Another criticism raised the by-now-familiar problem of the use of posed facial displays. Thus, Frid-

73. P. Ekman, "An Argument for Basic Emotions," *Cognition and Emotion* 6 (3–4) (1992): 169–200. The list of basic emotion characteristics Ekman proposed at the time were: 1. Distinctive universal signs. 2. Presence in other primates. 3. Distinctive physiology. 4. Distinctive universals in antecedent events. 5. Coherence among emotional response. 6. Quick onset. 7. Brief duration. 8. Automatic appraisal. 9. Unbidden occurrence. Ekman's reference to "automatic appraisal" is worth noting. He first posited the existence of an "automatic appraisal mechanism" in 1977 when, without attempting to integrate the different theories of appraisal offered by Tomkins, Mandler, and Lazarus, he proposed that an automatic appraisal mechanism selectively attended to those internal and external stimuli that, operating without the individual's consciousness, immediately triggered the subcortical affect programs, thereby setting off emotion-specific changes in expression and physiology. He contrasted automatic appraisal with the more extended appraisal processes that had interested the cognitivists (P. Ekman, "Biological Contributions to Body and Facial Movement," 58–59). In 1984 he suggested that the conflict between Lazarus and Zajonc that had recently erupted over the role of cognition in emotion was due in part to the fact that the two men were focusing on these two different types of appraisal (P. Ekman, "Expression and the Nature of Emotion," in *Approaches to Emotion*, eds. P. Ekman and L. Scherer [New York, 1984], 338). In 1992 he explicitly aligned his views with those of Zajonc, Öhman, LeDoux, and others who minimized cognition in automatic appraisal and emphasized the relatively hardwired, species-typical character of emotional reactions. He drew attention in this regard to what he detected as a "major shift" in Lazarus's position as the latter attempted to incorporate evidence concerning the basic emotions into his work, citing Lazarus as stating that "'I distinguish between two modes of appraisal: one automatic, unreflective, and unconscious or preconscious, the other deliberate and conscious'" (P. Ekman, "An Argument for Basic Emotions," 188). Lazarus's evolving views about appraisal are discussed in my chapters 3 and 4.

74. This has not prevented emotion theorists and researchers from routinely citing the 1983 Ekman, Levenson, and Friesen study, and a related study of 1990, as providing good evidence for the existence of signature patterns of autonomic responses associated with specific basic emotions. See for example Paul E. Griffiths, *What Emotions Really Are*, 83; Craig DeLancey, *Passionate Engines: What Emotions Reveal about Mind and Artificial Intelligence* (Oxford, 2002), 11; Jesse Prinz, *Gut Reactions: A Perceptual Theory of Emotion* (Oxford, 2004), 73.

lund noted that the autonomic changes observed in Ekman et al.'s 1983 experiment might have occurred simply as a consequence of the subjects' preparation for intended actions regardless of their emotional state. Such intentional actions might include deceptive ones, suggesting that any associated autonomic changes might be more a function of the facial displays than putative internal emotional states. The appropriate control procedure would thus have required using trained actors briefed on the experiment and told to feign expressions without "real" emotion, in order to see if their poses produced similar physiological signs. Indeed, as Fridlund observed, Ekman et al.'s 1983 data even suggested this explanation, because the latter observed similar autonomic differences when the subjects were merely requested to perform certain facial actions without their being cued for emotion.

Fridlund therefore offered three possible explanations for the 1983 findings: 1. Like Stanislavskian or method actors who become caught up in their role, the actors in the experiment actually became emotional while performing the "directed facial action" mimicry task, thereby causing the associated physiological changes. Unfortunately, however, self-ratings of emotion for the requested facial action task were not reported. 2. The observed autonomic changes might have resulted simply from the actors being *asked* to make particular faces. 3. The autonomic changes might not have related to emotion but to facial displays as such. As Fridlund observed in this regard, this was indeed Ekman et al.'s interpretation of the directed facial action data, a finding that was compatible with Fridlund's view that facial displays served social motives and might occur regardless of a person's emotional condition. He concluded that because the study of induced emotion had not included a "pretend" control group, there was no evidence that emotion per se related to autonomic functioning. From his perspective, the implication could be stated simply: "whether an organism *is* angry—whatever one means by the term—should have less bearing on autonomic functioning than the fact that it *acts* angry" (*HFE*, 171).

More generally, Fridlund suggested that all experiments that requested subjects to perform simulated facial displays were problematic, because these requests were implicitly confounded with suggestions to subjects about how they should act. The experimenter functioned in this regard like a director, and the subject like an actor who performs a role. As Fridlund put it, "it is the role or 'set' taken in the given social context that determines the emotion, not the displays themselves, whether facial or otherwise" (*HFE*, 179). He noted that such a performative analysis was especially applicable to Ekman et al.'s 1983 experiment in which many of the subjects were indeed actors, and the directed facial actions were obtained by using a "coach" and a mirror. It is not obvious that subsequent experiments using actors to pose expressions have been able to avoid the problem of such confounds.[75]

Another topic that has preoccupied Ekman over the years is the question of lying. As I have observed, his neurocultural theory was and remains premised on the idea that a distinction can be made between the genuine, spontaneous, universal, and involuntary expressions we make when we are authentically emotionally aroused, and the facial displays we make when we deliberately feign, lie, or manage our expressions in order to conform to the public requirements of facial behavior and display. The claim has been that under certain conditions information from unmanaged emotions may leak out onto the face in the form of micro movements detectable by the naked eye by well-trained observers, a position that implies the possibility of developing new techniques for lie detection and surveillance.

An important moment in the development of Ekman's views in this connection occurred when he introduced a distinction between the so-called "Duchenne" smile versus the "non-Duchenne" smile, a distinction that goes back to the work of the nineteenth-century scientist Duchenne de Boulogne. As already discussed in chapter 1, a Duchenne or true smile of happiness only occurs when two different muscles contract simultaneously: the *zygomatic major* muscle pulling the lip corners upward toward the cheek bone, and the *orbicularis oculi* muscle, which raises the cheek and gathers the skin inward from around the eye socket in a wincing movement. Since contractions of the *orbicularis oculi* muscle are hard to control

75. Ekman argued that in his 1983 experiment he had eliminated the risk of confounding the results with extraneous affects, such as frustration or embarrassment, because he used trained actors and scientists from his laboratory who were comfortable having their faces filmed (Ekman et al., "Autonomic Nervous System Activity Distinguishes Among Emotions," 1208). Cf. Ekman, "Expression and the Nature of Emotions," 324–25. Ekman seemed to think that because the scientists and trained actors who served as his experimental subjects in his 1983 experiment were so accustomed to being filmed, they could be relied upon to represent or pose authentically pure emotional expressions on demand by performing as if no one was watching. He made the same claim in Levenson, Ekman, and Friesen, "Voluntary Facial Expression Generates Emotion-Specific Autonomic Nervous System Activity," 27 (1990): 363–84, which also used a mix or actors and emotion researchers as experimental subjects. But this merely asserted what had to be proved. Indeed, it is not clear what could count as evidence on this point. The facial expressions and other results obtained by the use of trained actors and scientists in such an experiment would have to be compared to the authentic expressions and displays produced by people whose status as individuals uncontaminated by social conventions (or indeed suggestions) would have to be beyond question. But how could that condition be achieved?

voluntarily, the claim is that voluntary or simulated smiles never exactly match true or involuntary smiles of enjoyment. On this basis, Ekman declared that there were measurable differences between felt and false smiles and that data from studies that failed to recognize the distinction between these two kinds of smile and instead treated smiles as a single class of behavior were therefore disqualified. (The disqualification of course also applies to any of Ekman's own earlier studies that had ignored the distinction, although he was not always completely frank about this fact.)[76]

Over the years Ekman and his colleagues have published several influential papers offering a theoretical rationale for those claims and some supporting evidence. Thus, in Ekman and Friesen's paper "Felt, False, and Miserable Smile" (1982), the authors distinguished "felt" smiles from "false" or "miserable" smiles:

Felt smiles were defined as "all smiles in which the person actually experiences, and presumably would report, a positive emotion." These are the involuntary, universal smiles of felt emotion that occur when people are genuinely happy. They were said to depend on the action of both the *zygomatic major* and the *orbicularis oculi* muscles in the solitary condition when the display rules for managing social expression in public are not in operation.

According to the authors, *false smiles* are smiles that are deliberately made to convince another person that positive emotion is felt when it is not: they are the smiles we perform as forms of address to others in accordance with cultural conventions or display rules. Ekman and Friesen further distinguished between two kinds of false smiles: in "phony smiles" nothing much is felt but an attempt is made to appear as if positive feelings are being experienced; in "masking smiles" strong negative emotion is felt but an attempt is made to conceal these negative feelings by appearing to feel positive. Ekman and Friesen observed that sometimes people may succeed in producing a false smile which cannot be distinguished from a felt smile. Phony smiles are more successful in this regard than masking ones, because in phony smiles no conflicting emotion interferes, whereas masking smiles conceal negative emotions and hence are likely to show leakage if the negative emotions are so strong as to leak out on the face in the form of subtle micro-movements detectable by FAC.

76. See P. Ekman, "Facial Expressions of Emotion: New Findings, New Questions," *Psychological Science* 3 (1992): 37.

Miserable smiles are different again: these deliberate smiles are not attempts to disguise negative emotions but rather serve to make it clear to the self and others that the person is unhappy but that negative feelings will be contained.[77]

In support of such views, in a paper of 1988, Ekman, Friesen, and O'Sullivan used FACS to examine the incidence of the Duchenne smile and the masking smile when subjects were truthfully describing positive emotions and when they feigned enjoyment to conceal strong negative feelings. The authors claimed that, as predicted, the Duchenne smile occurred "more often when people were actually enjoying themselves as compared with when enjoyment was feigned to conceal negative emotions."[78] In another experiment, published in 1990, subjects watched very short, silent color films intended to evoke positive or negative emotions while alone in the viewing room; the subjects believed the purpose of the experiment was to gauge their physiological responses, as recorded by EEG measuring devices, and were unaware that their faces were being videotaped by a hidden camera. The subjects' secretly recorded facial expressions were again scored with FACS, and the results were said to confirm that there were more Duchenne smiles during the positive films than during the negative films and that only Duchenne smiles were correlated with the subjects' experience of enjoyment.[79]

Many other experiments by Ekman and like-minded researchers, based on the distinction between Duchenne and non-Duchenne smiles, have been reported in the literature. But almost from the start several scientists have contested the results of such experiments and the premises held to be supported by them. Thus, in a review of the findings in 1997, Fernández-Dols and Ruiz-Belda commented: "Unfortunately, no necessary or sufficient link between happiness and Duchenne smiles has been substantiated by other researchers."[80] In their discussion, Fernández-Dols and Ruiz-Belda approached the topic of smiling from the perspective of

77. P. Ekman and W. V. Friesen, "Felt, False, and Miserable Smiles," *Journal of Nonverbal Behavior* 6 (1982): 238–52.

78. P. Ekman, W. V. Friesen, and M. O'Sullivan, "Smiles When Lying," *Journal of Personality and Social Psychology* 54 (1988): 414–20.

79. Paul Ekman, Richard J. Davidson, and Wallace V. Friesen, "The Duchenne Smile: Emotional Expression and Brain Physiology II," *Journal of Personality and Social Psychology* 58 (2) (1990): 342–53.

80. José-Miguel Fernández-Dols and María-Angeles Ruiz-Belda, "Spontaneous Facial Behavior During Intense Emotional Episodes: Artistic Truth and Optical Truth," in Russell and Fernández-Dols, *The Psychology of Facial Expression*, 262–63.

the sociality of emotional expressions, a perspective that had recently emerged as the central concern of Fridlund and others. According to such developing views, if it was the case, as the evidence was starting to suggest, that the social context determined smile behavior, including the production of so-called authentic or Duchenne smiles, then Ekman's distinction between the genuine emotional expressions we make when we are alone, and the phony or false expressions we make when we are with others because of the intervention of "display rules," was in jeopardy.

Curiously, it has hitherto has gone unnoticed that Ekman himself had begun to undermine that distinction in his own analysis of smiles. This is because in his and Friesen's 1982 discussion of miserable smiles—defined, as we have seen, as the kind of smiles people make when they are feeling miserable but wish to cover up their negative feelings by communicating the message to others that the negative feelings will be contained—the authors mentioned that they had first observed such smiles long before, in their earlier Japanese–American study. As they stated:

> We first began to think about miserable smiles when we noted (Ekman, 1972) that some subjects smile when watching films intended to induce stress. Since they were alone, and did not think anyone was watching or recording their facial expressions, it seemed unlikely that these could be masking smiles. The smiles did not look at all convincing. The people looked unhappy while they smiled.[81]

Indeed, when we go back to the 1972 report of that famous study of the spontaneous expressions of Japanese and American students' responses to stress films, we find the data showing not just negative expressions in response to the stress films, as expected, but some positive emotion faces in both American and Japanese students.[82] In their 1982 paper, Ekman and Friesen now identified those positive emotion faces as involving miserable smiles. The implication was that, according to Ekman himself, *miserable smiles—defined as intentional forms of communication and as such governed by display rules—occurred when the viewing subjects were* alone, yet the whole point of the famous Japanese–American experiment was to show the absence of display rules governing facial expression when subjects were in the alone or solitary condition. Similarly, in 1982 Ekman's former student and coauthor Joseph C. Hager commented on Ekman, Friesen, and Ancoli's 1980 experiment in which subjects were alone in the

81. Ekman and Friesen, "Felt, False, and Miserable Smiles," 249.

82. Ekman, "Universals and Cultural Differences in Facial Expressions of Emotion," 256.

viewing room and were not aware of being videotaped by observing that the absence of an audience "probably tended to reduce the influence of social display rules on muscle movements," but that "*these rules can operate habitually even when a person is alone*" (my emphasis).[83]

These are rather astonishing admissions concerning facial displays in the solitary and hence ostensibly display-rule-free condition, and I shall have more to say about their implications in later chapters. For the moment it suffices to note the fragility of Ekman's distinction between the authentic or spontaneous, involuntary signs of emotional arousal when we are in the alone state and the faces we make in public when we are influenced by display rules. In fact, these statements by Ekman and Hager in 1982 undo that distinction altogether, leaving it unclear how one might then determine when such solitary display rules are in effect and when they are not.[84] But I also want to stress that in making these statements Ekman and Hager do not seem to have been aware of having committed self-inflicted wounds. Not for the first time, they appear not to have realized that they had fallen into incoherence, but instead continued to adhere to the same neurocultural formulations that their own passing comments had essentially called into question. Thus, in spite of emerging criticisms, throughout the 1980s and 1990s Ekman and his colleagues continued to develop a formidable research program based on his assumptions and methods, including the use of his *Pictures of Facial Affect* and Facial Action Coding System, a research program that was so persuasive that it swept many of the most active and prominent affect researchers into its fold.

"HAVE NO FEAR"[85]

One scientist who has deployed Ekman's theoretical assumptions and methods to great effect is Antonio Damasio, arguably the best-known neuroscientist in the emotion field today. Damasio accepts the Tomkins–Ekman paradigm, at least for the basic emotions. He regards affective responses as biologically determined, adaptive processes that depend on innately set devices with a long evolutionary history. According to him,

83. Joseph C. Hager, "Asymmetries in Facial Expression," in *Emotion in the Human Face*, 2nd ed., ed. P. Ekman (New York, 1982), 335–36.

84. A point stressed by Fridlund regarding a similar admission in a later text by Ekman to the effect that display rules function in the alone condition. On this point, see chapter 5.

85. Antonio Damasio, *The Feeling of What Happens: Body and Emotion in the Making of Consciousness* (New York, 1999), 62; hereafter abbreviated as *FWH*.

although culture and learning introduce individual variations, emotions are fundamentally stereotyped and automatic responses of the body and face which can occur automatically, without conscious deliberation. That is why, he explains, Darwin was able to catalog emotional expressions in humans and animals, and that is why in different parts of the world and across different cultures, emotions are so easily recognized. "The thing to marvel at, as you fly high above the planet," Damasio has observed, "is the similarity, not the difference," in facial expressions among people, a view he sees as having been given "immeasurable support" by the work of Ekman. The biological function of emotions is thus to produce specific reactions to events in a fast and exquisitely reliable, automatic way. "For certain classes of clearly dangerous or clearly valuable stimuli in the internal or external environment," Damasio writes, "evolution has assembled a matching answer in the form of emotion. This is why, in spite of the infinite variations to be found across cultures, among individuals, and over the course of a life span, we can predict with some success that certain stimuli will produce certain emotions. (This is why you can say to a colleague, 'Go tell her that; she will be so happy to hear it')" (*FWH*, 53–54). (The reader will recognize the similarity between Damasio's claim and DeLancey's suggestion, mentioned in the introduction (n. 15), that someone jumping for joy at receiving good news is performing a purely "expressive" or noncognitive reaction. But one might object to Damasio in the same way I earlier objected to DeLancey's claims, namely, that the person won't be happy or jump for joy unless she comprehends the news, and if she comprehends it she is doing cognition.)

In 1994 Damasio teamed up with Ralph Adolphs to study the case of a young woman, SM, whose brain scan revealed that she suffered from an extremely rare genetic disease producing in her case complete, bilateral damage to the amygdala. Since the work of Joseph LeDoux and others, the amygdala has been implicated in rapid emotional responses, especially the emotion of fear.[86] Adolphs and his colleagues reported that SM was normal in every way except for one strange symptom: she was unusually—that is to say, inappropriately, even excessively—forthcoming with people. Indeed, she so lacked normal reserve and reticence that she had often been taken advantage of by those she trusted. When tested on her ability to judge prototypical facial expressions of six basic emotions in pictures taken from Ekman and Friesen's *Pictures of Facial Affect* (1976), SM was unable to identify the expression of fear, as was demonstrated by the fact that

86. Joseph LeDoux, *The Emotional Brain: The Mysterious Underpinnings of Emotional Life* (New York, 1996).

her ratings intensity for fearful faces was low compared to those of normal subjects, although she had no difficulty identifying familiar faces.[87]

In a subsequent study, Adolphs and his colleagues showed that SM was unable to draw a face representing fear; she complained that "*she did not know what an afraid face would look like,* and that she was unable to draw any depiction of it," even though she could depict other emotions. Moreover, tests also showed that although SM did not lack the concept of fear, she seemed to lack the capacity to experience fear in a normal way, since she did not appear to feel frightened given the appropriate stimulus.[88] Damasio stated in his commentary on the case that SM's fearlessness, the result of bilateral damage to her amygdala, had prevented her from learning "the significance of the unpleasant situations that all of us have lived through. As a result she has not learned the telltale signs that announce possible danger and possible unpleasantness, especially as they show up in the face of another person or in a situation" (*FWH*, 66).

That this was true appeared to be demonstrated in yet another study. The experiment called for SM and two other patients with complete bilateral amygdala damage to rate 100 slides of facial expressions, again selected from Ekman and Friesen's collection of such posed expressions, slides that had previously been rated by a group of normal individuals as indicating various degrees of "trustworthiness" and "approachability." As Damasio reported, these normal subjects had been asked an apparently simple question: "How would you rate this face on a scale of one to five, relative to the trustworthiness and approachability that the owner of the face inspires? Or, in other words, how eager would you be to approach the person with this particular face if you needed help?" (*FWH*, 66). Fifty faces that were judged by normals as inspiring trust and fifty faces that were judged untrustworthy were then shown to SM and the two other patients. The scientists reported that although the latter were quite capable of judging trustworthy faces normally, they were severely impaired in their ability to judge untrustworthy or dangerous faces.[89] As Damasio

87. R. Adolphs, D. Tranel, H. Damasio, and A. Damasio, "Impaired Recognition of Emotion in Facial Expressions Following Bilateral Amygdala Damage to the Human Amygdala," *Nature* 372 (December 15, 1994): 669–72. The patient was originally identified as "SM," and this is how she is referred to in most subsequent publications, although in his book *The Feeling of What Happens*, Damasio calls the patient "S." I shall refer to her as SM.

88. Ralph Adolphs, Daniel Tranel, Hanna Damasio, and Antonio R. Damasio, "Fear and the Human Amygdala," *Journal of Neuroscience* 15 (9) (September 1995): 5879–87.

89. Ralph Adolphs, Daniel Tranel, and Antonio R. Damasio, "The Human Amygdala in Social Judgment," *Nature* 393 (June 1998): 470–73. See also Ralph Adolphs and Daniel

observed, they "looked at faces that you or I would consider trustworthy and classified them, quite correctly, as you and I would, as faces that one might approach in case of need. But when they looked at faces of which you and I would be suspicious, faces of persons that we would try to avoid, they judged them as equally trustworthy . . . Immersed in a secure Pollyanna world, these individuals cannot protect themselves against simple and not-so-simple social risks and are thus more vulnerable and less independent than we are" (*FWH*, 67).

Several comments are in order. The first point to notice is that in Damasio's narrative of the case, the concepts of "trustworthiness" and "untrustworthiness" have been stripped of all context in order to treat these traits as objective, identifiable features of persons that are immediately, universally, and unambiguously readable in the human face. The aim of Damasio's and his colleagues' experiment was not to assess the meaning certain facial expressions might have had for SM but to establish the difference between abnormal and normal subjects on the basis of an ostensibly objective standard. The posed facial expressions drawn from Ekman's *Pictures of Facial Affect*, with which SM was tested, were taken to provide that objective standard because they came with a predetermined "correct" estimation of danger. The assumption underlying the research project was that dangerousness or what is frightening inheres unambiguously in certain facial expressions that everyone has the ability to decipher, because that ability is an evolved skill shared by everyone with normal brain function.[90] Russell has shown that when forced-choice

---

Tranel, "Emotion Recognition and the Human Amygdala," in *The Amygdala: A Functional Analysis*, 2nd ed. (Oxford, 2000), 587–630. In these experiments on SM, ratings of approachability and trustworthiness were analyzed for the fifty faces to which normal controls had assigned the most negative ratings, and for the fifty most positive faces. Subjects with bilateral amygdala damage rated the fifty most negative faces more positively than did either normal controls or patients with unilateral amygdala damage. All subject groups gave similar ratings to the fifty most positive faces. On the basis of the results, the researchers suggested that "the human amygdala triggers socially and emotionally relevant information in response to visual stimuli. The amygdala's role appears to be of special importance for social judgment of faces that are normally classified as unapproachable and untrustworthy, consistent with the amygdala's demonstrated role in processing threatening and aversive stimuli" (Adolphs et al., "The Human Amygdala in Social Judgment," 472).

90. As two recent critiques of Damasio's report of the case of SM (or "S") and other neuroscientific studies of the relationship between facial expression and emotion make clear, when they emphasize the bracketing, or loss of the social-interpretive dimension of the emotions, in Ekman's standardized, image-based approach: Daniel Gross, *The Secret History of Emotion: From Aristotle's Rhetoric to Modern Brain Science*

response formats and other problematic experimental methods are abandoned, the results obtained fail to support the Tomkins–Ekman position on the universal recognition of facial expressions of emotion. Ignoring such criticisms, Damasio employed the Tomkins–Ekman paradigm in order to argue that SM can't feel fear because she has an abnormal amygdala and therefore makes errors when judging the dangerous faces she encounters. Her emotional deficit is that she cannot conform to a fixed standard or norm.

What interests me here is the kind of scientific object that fear is imagined to be when analyzed in these terms. The issue is not whether the amygdala plays a role in the fear response—it clearly does—but what kind of role it plays and how that role is to be conceptualized. Among the questions that arise are: Did Damasio and Adolphs make unwarranted assumptions in their analysis of the case? To what extent did their commitment to Ekman's taxonomic approach to the emotions and their use of the latter's pictures of emotional expression predetermine their findings? How specific is the amygdala's role in the fear response? Does other work on the amygdala show that its part in fear processing is somewhat different from what Damasio and Adolphs first proposed?

The answer to the last question now appears to be yes. Subsequent studies of SM by Adolphs in collaboration with Russell (the same investigator who has queried the validity of Ekman's cross-cultural judgment studies) and others have suggested that the amygdala is involved not in making categorical fear judgments but in something different, a judgment of arousal levels based on attention to features such as wide-open eyes.[91] Tests have shown that SM fails to look normally at the eyes in *all* fa-

---

(Chicago and London, 2006), 28–39; and John McClain Watson, "From Interpretation to Identification: A History of Facial Images in the Sciences of Emotion," *History of the Human Sciences* 17 (1) (2004): 29–51. The latter rightly observes that FACS images "enable researchers to introduce reliability and consistency into their investigation but it is important to remember that these advantages are procured only by bracketing the conceptual quandaries so thoroughly detailed by earlier researchers" (45).

91. Already in 1997, in a discussion of the existing literature on the amygdala's role in the evaluation of emotional stimuli, Elizabeth Phelps raised questions about some of the more extreme claims being made about amygdala-damaged patients, pointing out that such patients are different but that their social deficits are subtle and not as dramatic as had been suggested. She proposed that the amygdala seemed to be activated whenever stimuli were arousing. E. A. Phelps and A. K. Anderson, "Emotional Memory: What Does the Amygdala Do?," *Current Biology* 7 (1997): R312–14. See also Adam K. Anderson and Elizabeth A. Phelps, "Is the Human Amygdala Critical for the Subjective Experience of Emotion? Evidence of Intact Dispositional Affect in Patients with Amygdala Lesions," *Journal of Cognitive Neuroscience* 14 (5) (2002): 709–20; and

cial expressions and that if she is explicitly instructed to look at the eye region when performing an emotion detection task, her recognition of fear is normal. Her selective impairment in identifying fear thus appears to be due to the fact that her defective amygdala is unable to direct her attention to the wide-open eyes that are thought to characterize the expression of fear. Furthermore, neuroimaging studies have revealed that humans presented with pictures of fearful faces do not report feeling "afraid," yet amygdala activity is nevertheless altered, suggesting that reported emotion and amygdala activation should not be equated. (Indeed, amygdala activation is associated not just with negative emotions, such as fear, but with positive emotions as well.)

The strong inference from these and related experiments is that, rather than functioning as the site for the production of discrete emotional states, such as fear, the amygdala modulates the vigilance and arousal levels required to attend to especially ambiguous stimuli of relevance to the organism (for example, fearful faces that provide information about the presence of a threat but not the source of that threat). At the very least these experiments suggest that the amygdala is not the locus of a discrete fear "entity" but that both vigilance and emotions are processes set in motion by amygdala activation.[92] Indeed, in a 2013 review of the evidence,

---

Elizabeth A. Phelps, "Emotion and Cognition: Insights from Studies of the Human Amygdala," *Annual Review of Psychology* 57 (2006): 27–53.

92. In a very large literature in an ongoing debate about the role of the amygdala in the recognition and experience of fear, involving experiments using standardized emotional stimuli, such as Ekman's *Pictures of Facial Affect* as well as moving images of actors performing whole-body and other emotional gestures and actions, see Ralph Adolphs, James A. Russell, and Daniel Tranel, "A Role for the Human Amygdala in Recognizing Emotional Arousal from Unpleasant Stimuli," *Psychological Science* 10 (1999): 167–71; R. Adolphs, D. Tranel, S. Hamann, A. W. Young, A. J. Calder, E. A. Phelps, A. Anderson, G. P. Lee, and A. R. Damasio, "Recognition of Facial Emotion in Nine Individuals with Bilateral Amygdala Damage," *Neuropsychologia* 37 (1999): 1111–17; R. Adolphs, F. Gosselin, T. W. Buchanan, D. Tranel, P. D. Schyns, and A. R. Damasio, "A Mechanism for Impaired Fear Recognition After Amygdala Damage," *Nature* 433 (January 6, 2005): 68–72; Ralph Adolphs and Michael Spezio, "Role of the Amygdala in Processing Visual Stimuli," *Progress in Brian Research* 156 (2006): 363–78; D. Tranel, G. Gullickson, M. Koch, and R. Adolphs, "Altered Experience of Emotion Following Bilateral Amygdala Damage," *Cognitive Neuropsychiatry* 11 (3) (2006): 219–32; Michael L. Spezio, Po-Yin Samuel Hang, Fulvia Catelli, and Ralph Adolphs, "Amygdala Damage Impairs Eye Contact During Conversations with Real People," *Journal of Neuroscience* 27 (15) (2007): 3994–97; Ralph Adolphs, "Fear, Faces, and the Human Amygdala," *Current Opinion in Neurobiology* 18 (2008): 166–72; Matthias Gamer and Christian Buchel, "Amygdala Activation Predicts Gaze toward Fearful Eyes," *Journal of Neuroscience* 29 (28) (2009): 9123–26; Jorge L. Armony, "Current Research in Behavioral Neuroscience:

Whalen et al. have concluded that "the human amygdala is responsive to the potentially predictive value of all facial expressions, in general, consistent with data showing that increased amygdala activation is not restricted to threat-related information."[93] Moreover, in a dramatic development, the same year, Justin S. Feinstein et al. showed that the patient SM, and other amygdala-damaged patients, do experience fear and panic when inhaling thirty-five percent carbon dioxide. These findings overturn the conclusions drawn from the studies of SM carried out over a period of more than twenty years that the amygdala, which SM lacks, is essential to the experience of fear.[94]

On the basis of these newer findings, some scientists involved in research on the human amygdala have begun to question the general taxonomic-categorical approach to the emotions associated with the Tomkins–Ekman paradigm.[95] But this is not a direction Damasio and

---

The Role(s) of the Amygdala," *Emotion Review* 5 (1) (2013): 104–15; Andrea S. Heberlein and Anthony P. Atkinson, "Neuroscientific Evidence for Simulation and Shared Substrates in Emotion Recognition," *Emotion Review* 5 (1) (2013): 162–77. In a 2007 review of some of the more recent results, Heberlein and Adolphs commented: "[T[he amygdala cannot be conceived as a simple 'fear generator' or even as a 'fear processor'; it plays roles in both the experience and the recognition of a range of negative emotional stimuli, as well as in certain highly arousing positive stimuli . . . [C]onsistent findings of amygdala involvement in processing fear-related stimuli do not mean that it is *dedicated* to fear-related stimuli . . . [T]he amygdala, or parts thereof, acts as a detector of salience or relevance" (Andrea S. Heberlein and Ralph Adolphs, "Neurobiology of Emotional Recognition: Current Evidence for Shared Substrates," in *Fundamentals of Social Neuroscience*, eds. E. Harmon-Jones and P. Winkielman [New York, 2007], 38–39). Of course, the introduction of the notion of "salience" raises the familiar epistemological problem of the "frame" or the assessment of relevance—in other words, it raises the problem of meaning.

93. Paul Whalen, Hannah Raila, Randi Bennett, et al., "Neuroscience and Facial Expressions of Emotion: The Role of the Amygdala-Pre-Frontal Interactions," *Emotion Review* 5 (1) (2013): 79.

94. Justin S. Feinstein et al., "Fear and Panic in Humans with Bilateral Amygdala Damage," *Nature Neuroscience* 16 (3) (2013): 270–72.

95. See Paul J. Whalen, "Fear, Vigilance, and Ambiguity: Initial Neuroimaging Studies of the Human Amygdala," *Current Directions in Psychological Science* 7 (1998): 177–88; Paul J. Whalen, Scott L. Rauch, Nancy L. Etcoff, Sean C. McInerney, Michael B. Lee, and Michael A. Jenike, "Masked Presentations of Emotional Facial Expressions Modulate Amygdala Activity without Explicit Knowledge," *Journal of Neuroscience* 18 (1) (1998): 411–18; M. Davis and P. J. Whalen, "The Amygdala: Vigilance and Emotion," *Molecular Psychiatry* 6 (2001): 13–34; Paul J. Whalen et al., "Human Amygdala Responsivity to Masked Fearful Eye Whites," *Science* 306 (2004): 2061. See also Luiz Pessoa and Ralph Adolphs, "Emotion Processing and the Amygdala: From a 'Low Road' to 'Many Roads' of Evaluating Biological Significance," *Nature Reviews* 11 (December 2010): 773–82. In

Adolphs seem to want to pursue. In line with some of the newer findings about SM, they have revised their understanding of the mechanism by which bilateral amygdala damage compromises the recognition of fear. But they do not appear to have altered their general commitment to the Tomkins–Ekman paradigm.[96] I can think of several reasons for the continued success of that paradigm. Its ostensibly objective approach to the affects; its solidarity with evolutionary theories of the mind; the agreement between its assumptions about the independence of the affect system and cognition and contemporary presuppositions about the modularity and encapsulation of brain functions; the congruence between its image-based approach to the emotions and neuroimaging technologies such as PET and fMRI; the promise it holds for surveillance experts keen to find ways of detecting liars as easily as a blood test can detect DNA—all these and other factors help explain why the "basic emotions" view is so entrenched in contemporary thinking.[97]

Especially important in this connection is the convenience of Ekman's methods in facilitating research. His photographs of posed expressions are so easy to use that even his critics have continued to employ them in

---

a 2007 paper, Whalen draws attention to experiments suggesting that the stimuli to which the amygdala responds do not even have to be biologically relevant to the organism, but that uncertainty or unpredictability is enough to active the amygdala (Paul J. Whalen, "The Uncertainty of It All," *TRENDS in Cognitive Sciences* 11 [2007]: 499–500).

96. In Ralph Adolphs, "Perception and Emotion: How We Recognize Facial Expressions," *Current Directions in Psychological Science* 15 (5) (2006): 222–26, the author accepts the idea that the amygdala's role in recognizing facial expressions encompasses a broader, more abstract, or perhaps a more dimensional rather than a categorical aspect of the emotions, of which fear is only one instance. But this leaves open the possibility that he is still committed to the idea of the existence of a set of "basic emotions" according to the Tomkins–Ekman paradigm. Cf. Ralph Adolphs, "Fear, Faces, and the Human Amygdala," *Current Opinion in Neurobiology* 18 (2008): 166–72. For Damasio's recent views on emotion and the affects, see his *Looking for Spinoza: Joy, Sorrow, and the Feeling Brain* (New York, 2003); and *Self Comes to Mind: Constructing the Conscious Brain* (New York, 2010). In the latter text, Damasio introduces an idiosyncratic distinction between "emotions," defined in Ekman-like terms (although in this particular text Ekman is not mentioned) as discrete, automatic, bodily responses to "emotionally competent stimuli," and "feelings," defined as "perceptions of what happens in the body and mind when we are emoting" (116–17). But he makes no reference to the case of SM.

97. The compatibility between Ekman's hypostatization of the facial image and the hypostatization of the image in new imaging technologies such as PET and fMRI is striking. In this connection, see Joseph Dumit, *Picturing Personhood: Brain Scans and Biomedical Identity* (Princeton, 2004), in which the author suggests that PET and fMRI reinforce the notion that we are what our brain images tell us we are.

their own experiments—an extraordinary fact, when you think about it.[98] Moreover, I sense that Ekman's critics face another difficulty, which is that the moment one abandons the basic emotions approach in favor of some kind of intentionalist interpretation of the kind associated with Freud, appraisal theorists, and social constructionists, one finds oneself forced to provide thick descriptions of life experiences of the kind that are familiar to anthropologists and indeed novelists but are widely held to be inimical to science.[99] At the same time, one is obliged to engage with an array of tremendously difficult questions about the nature of intentionality, including the intentionality of nonhuman animals, that have traditionally belonged to the domain of philosophy.

Quite apart from those considerations, however, it is precisely Ekman's *non- or anti-intentionalism* that makes his work particularly attractive at the present time, at any rate in certain quarters. For, once one imagines that emotions are non-intentional states that are simply triggered by various stimuli, and once one imagines that, as inherited patterns of response, under the right conditions they will inevitably express themselves on the face—which is what it means for them to be universal—one is likely to conclude that the inner truth about a person will be detectable by properly trained observers, which is to say, one will conclude that there is an important sense in which the body cannot lie. That is why since 9/11 Ekman's research program has been of interest to American intelligence and security agencies.[100]

98. Another now widely used set of standardized pictures for studying emotion is the International Affective Picture System (IAPS), developed by the National Institutes of Health. It comprises hundreds of color photographs ranging from the pictures of the most mundane, everyday objects and scenes, to rare or exciting ones, including mutilated bodies and erotic nudes. The normative judgments of such images is based on three dimensions of valence, arousal, and dominance, meaning that observers in the studies performed to standardize the images are asked to rate how pleasant/unpleasant, calm/excited, and controlled/in control they felt when looking at each picture. In other words, the IAPS does not register categorical emotion judgments.

99. In his novel *Your Face Tomorrow*, vol. 1, *Fever and Spear*, trans. Margaret Jull Costa (New York, 2005), Javier Marías offers a fascinating fictional account of what might be involved in the reading of faces and facial behavior. Indeed, it is striking that this topic is thematized in his book at a time when, post 9/11, these issues are of such urgent concern in contemporary society and psychology. Suffice it to say that Marías's position is entirely on the side of narrative-thick description and the observation over time of multiple aspects of human behavior, not just the observation of isolated faces, much less static depictions of these.

100. Ekman has been centrally involved in the creation of the Screening of Passengers by Observational Techniques (SPOT) program, funded by the Department of

Whereas for critics of Ekman's approach to the emotions such as Fridlund, it does not follow that what is hidden in deception is destined to "leak out" from unmanaged behavior. Rather, humans and nonhuman animals produce facial behaviors or displays when it is strategically advantageous for them to do so and not at other times, because displays are dynamic and often highly plastic social and communicative signals. Deception is thus omnipresent in nature and potentially highly advantageous for the displayer, not something that covers over the hidden truth of authentic feelings (*HFE*, 137–39). But this way of thinking introduces a degree of complexity and uncertainty that contrasts to its disadvantage with the reassuring idea that the truth of our emotions is bound to reveal itself.

Homeland Security after 9/11. Starting with a test program in 2004, SPOT employs Behavioral Detection Officers (BDOs) at airports for the purpose of detecting behavioral-based indicators of threats to aviation security. The claim is that judgments of credibility can be made by observing facial and nonverbal expressions and cues indicative of stress, fear, and deception. The Department of Homeland Security has been criticized for failing to scientifically validate the SPOT program before operationally deploying it, but even though the results have been completely disappointing, the project is still in operation. For a discussion of SPOT, which has spent nearly $1 billion, see the Hearing Before the Subcommittee of Investigation and Oversight, Committee on Science, Space, and Technology, House of Representatives April 6, 2011, available at http://science.house.gov. An especially devastating assessment of the scientific support for SPOT is given in testimony before the Subcommittee by Maria Hartwig, Associate Professor, Department of Psychology, John Jay College of Criminal Justice.

[CHAPTER THREE]

RICHARD S. LAZARUS'S APPRAISAL THEORY I

Emotions as Intentional States

> More recently, especially among social and humanistic psychologists, the contextualist perspective has been adopted. The root metaphor is the historical act in all its complexity . . . The phenomena of interest will be intentional acts, acts designed to transform or transfigure the world; and the acts are performed by persons as agents, not as mechanical automata . . . Unlike many theorists in the recent academic tradition, I do not treat "emotion" as detachable from social contexts.[1]

These remarks by Theodore Sarbin in 1986 can serve as the starting point for an analysis of the contribution made by the American psychologist Richard S. Lazarus (1922–2002) to the study of emotions between the 1960s and the early years of the twenty-first century. Sarbin made his comments as an intervention in the well-known debate between Lazarus and Robert Zajonc about the role of cognition in the emotions. That debate had started six years earlier, in 1980, when, without naming Lazarus, who nevertheless understood that he was one of the main targets of the attack, Zajonc published the first in a series of papers contending that the affects did not depend on cognition, as several psychologists had argued, but could occur without extensive cognitive coding. Lazarus responded to Zajonc by insisting that emotions flow from cognitive processes, and the battle was joined. It lasted for many years. Indeed, in an important sense it is still not over.

What was the controversy between Zajonc and Lazarus really about? It has more than once been suggested that it was not about very much, that the dispute between the two men was not an empirical or even a theoretical one but was merely verbal or definitional: the two men were in fact describing the same thing, namely, the speedy, apparently automatic, involuntary response to emotionally salient stimuli, but one of them, Zajonc, called that response "noncognitive" whereas the other, Lazarus, called it "cognitive."[2] But this interpretation risks glossing over fundamental

1. Theodore R. Sarbin, "Emotion and Act: Roles and Rhetoric," in R. Harré, ed., *The Social Construction of Emotions* (Oxford, 1986), 83.

2. H. Leventhal and K. R. Scherer, "The Relationship of Emotion and Cognition: A Functional Approach to a Semantic Controversy," *Cognition and Emotion* 1 (1987): 3.

issues. Setting aside the question of the validity of Sarbin's adherence to a "dramaturgical" or "role-playing" theory of emotional action, his remarks have the merit of zeroing in on the most central question at stake in the dispute—the question of *intentionality*.[3]

In Lazarus's theory of the emotions, intentionality was a fundamental concern. It was his commitment to the idea that emotions involve the cognition of objects—which is to say that they involve questions of meaning—that makes the idea that they are intentional states a central issue in his work. The claim that emotions are intentional states has had a distinguished history: in the philosopher Franz Brentano's phenomenology; in Sigmund Freud's psychoanalysis, in which emotional states can be unconscious because repressed (or rather, the original object of the emotion may be repressed and hence unavailable for consciousness, since Freud tended to think that emotions themselves were in some sense always conscious); in Wittgenstein's remarks on the philosophy of psychology, in which he repeatedly questioned William James's non-intentionalist reduction of emotions to bodily sensations; in Magda Arnold's post-Brentano and post-Freudian approach to the emotions based in part on Sartre's intentionalist arguments; in Anthony Kenny's Wittgenstein-inspired discussion of emotion; in Robert Solomon's existential-phenomenological discussion of affect; in Sarbin's social-constructionist approach to emotion; and in the work of more recent emotion theorists, such as that of Alan Fridlund, who has offered a "Behavioral Ecology" or "Communicative" theory of the emotions in which the intentionality of animal action is a given; in Peter Goldie's treatment of the emotions; and so on.

Yet Lazarus's approach to the topic was often tentative and confused. In particular, as time went on he had to confront the question of what kind of claim it is that emotions are intentional states or actions. Is the claim fundamentally a constitutive-conceptual one? Could one say in a Wittgensteinian spirit that it belongs to the very "grammar" of the emotions that they are intentional states—that emotions are necessarily about some object or event, so that the meaning of the various cognized objects or situations or events to which emotions are directed is therefore one of its constitutive or essential features? Or is the claim that emotions are intentional

3. In their interesting discussion of Carl Stumpf's Brentano-influenced intentionalist account of the emotions, Reisenzein and Schönpflug have rightly observed that the concept of intentionality has tended to be a crucial missing term in the post–World War II controversy over the cognition-emotion relationship. Rainer Reisenzein and Wolfgang Schönpflug, "Stumpf's Cognitive-Emotional Theory of Emotion," *American Psychologist* 47 (1) (1992): 34–45.

states an empirical assertion about the causal link between emotions and their objects, so that what is required is scientific evidence about that link and about the causal mechanisms involved? Are those different kinds of claims incompatible or can one adopt both a conceptual or "grammatical" *and* an empirical position with respect to the nature of emotion? And if the claim is an empirical one, what methods—introspective, behavioral, physiological, ethological—are best suited to determining the causal link between cognition and affect? We will see that Lazarus did not find it easy to answer these questions. Indeed, his life's work can be regarded as a kind of litmus test of the difficulties involved in formulating a cogent position on the cognitive-intentional nature of the emotions. His arguments proved hard to evaluate precisely because it was not clear to others and even to himself at what level he was operating.

Exacerbating Lazarus's situation were several other factors. In the first place, the concept of intentionality involves some of the most intractable problems in the philosophy of mind, so it is hardly surprising that as a busy research scientist without philosophical training, Lazarus could not master the problems involved. Wittgenstein's famous remark that "in psychology there are experimental methods and *conceptual confusions*" is especially pertinent in this regard, for the controversy over the nature of the emotions raised deep conceptual questions that not only confounded psychologists but troubled philosophers as well. A further complicating factor was that the post–World War II "cognitive revolution," which challenged behaviorism by making it respectable for American psychologists to study mental processes once again, coincided with the cybernetic revolution, with the result that cognitive psychology was quickly captured by computational assumptions. Although the assimilation between cognitive psychology and information theory was not inevitable, their historical convergence, as Shanker has superbly demonstrated, can be traced to shared epistemological commitments of a Cartesian, dualistic-mechanist kind. The result was that cognitive psychologists were tempted to explain the operations of the mind in subpersonal computational-mechanical terms, which tended inevitably to marginalize the whole person and deflect attention from the meaningful mental contents that were the ostensible focus of psychological interest.[4]

One of Lazarus's more interesting objections to Zajonc's noncognitive theory of emotion, an objection inspired in part by Hubert Dreyfus's well-known critique of artificial intelligence, was that it was indebted to a false

4. Stuart Shanker, *Wittgenstein's Remarks on the Foundations of AI* (London and New York, 1998).

picture of how the mind works because it mistakenly assumed that cognition and meaning could be built up out of meaningless "bits" of information by hypothetical information-processing systems in the mind-brain. But we shall see that Lazarus himself was unable to resist some of the same computational assumptions, with the result that he risked losing control of his most important insights. In particular, his theory about the role of cognitive appraisal in emotions rested on the presupposition that there exists a gap or separation between the person and the world, which it is the function of cognition or appraisal to close, with the result that our emotional evaluations of objects are not direct and immediate but indirect and mediated—as if the meaning of objects for the individual had to be tacked on to neutral percepts of some kind.

In fact, as I will argue, Lazarus's entire picture of appraisal as involving inner cognitions intervening causally between the person and the world—a picture he shared with information-processing theorists—was a mistake, one that led to several dead ends. As I noted in the introduction, an "embodied world-taking cognitivism" of the kind recently advocated by Phil Hutchinson, and inspired by Wittgenstein and the philosopher John McDowell, might have saved Lazarus from some of these errors. Hutchinson accepts the Kantian thought that "intuitions [or perceptions] without concepts are blind" in order to insist, in the terms brilliantly developed by McDowell, that our perceptions are conceptual or cognitive "all the way out," which is to say that they are inherently conceptual. The resulting account of emotions is a cognitivism that emphasizes the ways in which humans and other animals are alive to aspects of the world—not to the disenchanted world of the modern natural sciences that stands external to minds, but to the cognized, conceptualized world. In Hutchinson's case, it is a cognitivism that, because it does not make the attribution of propositional attitudes or speech central to a theory of the intentionality of the emotions, is not vulnerable to the charge that it is unable to account for emotions in nonhuman animals.[5] But Lazarus himself was far from being able to articulate such a position. Instead, he adopted views that tended to undercut some of his best insights and, in an otherwise laudatory impulse to relate his ideas to those of his contemporaries and critics, offered

5. Phil Hutchinson, *Shame and Philosophy: An Investigation in the Philosophy of Emotion and Ethics* (Basingstoke and New York, 2008); Hutchinson, "Emotion-Philosophy-Science," in *Emotions and Understanding: Wittgensteinian Perspectives*, eds. Ylva Gustafsson, Camilla Kronqvist, and Michael McEachrane (Basingstoke and New York, 2009), 60–80; John McDowell, *Mind and World* (Cambridge, 1994).

several concessions and accommodations that also had the overall effect of undermining the coherence of his approach.

In order to elucidate these and related points, in this chapter I shall discuss Lazarus's early work on subliminal perception or "subception," his subsequent experiments on stress and emotion, and his early theorizations of appraisal. In the following chapter I will turn to the debate generated in the 1980s by Zajonc's noncognitive claims about affect in order to assess Lazarus's response. In that chapter I will also consider Lazarus's later writings, focusing on certain criticisms made by his peers that throw light on the continued difficulties Lazarus faced in his effort to establish a science of emotions defined as intentional states.

SUBCEPTION, PERCEPTUAL DEFENSE, AND THE "NEW LOOK" IN PSYCHOLOGY

"Subception" is the name Lazarus coined to describe an individual's automatic discrimination of threat or danger without conscious awareness of the stimulus. An early experiment on that topic, published with Robert McCleary in 1951 when Lazarus was starting his career at Johns Hopkins University, involved conditioning subjects to a set of nonsense syllables associated with painful electric shocks and, in a later session, comparing the subjects' psychogalvanic skin responses to the shock syllables and to syllables that had not been associated with a shock. Lazarus and McCleary reported that even when subjects incorrectly identified which syllables had been tachistoscopically flashed on a screen, because the speed used was below the recognition threshold, they nevertheless produced larger psychogalvanic skin responses to the shock syllables than to the non-shock ones. The authors concluded that subjects could discriminate between the threat and non-threat syllables even in the absence of conscious awareness of the presented stimulus. In other words, some sort of judgment of the stimulus had occurred at an unconscious level, as indicated by the differential skin responses.[6] As Lazarus subsequently observed of the implications of his findings: "In effect, people can instantly 'recog-

6. Richard S. Lazarus and Robert A. McCleary, "Autonomic Discrimination Without Awareness: A Study of Subception," *Psychological Review* 58 (1951): 113–22. The authors coined the term "subception" in an earlier paper: Robert A. McCleary and Richard S. Lazarus, "Autonomic Discrimination Without Awareness: An Interim Report," in *Perception and Personality: A Symposium*, eds. Jerome S. Bruner and David Krech (Durham, NC, 1949 and 1950), 171–79.

nize' that a situation is dangerous or benign and act accordingly without having much information about it and/or without consciously having examined the cognitive premises on which the appraisal is based."[7]

Lazarus and McCleary's experiment belonged to the movement launched by Jerome Bruner, George Klein, and others in postwar American psychology known as the "New Look" in perception. This was a movement that challenged the classical view of perception as involving the subject's passive registration of external stimuli by emphasizing instead the active, dynamic role of individual motivation and personality in the perceptual process.[8] The "New Look" owed a debt to Freudian ideas about unconscious motivation although, owing to hostility to psychoanalytic concepts among most American psychologists, Lazarus deliberately avoided the term "unconscious" when reporting his results and introduced instead the word "subception" (from *sub*conscious per*ception*).[9]

Although the phenomenon of perceptual defense was hardly new, the experiments performed on the topic in the postwar period generated a huge controversy that lasted for many years.[10] As Norman Dixon showed in an interesting book published in 1971 at a time when perceptual defense was still the topic of much dispute, the problems associated with this work were indeed in part technical and methodological. Other factors also influenced the course of the debate—for example, many psychologists rejected the idea of unconscious or subconscious discrimination be-

7. Richard S. Lazarus, "A Cognitivist's Reply to Zajonc on Emotion and Cognition," *American Psychologist* (February 1981): 223. Or, as he also later noted in a comment that brings out the intentionalism of his views about perceptual defense, when people appraise a situation unconsciously, the emotions experienced or displayed seem to make little or no sense "because the intentional premises are hidden." Richard S. Lazarus, "Cognition and Motivation in Emotion," *American Psychologist* 46 (April 1991): 363.

8. On the "New Look" in American psychology, see Jerome S. Bruner and Cecile C. Goodman, "Value and Need as Organizing Factors in Perception," *Journal of Abnormal and Social Psychology* 42 (1947): 33–44; Jerome S. Bruner and George S. Klein, "The Function of Perceiving: New Look Retrospect," in *Perspectives in Psychological Theory*, eds. S. Wapner and B. Kaplan (New York, 1960), 61–77.

9. As Lazarus observed, "psychologists in those days were simply not ready to assimilate comfortably the idea of unconscious processing, especially if it contained the dynamic implication of Freudian thought that sometimes ideas are kept out of awareness as an ego-defense." Lazarus, "Cognition and Motivation in Emotion," 360.

10. For a discussion of the bitter and sustained controversy over perceptual defense, see for example M. H. Erdelyi and B. Goldberg, "Let's Not Sweep Repression Under the Rug: Toward a Cognitive Psychology of Repression," in *Functional Disorders of Memory*, eds. J. F. New Kihlstrom and F. J. Evans (New York, 1979), 378–79.

cause they feared its commercial exploitation in the form of subliminal advertising. But the most basic objection stemmed from the very idea of intentional defense—the idea that the mind could intentionally block out threatening stimuli by making unconscious evaluations of their affective meaning. Dixon commented in this regard that

> Few findings in psychology have generated more heat than the discovery that recognition thresholds depend upon the emotional connotation of that which is recognized. To psychologists years hence, this academic pother may well seem rather odd since emotional connotation is but a parameter of meaning, and few would deny, even nowadays, that the ease or difficulty with which one recognized somebody or something depends upon their meaning or personal significance . . . Why, then, all this fuss over perceptual defence?[11]

One reason for the fuss, as Dixon explained, was that the notion of perceptual defense, or subception, seemed paradoxical because it implied that the mind simultaneously knows and does not know the same thing. Numerous critics voiced that objection (*Subliminal Perception*, 221).[12] As

11. Norman F. Dixon, *Subliminal Perception: The Nature of a Controversy* (London, 1971), 179. Cf. W. P. Brown, "Conceptions of Perceptual Defence," *British Journal of Psychology, Monograph Supplements* (Cambridge, 1961). As Dixon noted, "far from being an unlikely phenomenon, perceptual defence was merely a laboratory demonstration of processes akin to those underlying hysterical blindness and psychogenic deafness" (*Subliminal Perception*, 180). Yet the notion of perceptual defense, or subliminal perception, proved immensely controversial. Of the great number of contradictory results produced during the debate over subliminal perception, Dixon wrote: "Truly, we are in deep waters" (*Subliminal Perception*, 188). He went on to observe that each new methodological departure only served to introduce new problems, and added: "One begins to suspect that there is something inherent in the logic of the situation which always precludes a satisfactory solution" (*Subliminal Perception*, 186). Commenting on the possibility that demand characteristics might have influenced the results of some experiments, Dixon observed that "some experimenters, some academic departments, and even some countries, invariably provide positive evidence for subliminal effects, while others with almost equal monotony do not. This suggests that other things being equal, differences in the belief system of the experimenter concerned is a crucial and deciding variable, and that this communicates itself in some way to the experimental subject, and makes the subject receptive or resistant to subliminal effects" (*Subliminal Perception*, 242).

12. Leo Postman, Jerome S. Bruner, and Elliott McGinnies had already voiced that concern in "Personal Values as Selective Factors in Perception," *Journal of Abnormal and Social Psychology* 43 (1948): 142–54. "One may inquire at this point," they wrote, "How does the subject 'know' that a word should be avoided? In order to 'repress,' he must first recognize it for what it is. We have no answer to propose . . . Of only one

the eminent perception researcher James Gibson objected: "This seems to imply unconscious defence mechanisms governing perception as well as motivated behavior—wishful perceiving. But to say that one can perceive in order to *not* to perceive is a logical contradiction" (cited in Dixon, *Subliminal Perception*, 228–29). Of course, the psychoanalytic concepts of the unconscious and repression were meant to solve that paradox or contradiction. But many theorists rejected Freud's solution as unsatisfactory because it seemed to merely reproduce the problem, not solve it, by implying the existence of an inner ego or homunculus capable of deciding which stimuli are acceptable and which must be defended against and repressed because they are threatening to the individual.[13]

In an attempt to solve the paradox of perceptual defense, investigators proposed many alternative solutions. These alternatives often took the form of neurophysiological theories in which the unconscious discrimination of stimuli was viewed as a consequence of a peripheral or central nervous system "gating" or "filtering" of sensory inputs and flows.[14] Those physiological approaches were themselves sometimes informed by the new cybernetic ideas, with the result that subception effects were treated as the consequence of hypothetical information-processing mechanisms capable of discriminating between different system inputs. The dualism

---

thing we can be fairly sure: reactions occur without conscious awareness of what one is reacting to. Psychological defense in perception is but one instance of such 'unconscious' reaction" (153).

13. On the dilemmas of Freud's solution to the problem of self-deception, see Simon Boag, "Realism, Self-Deception, and the Logical Paradox of Repression," *Theory and Psychology* 17 (2007): 421–47. As Walter Benn Michaels has remarked, the paradox is at once psychological and epistemological: it is psychological in that according to Freud's theory of repression it involves the refusal to know something based on the theory of repression; and it is epistemological because it involves the paradox of self-deception according to which the subject simultaneously does and doesn't believe something, an impossibility formulated by Wittgenstein in his discussion of Moore's paradox. See Walter Benn Michaels, "The Death of a Beautiful Woman: Christopher Nolan's Idea of Form," www.electronicbookreview.com, October 1, 2007.

14. Dixon proposed just such a solution: "Subliminal perception can be envisaged as very largely a physiological phenomenon, and needs to be explained in physiological terms. It is concerned with neural processes initiated by a physical stimulus and terminating in a physiological response, whether this involves activity of the speech muscles or activity of the autonomic nervous system. By definition, there is no psychological (mental) content to this process, and, therefore, no solely psychological explanation of the phenomenon can possibly be adequate. To talk of unconscious feelings, memories, wishes, and fantasies is a contradiction in terms, and adds little by way of explanation" (*Subliminal Perception*, 244).

inherent in such approaches, according to which what is mental must be conscious and what is not conscious must therefore be neurophysiological or corporeal, is revealed in Dixon's remark that

> If only conscious minds can "discriminate" and "recognize," then obviously it is absurd to talk of discrimination without awareness or perceptual pre-recognition, but if we take these words to signify not the consciousness of the act, but merely the operation it performs the difficulty evaporates . . . [I]t is not difficult to build a machine which is programmed to "discriminate" between different inputs, and to classify the resultant discriminanda according to some prearranged principle. Nor is it difficult to conceive of this classification having an effect on subsequent read out. (*Subliminal Perception*, 229)

In this formulation, the concept of subception as an unconscious mode of meaning-making is transformed into a concept of subception as a mechanical response—as if, unknown to the subject himself or herself, subpersonal processes or "programs" do the perceiving and judging that have been denied the mind and indeed the person.

Of course, as Dreyfus argued in his 1972 critique of cybernetics and artificial intelligence which appeared a year after Dixon's book on subliminal perception, the latter's widely shared belief that a machine could easily be programmed to discriminate stimulus inputs turned out to be naïvely optimistic. Dreyfus suggested that it was impossible to derive meaning from bits of meaningless information, so that attempts to frame issues of meaning in terms of information theory were bound to fail because they begged all the crucial epistemological, psychological, and ontological questions—a prediction that was soon borne out when efforts to build intelligent machines faltered or failed.[15] Nevertheless, for many researchers in cognitive science the attraction of cybernetics and information-processing theory was very great.

A defining moment came in 1974 when the psychologist Matthew Erdelyi in a "New Look at The New Look" reformulated subception in strictly information-processing terms. Erdelyi argued that the paradox of perceptual defense, according to which the subject has to know something in order to defend against knowing it, could be solved by treating cognition not as a single, unified phenomenon but as a multicomponent filtering and encoding process involving several parallel subsystems operating selec-

15. Hubert Dreyfus, *What Computers Can't Do: The Limits of Artificial Reason* (New York, 1972).

tively at different levels and with different outcomes.[16] On this hypothesis, there was no paradox about subception because affective and other signals could be managed by hypothetical information-processing systems running independently of conscious processing and before conscious cognition could kick in. In the next chapter, we shall see that Erdelyi's cybernetic model of unconscious defense contained the seeds of Zajonc's theory of emotion according to which affective signals are handled entirely separately from cognition because separate information-processing or neural pathways are involved.[17]

It could be argued that Erdelyi's cybernetic hypothesis was no more of a solution to the paradox of subception than Freud's psychoanalytic approach, since it merely restated the problem of stimulus recognition and defense in computational terms while begging every conceivable question about how the information-processing system could, by performing the necessary filtering processes, "recognize" contextual cues and create semantic meaning from "information" held in raw storage. But for many psychologists Erdelyi's proposal had the irresistible advantage that, unlike psychoanalytic or quasi-psychoanalytic theories of ego defense of the kind Lazarus and other New Look theorists had put forward to explain the phenomenon of subception, it avoided any appeal to unconscious intention by reformulating mentation in strictly computational-corporeal terms.

In those early years Lazarus was gripped by the idea that affective and motivational factors played a dynamic role in perception, and he tended to stick to ego-psychology explanations of the phenomenon of psychic defense. He was not tempted by approaches that reduced perceptual and emotional processes to the workings of neurophysiological or information-processing systems (even if his own formulations sometimes

16. Matthew Hugh Erdelyi, "A New Look at the New Look: Perceptual Defense and Vigilance," *Psychological Review* 81 (1974): 1–25.

17. As Erdelyi himself noted of his 1974 contribution: "There is probably no defense process discussed by Freud that does not naturally lend itself to a computer articulation. . . . We have already seen that certain logical and philosophical objections to defense processes collapsed with the advent of the computer analogy. If the computer could be programmed to simulate perceptual defense effects, then all the paradoxes that supposedly rendered the process impossible must be in error. Of if a persuasive philosopher such as Sartre proves that a system cannot lie to itself (and that, therefore, there cannot be self-deception and, therefore, no repression), the *idiot savant* computer can set matters straight immediately by showing that it can deceive itself and thereby undeceive us brighter human savants." Erdelyi and Goldberg, "Let's Not Sweep Repression Under the Rug: Toward a Cognitive Psychology of Repression," 390.

strayed in that direction). Instead, after leaving Johns Hopkins for Clark University in Massachusetts, in 1957 he moved to the Department of Psychology at the University of California at Berkeley, where he would spend the remainder of his career. During the 1960s at Berkeley he carried out an important series of experiments on stress, in the course of which he extended and refined his views about the part played by unconscious defense in psychic life. But it was during his efforts to theorize the cognitive appraisal processes involved in stress that problems in his theorizing began to surface.

LAZARUS'S EXPERIMENTS ON STRESS

In the 1950s Lazarus took up the problem of stress in an impressive laboratory program of research that lasted until 1972. The concept of stress had come into prominence in World War II when pilot error and other forms of impairment due to fatigue had become a topic of concern in the military. Physicians in the postwar period concerned with caring for survivors of the concentration camps were also dealing with the phenomenon of stress in the form of post-traumatic symptoms and the "survivor syndrome."[18] So were doctors and others investigating the effects of natural disasters and other life-threatening conditions, although during the early postwar years there was not much contact between the clinicians and the scientists who were carrying out experiments in the laboratory.[19] By the time Lazarus published his first book on stress in 1966, stress had garnered so much attention that he complained that it was virtually impossible for him to do justice to or even fully cite the huge literature that had already been published on the topic.[20]

18. Ruth Leys, *From Guilt to Shame: Auschwitz and After* (Princeton, 2007).

19. For the history of the stress concept, see especially *Stress, Shock, and Adaptation in the Twentieth Century*, eds. David Cantor and Edmund Ramsden (Rochester, NY, 2014); and Cary L. Cooper and Philip Dewe, *Stress: A Brief History* (Maiden, MA, 2004).

20. Richard S. Lazarus, *Psychological Stress and the Coping Process* (1966), viii–ix; hereafter abbreviated *PS*. Lazarus suggested that, although the issues encompassed by the word "stress" had often been considered under the rubric of "emotion," the word "stress" was adopted because it connoted the negative aspects of emotion, such as fear, anger, and depression, and emphasized disturbances in or failures of psychological and biological adaptation, an emphasis not found in the concept of emotion. He also suggested that in the era of behaviorism, the engineering concept of stress might have been more acceptable to psychologists than the word "emotion," with its implied reference to states of mind (*PS*, 10). See also Richard S. Lazarus, "From Psychological Stress to the Emotions: A History of Changing Outlooks," *Annual Reviews* 44 (1993): 1–22.

The general thrust of Lazarus's work on stress, which he offered as a continuation of the "New Look" in psychology, was to reject the dominant tendency to regard the stress threat as objectively "out there as an attribute of the stimulus" and to argue instead that the meaning of a stimulus depended on the contributing role of personality and individual differences. According to him, in a critique aimed at the majority of his fellow researchers who regarded stress as a purely external, objective event, the stressful nature of a stimulus could not be treated as objective or predetermined by the experimenter because its significance was also a function of a "person's appraisal of the meaning of the stimulus for the thwarting of motives of importance to him"—which is to say, on "the interpretation the person makes concerning its future significance."[21]

Since apparently similar stressful conditions meant different things to different individuals—since a situation that appeared threatening to one subject might not be threatening to another—Lazarus suggested that the role of the *person* in the stress response had to be taken into account. Instead of assuming, as most stress researchers did, that subject populations were homogeneous, Lazarus therefore emphasized the importance of individual differences. Coping styles and strategies—which is to say, cognitive styles—might vary depending on a person's specific motivational history, with the result that the stressor or stimulus condition could not be isolated from the question of individual meaning. In short, the accent of Lazarus's work on stress fell on the connotation of the situation or event for the individual and on the defensive appraisals he or she deployed in order to cope with danger. Lazarus suggested that the entire field of stress research had to start over so that the role of personality factors and individual differences could be considered.[22] Over time, he

21. Richard S. Lazarus, "A Laboratory Approach to the Dynamics of Psychological Stress," *American Psychologist* 19 (1964): 404.

22. Richard S. Lazarus, *The Life and Work of an Eminent Psychologist: Autobiography of Richard S. Lazarus* (New York, 1998), chapters 6–8; Richard S. Lazarus, James Deese, and Sonia Foster, "The Effects of Psychological Stress Upon Performance," *Psychological Bulletin* 49 (4) (1952): 293–371; Richard S. Lazarus and C. W. Eriksen, "Effects of Failure Stress Upon Skilled Performance," *Journal of Experimental Psychology* 43 (1952): 100–105; Richard S. Lazarus and J. Keenan, "Anxiety, Anxiety-Reduction and Stress in Learning," *Journal of Experimental Psychology* 46 (1953): 55–61; Richard S. Lazarus and Robert W. Baker, "Psychology," *Progress in Neurology and Psychiatry* 11 (1956): 253–71; Richard S. Lazarus, "Motivation and Personality in Psychological Stress," *Psychological Newsletter* 8 (1957): 159–93; Richard S. Lazarus, "A Program of Research in Psychological Research," in *Festschrift for Gardner Murphy*, eds. J. G. Peatman and E. L. Hartley (New York, 1960), 313–29; William Vogel, Robert W. Baker, and Richard S. Lazarus, "The Role of Motivation in Psychological Stress," *Journal of Abnormal and Social Psychology* 56

shifted attention from the concept of stress to that of emotion, because he regarded emotion as the more embracing term.[23]

Lazarus's aim was to devise experiments in which the relationship between specific stressor conditions and individual responses could be operationalized, manipulated, and controlled. In a typical stress experiment, the scientist induced reactions by subjecting individuals to stressful situations under certain conditions, such as having them perform tasks under time pressure, or exposing them to painful-stressful stimuli such as small electric shocks or taboo words. The subjects' responses were assessed by measuring changes in psychogalvanic skin reflexes, heart rate, and other gauges of autonomic arousal, as well as by evaluating subjective reactions, moods, and personality characteristics through the use of questionnaires and interviews. One of the challenges Lazarus hoped to meet in designing his experiments was to deploy stress stimuli that were as realistic as was possible within the confines of the laboratory in order to fully engage his experimental subjects—college students, for the most part. That is why starting in 1959 he began exploiting the stress-inducing qualities of *Subincision*, a short, silent, black-and-white anthropological film, showing naked Aborigines in the Australian bush undergoing puberty initiation rites. The film included scenes of adolescents, who appeared to volunteer for the procedure, repeatedly wincing and writhing in pain as older men in the tribe ritualistically cut their penises with sharp stones.[24] We have al-

---

(1) (1958): 105–12; William Vogel, Susan Raymond, and Richard S. Lazarus, "Intrinsic Motivation and Psychological Stress," *Journal of Abnormal and Social Psychology* 58 (2) (1959): 225–33.

23. Richard S. Lazarus, "Emotions and Adaptation: Conceptual and Empirical Relations," in *The Nebraska Symposium on Motivation*, ed. W. J. Arnold (Lincoln, NE, 1968), 175–76.

24. *Subincision*, which was made by the anthropologist Géza Róheim in the 1940s, had already been used to study castration and mutilation fantasies experimentally. See J. B. Schwarz, "An Empirical Test of Two Freudian Hypotheses Concerning Castration Anxiety," *Journal of Personality* 24 (1956): 318–37; and A. Aas, *Mutilation Fantasies and Autonomic Response* (Oslo, 1958). For a contemporaneous discussion of the meaning of the subincision ritual, see G. Róheim, "The Symbolism of Subincision," *American Imago* 6 (1949): 321–28; J. E. Cawe, N. Djagamarra, and M. G. Barrett, "The Meaning of Subincision of the Urethra to Aboriginal Australians," *British Journal of Medical Psychology* 39 (1966): 246–53; and Ashley Montagu, *Coming Into Being Among the Australian Aborigines: The Procreative Beliefs of the Australian Aborigines* (London, 1974). Lazarus came to recognize that the content of the film was very complex, so it was not clear what aspects of it were most important to each individual viewer. Themes included fantasies of castration, sexual exposure, mutilation, and pain, all of which were confounded. He eventually concluded that the film was too complex to serve as a useful

ready encountered this film in the work of Ekman, who borrowed it from Lazarus (see chapter 2, n. 47). Lazarus's rationale for using *Subincision* was that it provided a more vivid and immersive, if vicarious, experience of stress or threat to experimental subjects than did the tamer and more deceitful stimuli hitherto used in experiments of this kind.

As his research program advanced, in an elegant series of experiments he and his co-researchers sliced *Subincision* into short clips so as to be able to assess as precisely as possible the influence of the film content. Autonomic changes in the viewing subjects were monitored continuously during film screenings; fluctuating experiential reactions were assessed by questionnaires, post-film interviews, and other methods; and a "neutral" film was used as a control. In some of the experiments, other stress films, such as workshop accident films, were also employed to manipulate cognition and evaluate stress responses. Lazarus and his research group were able to demonstrate "huge main effects" of the contents of *Subincision* on their viewing subjects, as evidenced by large increases in autonomic arousal and related measures. Thus, the stressor film produced much higher levels of response than did the control film. But the stressor film also produced more individual variability in mood changes, coping mechanisms, and patterns of autonomic arousal.

So on the one hand, Lazarus and his team found large, consistent film effects on autonomic indices and subjective reactions. But on the other hand, they found many individual differences. Indeed, while some subjects expressed extreme distress, including disgust and horror, at viewing the scenes of genital cutting, others denied being disturbed. In an attempt to discover which scenes were the most upsetting, Lazarus and his team tried to pinpoint the parts of the film where the greatest autonomic arousal tended to occur. Here, too, they detected a peak reaction at the initial scenes of cutting. Nevertheless, some subjects showed little response to those scenes but reacted strongly to other moments in the film. Lazarus thought one of his most interesting findings was that those who denied being disturbed by the stressor film nevertheless registered autonomic changes of great magnitude, so there was a "baffling" discrepancy between subjective report and the psycho-physiological reaction, a discrepancy whose significance was not clear.

The discovery that some viewers denied that they were upset by the movie suggested to Lazarus that they were unconsciously using defensive

---

instrument of further research, and his turn away from laboratory to clinical research seems to have been motivated in part by this insight.

coping strategies. This led him to recognize that *how* the film content was presented could directly influence outcomes. Thus, Lazarus and his team found that if they added a soundtrack to the film clips that took a distant, intellectual, and detached anthropological view of the proceedings, or one using an official-sounding travelogue voice that took a denying attitude toward the scenes (as if the cutting were a happy experience for the boys), the threat of the film, as measured by autonomic changes and subjective responses, was reduced. But if a "trauma" soundtrack was added, emphasizing the horror of the situation, the dread of the Aborigine boys, and the harmful consequences that would befall some of them, the stressful response was increased. At the same time, individual differences in viewing responses continued to manifest themselves, so that the experiments revealed the existence of both typical changes and individual variations due, as Lazarus proposed, to personality differences and other motivating factors. Thus, while he did not deny that there were classes of conditions that served as stressors for most people, he also proposed that the more complex the content of the stressor or the situation, the more likely it was that individual differences relating to motives, needs, and personality would produce different areas of sensitivity. In a series of experiments conducted first in Japan and then back in the United States, Lazarus extended his studies to examine the role of cultural differences in emotional responses.[25]

25. Joseph C. Speisman, Janet Osborn, and Richard S. Lazarus, "Cluster Analyses of Skin Resistance and Heart Rate at Rest Under Stress," *Psychosomatic Medicine* 23 (4) (1961): 323–43; Richard S. Lazarus, Joseph C. Speisman, Arnold M. Mordkoff, and Leslie A. Davison, "A Laboratory Study of Psychological Stress Produced by a Motion Picture Film," *Psychological Monographs* 34 (1962): 144–56; Richard S. Lazarus, Joseph C. Speisman, and Arnold M. Mordkoff, "The Relationship Between Autonomic Indicators of Psychological Stress: Heart Rate and Skin Conductance," *Psychosomatic Medicine* 25 (1) (1963): 19–30; Richard S. Lazarus, "A Laboratory Approach to the Dynamics of Psychological Stress," *American Psychologist* 19 (1964): 400–411; Richard S. Lazarus and Elizabeth Alfert, "Short-Circuiting of Threat by Experimentally Altering Cognitive Appraisal," *Journal of Abnormal and Social Psychology* 69 (1964): 195–205; Joseph C. Speisman, Richard S. Lazarus, Arnold M. Mordkoff, and L. A. Davison, "The Experimental Reduction of Stress Based on Ego-Defense Theory," *Journal of Abnormal and Social Psychology* 68 (4) (1964): 367–80; Richard S. Lazarus, Edward M. Opton Jr., Markellos S. Nomikos, and Neil O. Rankin, "The Principle of Short-Circuiting of Threat: Further Evidence," *Journal of Personality* 33 (1965): 622–35; Edward J. Malmstrom, Edward M. Opton Jr., and Richard S. Lazarus, "Heart Rate Measurement and the Correlation of Indices of Arousal," *Psychosomatic Medicine* 27 (1965): 546–56; Richard S. Lazarus and Edward M. Opton Jr., "The Study of Psychological Stress: A Summary of Theoretical Formulations and Experimental Findings," in *Anxiety and Behavior*, ed. Charles D. Spielberger (New York, 1966), 225–62.

Lazarus's central finding in these experiments was that what mattered in stress was the consciously or unconsciously cognized meaning of the situation for the individual. At first he deployed the familiar psychoanalytic concept of ego defense to explain how subjects coped with stress. But as his work progressed and as he began to read more widely in the literature on stress and the coping process, "appraisal" emerged as a key concept. He owed the term to the work of Magda Arnold (1903–2002), a pioneer of emotion research who in 1960 had published an important book on emotions that decisively influenced the development of Lazarus's theoretical ideas.[26] Thus, in his first reference to the appraisal concept in 1964, Lazarus and his coauthors cited Arnold for arguing persuasively that "an emotion implies an evaluation of a stimulus as either harmful or beneficial." They stated that "In our own view . . . a stimulus must be regarded by the person as a threat to his welfare in order for stress responses to be produced. Thus, the same stimulus may be either a stressor or not, depending upon the nature of the cognitive appraisal the person makes regarding the significance for him."[27] A year later, Lazarus and his team again acknowledged a debt to Arnold by observing that the short-circuiting of threat by defensive processes was accomplished by "modification of appraisal—that is, the interpretation, the recognized meaning—of the events in the stimulus film. The theoretical sequence posited by this view is similar to the conceptual statement put forward by Arnold (1960) and is consistent with Schachter and Singer's (1962) emphasis on cognitive processes as critical in determining emotional reaction."[28] From then on, Lazarus repeatedly credited Arnold with being the first to provide a "truly systematic use of the concept [of appraisal] in a serious theoretical treatment of the field of emotion."[29]

26. For biographical information about Arnold and discussions of her contributions, see *Cognition and Emotion* 20 (7) (2006), a special issue devoted to Arnold, edited by Stephanie Shields and Arvid Kappas.

27. Speisman, Lazarus, Mordkoff, and Davison, "The Experimental Reduction of Stress Based on Ego-Defense Theory," 367.

28. Lazarus, Opton, Nomikos, and Rankin, "The Principle of Short-Circuiting of Threat: Further Evidence," 631.

29. Lazarus, *The Life and Work of an Eminent Psychologist*, 169. Lazarus added that Arnold's book influenced him to abandon the term "perception," which he had originally used "in its broadest sense to imply personal meaning," and instead to employ the term "appraisal" systematically "because it clearly denoted evaluation" (169). This suggests that it was the substitution of Arnold's term "appraisal" for that of "perception" that played a role in Lazarus's mistaken separation of these two functions or processes, discussed below.

What is important to grasp about Arnold's influence on the development of Lazarus's views is what she got right about appraisal—and what she got wrong. The first thing to note in this regard is that Arnold presented her general position on emotion as a phenomenological one. As Reisenzein has helpfully observed, her debt to the phenomenology of Brentano and Husserl was largely indirect, through the influence of their intellectual heirs and successors, such as Scheler and especially Sartre.[30] In his little book *The Emotions: Outline of a Theory* (1948), Sartre had argued with reference to the writings of Husserl that if psychologists wanted to understand the emotions they would have to grasp the essence of human reality, not reduce the human being to the sum of isolated, externally observed psycho-neurological and behavioral "facts." The essence of human reality included, crucially, the meanings and values that constituted consciousness. According to Sartre, then, the essence of the emotions as organized kinds of consciousness is that they are directed toward the world of objects such that signification or meaning inheres in them. As Sartre wrote, emotion "*is* strictly to the extent that it signifies."

For Sartre, therefore, the relation of emotions to their objects is not an accidental or contingent one but a necessary or essential feature, and any properly phenomenological psychology of emotions has to start from that premise. It has to recognize that, since emotion does not exist as a corporeal phenomenon, because "a body cannot be affected, for want of power to confer meaning on its own manifestations," a phenomenological approach to the affects must accept the purposive-intentional character of emotional reactions. From this point of view, the essence of emotion is not the mere consciousness of an affective state but a consciousness *of* something—of a real (or imagined) object. "[T]he man who is afraid is afraid *of* something," Sartre observed, and the affected subject and the affective object are thus "bound in an indissoluble synthesis." According to Sartre, even in the case of apparently objectless feelings, such as "one of those indefinite anxieties which one experiences in the dark, in a sinister and deserted passageway, etc., one is afraid *of* certain aspects of the night, of the world."[31] (This is an important claim in the context of the later debates over the cognitive nature of emotions in which, as we saw in

30. Rainer Reisenzein, "Arnold's Theory of Emotion in Historical Perspective," *Cognition and Emotion* 20 (7) (2006): 920–51.

31. Jean-Paul Sartre, *The Emotions: Outline of a Theory* (New York, 1948), 51–52.

the introduction, theorists such as Paul Griffiths have routinely appealed to the objectless character of some emotions, such as some depressive states, to argue for the non-intentional nature of the basic affects.)

Arnold endorsed Sartre's views. She stated in this regard that "Sartre sees the 'signification' of emotion in its unique relation to reality. Emotion always has an object reference. A man is always afraid of *something*, angry at *something*. As long as the emotion lasts, it maintains that focus on the object. Emotional consciousness is an unreflective awareness of the object, not a person's awareness that he has an emotion" (her emphasis).[32] As she also affirmed: "In his insistence that emotion can be understood only as a special relationship of the human being to the external world, Sartre's theory is a much needed corrective. There is no doubt that any emotion refers to an object and clings to it as long as it remains an emotion. It is also true that emotional awareness is unreflective, immediate, object-bound. Nevertheless, it remains a reaction to the object, and therefore is active, not passive" (*EP*, 160). This last statement also brings out the importance Arnold attached to the unreflective nature of emotional judgments or appraisals. In her view many of the evaluations involved in appraisals are not deliberate or intellectual in nature but more like the rapid, intuitive judgments a person makes when performing a skill such as throwing a ball, driving a car, or playing the piano. A reflective judgment might follow the intuitive or non-reflective one.[33]

So far, so good. But there were other aspects of Arnold's approach to emotion that from the start compromised her analysis. In particular, in spite of her assertion, following Sartre's lead, that phenomenologically the appraisal of the object is immediate and unreflective, she nevertheless proceeded as if there was a gap between the person and the intended object, a gap that had to be filled by a cognitive appraisal process intervening between them. The result was that in Arnold's analysis perception and appraisal were split apart into two separate processes. "Both perception and emotion have an object," she wrote, "but in emotion the object is known in a particular way. To perceive or apprehend something means that I know what it is like as a thing, apart from any effect on me. To like or dislike it means that I know it not only objectively, as it is apart from me, but also that I estimate its relation to me, that I appraise it as desirable or undesirable, valuable or harmful for me, so that I am drawn toward it or

32. Magda Arnold, *Emotion and Personality* (New York, 1960), vol. 1, *Psychological Aspects*, 158–59; hereafter abbreviated *EP*.

33. See in this connection Randolph R. Cornelius, *The Science of Emotion: Research and Tradition in the Psychology of Emotions* (New York, 1996), 117.

repelled by it" (*EP*, 171). To which she added: "Before anything can have 'meaning' for us, it must be seen as a thing (must be *perceived*) and must also be seen in some relationship to us (it must be *appraised*). Meaning comes with appraisal" (*EP*, 171). On this model, perception gives me "objective" knowledge or a "simple apprehension" of an object in the world but not its meaning *for me*, because to evaluate its meaning *for me* I have to make a separate assessment of its personal significance. Or, as Arnold also stated: "To know or perceive something and to estimate its effect on us are two distinct processes, and appraisal necessarily presupposes perception . . . To estimate how [an object] affects us personally . . . seems to require a further step beyond perception . . . Following upon perception and completing it, appraisal makes possible an active approach, acceptance or withdrawal, and thus establishes our relationship to the outside world" (*EP*, 176).

One way Arnold had of describing the appraisal process was to suggest, in terms borrowed from William James, that what appraisal does is to turn "cold" perceptions, which lack emotional meaning, into "hot" ones, in which the object perceived becomes affectively significant for the individual. As she put it: "When we examine the features to be explained in emotion, we find that at the base of every emotion there is some kind of perception or awareness of an object, a person, or a situation, which in some cases becomes emotional, in other cases remains (in the words of William James) a 'cold perception.' Therefore, it stands to reason that the perception that arouses an emotion must be somehow different from the mere perception of an object as such, which does not arouse an emotion" (*EP*, 93).[34] But here, objections immediately arise. From what ostensibly neutral perspective can I perceive an item in the world as an objective "thing" in itself without regard to its meaning for me? How can I see an object "as such" before it acquires any personal meaning? It's not at all clear that these formulations, which imply a splitting apart of the perception of the object from its significance for the person, are compatible with Arnold's self-proclaimed phenomenological standpoint. Moreover, once such a gap between perception and personal meaning is posited, then it is natural to assume that the scientist's goal must be to discover the causal processes or mechanisms by which the gap is closed. But this suggests that for Arnold the relationship between emotions and their objects is not an inherent and necessary one, as Sartre had argued, but is merely

34. For useful recent discussions of James's views, about the import of which there is still no consensus, see "William James and His Legacy," a special section devoted to his work in *Emotion Review*, January 2014.

contingent—as if, for her, the relationship is the consequence of a merely empirical correlation or connection.

Lazarus made the same troubling moves. "The cognitive interpretation and the emotional effect of a stimulus is [*sic*] determined not only by the objective properties of the stimulus but also by factors within the psychological structure of the individual such as motivation and knowledge," he wrote in an early statement.[35] Or, as he also observed:

> The process of appraisal depends on two classes of antecedents. One consists of factors in the stimulus configuration, such as the comparative power of the harm-producing condition and the individual's counter-harm resources, the imminence of the harmful confrontation, and degree of ambiguity in the significance of the stimulus cue. The second class of factors that determine the appraisal are within the psychological structure of the individual, including motive strength and pattern, general beliefs about transactions with the environment, and intellectual resources, education, and knowledge.[36]

Lazarus posited the same distinction between perception and appraisal in the following statement:

> *For threat to occur, an evaluation must be made of the situation, to the effect that a harm is signified. The individual's knowledge and beliefs contribute to this. The appraisal of threat is not a simple perception of the elements of the situation, but a judgment, an inference in which the data are assimilated to a constellation of ideas and expectations.* If you change the background of cognition, the same situation will now have a very different significance, perhaps no longer signaling harm to the individual . . . While the objective nature of the stimulus configuration contributes importantly to the appraisal process, it is always in interplay with the psychological structure of the individual. (Lazarus, PS, 44, 55; his emphasis)

One can see why Lazarus wanted to say things like this. He wished to acknowledge the fact that features of a situation that do not bother one person do cause distress or emotional upset in another—as, for example, in the case of the viewers of the film *Subincision*, some of whom reported that they were not at all dismayed by the scenes of genital mutilation while

35. Lazarus, Opton, Nomikos, and Rankin, "The Principle of Short-Circuiting of Threat: Further Evidence," 631, note.

36. Lazarus and Opton, "The Study of Psychological Stress," 229.

others reported feeling disgust. It was as if the "same" film aroused quite different emotions in different people because of the different evaluations they placed on it. But in what sense was the film "stimulus" the "same" for both types of viewers? The problem in Lazarus's analysis begins when, in the passages just quoted, he differentiates between the "elements of the situation" or the "data" in the "stimulus configuration" on the one hand, and the meaning or emotional evaluation that is conferred on these same elements or data on the other—as if appraisal is a matter of imposing meaning on a percept that is not in itself meaningful to the individual subject until the appraisal process intervenes. In other words, Lazarus proceeded as if he believed that meaning has to be added, tacked on, to a supposedly impartial or unbiased and therefore not yet fully meaningful perception of some kind.

He used the same language as Arnold when he described the process of appraisal as one in which a person goes from a "cold" perception to a "hot" appraisal or cognition (*PS*, 52). He gave as an example the way an ordinary knife can suddenly appear threatening to an individual. The threat value of a knife, he wrote, "depends on inferences about the intentions of the holder. We deal with knives all the time, but it is only a particular pattern of cues (for example, the way it is held, the expression on the face, or the content of the words spoken) that suggest [*sic*] that there is any danger. And the significance of this pattern is established through experience" (*PS*, 395). What seems problematic here is Lazarus's idea that in this situation the knife first appears as a neutral and unthreatening object but then is subsequently inferred to be threatening on the basis of cues of a more personally meaningful kind.

I can put the difficulty in this way: however much Lazarus tried to emphasize the role of context and the framing process in emotion, he couldn't give up, or work his way around, the idea of a "given" state of affairs in perception, one that could in principle be objectively or neutrally described independently of the person actually encountering the scene. But *who* has this objective or neutral perception, and *to whom* does it belong? To be "objective" in the way Lazarus conceived it, the perception would have to be experienced from the perspective of no one in particular.[37] Moreover, by positing the need for an appraisal process ancillary or supplementary to perception and designed to mediate between the impersonally perceived object and the individually motivated

37. In his 1966 book on stress, in a section entitled "Examples Where Appraisal Deviated from the Objective Facts," Lazarus defines "the point of view of the objective facts" as facts judged "consensually" (*PS*, 109).

subject, Lazarus was proposing that emotional meaning is not immediate and direct, as Sartre had claimed, but is indirect because it is an "inference" made from an initially "neutral" percept when the latter's significance is subsequently or belatedly recognized. But this was to assert that the relationship between emotion and the intentional object was not a constitutive or essential one, as Sartre had argued, but was accidental or external—as if the significance of the object was not inherent in emotion but depended contingently on what the object happened to mean to a particular individual owing to his or her particular history and the mediation of appraisal. It is the concept of "mediation" itself that is problematic here, because it suggests that meaning, including affective meaning, is not directly apprehended but has to be added on to objects in the form of supplementary, personal interpretations.

It is interesting that Lazarus did not derive more help in thinking about these matters from another important text that he read at this time, the philosopher Anthony Kenny's short book *Action, Emotion and Will* (1963), which concerned the intentionality of emotion and action. Kenny's views reflected the influence of Wittgenstein rather than of Sartre, which is to say that his discussion was framed in the terms of ordinary language philosophy, or in what might be called "grammatical" terms, rather than in phenomenological-existential terms. Kenny was influenced not only by Wittgenstein but also especially by the philosopher Elizabeth Anscombe (a student of Wittgenstein), whose short book *Intention* (1959) has by now achieved the status of a classic but whose difficult and sophisticated arguments Kenny was attempting to work out in a more readable form.[38]

In his book, Kenny maintained that emotions could not be understood as intentional actions by virtue of their being related to some features of an agent's supposed inner mental states functioning as the causes of his or her affects. He made use of Wittgenstein's arguments against the idea of a private language and of private mental states to contend against the common picture of the emotions as inner psychic events that were only contingently connected to their objects. "Wittgenstein has shown that a

38. Anthony Kenny, *Action, Emotion and Will* (London, 1963; hereafter abbreviated as *AEW*. A few years later the philosopher Norman Malcolm offered a related Wittgensteinian critique of the attempt to explain intention and other mental concepts by appealing to internal "cognitive processes," focusing especially on Chomsky's hypothesis of the existence of an innate grammatical structure. See his "The Myth of Cognitive Processes and Structures," in *Cognitive Development and Epistemology*, ed. Theodore Mischel (New York, 1971), 385–92.

purely mental event, such as Descartes conceived emotion to be, is an *Unding* [absurdity]," he wrote:

> Any word purporting to be the name of something observable only by introspection, and merely causally connected with publicly observable phenomena, would have to acquire its meaning by a purely private and uncheckable performance. But no word could acquire a meaning by such a performance; for a word only has meaning as part of a language; and a language is something essentially public and shareable. If the names of emotion acquire their meaning for each of us by a ceremony from which everyone else is excluded, then none of us can have any idea what anyone else means by the word. Nor can anyone know what he means by the word himself; for to know the meaning of a word is to know how to use it rightly; and where there can be no check on how man uses a word there is no room for "right" or "wrong" use. (*AEW*, 13)

Kenny traced that mistaken causal view of the emotions back to the influence of Descartes, commenting that

> In arguing against the notion of the "private object" Wittgenstein chose as his example a sensation, pain; no doubt because it is in such cases that it seems most plausible to suggest that a word might acquire a meaning by a private ostensive definition. With the emotions, the Cartesian idea of a purely mental event runs into extra difficulty. Emotions, unlike pain, have objects: we are afraid *of* things, angry *with* people, ashamed *that* we have done such-and-such. This feature of the emotions, which is sometimes called their "intensionality," is misrepresented by Descartes, who treats the relation between a passion and its object as a contingent one of effect to cause . . . [I]t is well known how difficult it is to give any account on Cartesian principles of a causal relation between physical and spiritual events. Here I will remark merely that the intensionality of the emotions adds a further reason for denying that emotion-words acquire their meanings by some private ostensive definition. Many people are attracted by the idea that the meaning of the word "pain" is learnt by picking out a recurring feature of experience and associating it with the sound of the word. It is much less plausible to suggest that the meaning of "fear" is learnt in the same way, when we reflect how very different from each other fears of different objects may be. *What* is the feature of experience, recognizable by introspec-

tion without reference to context, which is common to fear of mice and fear of waking the baby, fear of overpopulation and fear of being overdressed, fear of muddling one's sequence of tenses and fear of hell? (*AEW*, 13–14)

Kenny went on to suggest that the same erroneous Cartesian assumptions informed recent attempts to explain emotions by identifying their psychological or physiological causes, and tried to show in some detail why such approaches were conceptually and empirically inadequate. It is not that he rejected experimentation out of hand. He regarded as useful the methods being used to measure the intensity of emotions as precisely as possible based not only on what people said but also on the measurement of bodily changes, including facial expressions and psychogalvanic skin reflexes when subjects were exposed to shocks or films or other stimuli (these were just the kind of methods Lazarus himself was employing). But he cautioned that too many psychologists mistakenly believed that by measuring these phenomena they were measuring emotion itself. In his judgment, they were not. His argument was not just that scientists had so far been unable to detect distinct physiological or behavioral patterns associated with specific emotional states, although that was one of his points. Thus, he noted with reference to recent experiments that the same physiological and behavioral patterns had been detected in quite different affective states (a finding that undermined the kind of claims Tomkins had just made to the effect that different emotions could be defined by distinct patterns of physiological-behavioral response, though Kenny did not mention Tomkins). But Kenny's objection to such approaches was also the conceptual one that in such experiments psychologists were obliged to treat certain affects, such as hope, love, vanity, or fear of a world war, which were not easily connected to precise bodily changes, as degenerate or aberrant cases of emotion—as if a chronic fear of war was different from sudden fright only in being less intense and therefore less objectively measurable.[39]

As Kenny saw it, emotions were not mere behaviors but *motivated actions* whose status, as such, experiments were in principle incapable of capturing. We might put it that for Kenny emotions relate to their objects

39. As we saw in chapter 1, Tomkins in 1962 had indeed proposed that there exists a number of discrete emotions or "affect programs" with signature physiological and behavioral characteristics. Kenny in 1963 did not cite Tomkins but relied on reports of experiments in Robert S. Woodworth's influential textbook *Experimental Psychology* ([1938] 1954, 2nd ed., with Harold S. Schlosberg), as well on more recent findings (*AEW*, 34ff).

not as a result of causal mechanisms but by virtue of conceptual necessity. He observed:

> The precisely measurable bodily phenomena which psychologists study are not identical with the feelings of which they are characteristic; for unlike the feelings they have a merely contingent connection with motivated behavior. There is a conceptual connection also between a feeling and its object, whereas the physiological processes studied by psychologists lack intensionality. Bodily changes may be the vehicle of an emotion, but they are not themselves emotion. As Aristotle said, a man who defines anger as a bubbling of the blood about the heart gives only the matter without the form. (*AEW*, 38)

According to Kenny's Wittgensteinian approach to the emotions, understanding another person's emotional state is not a matter of inference, as Lazarus appeared to suggest, but of the direct apprehension or perception of facial and gestural expressions and the mastery of the grammar or meaning of emotion terms. He therefore suggested that all efforts to attribute emotions to putative inner mental states linked causally and contingently to external objects were confused or mistaken.

It is impossible to do justice here to the further details of Kenny's difficult arguments. He has been described as one of the first philosophers in the analytic tradition to defend the role of cognition in emotion, which is presumably why his book attracted Lazarus's attention.[40] Yet, if we ask what Lazarus got out of Kenny's text, the answer is—not much. Lazarus cited Kenny's book in one of his earliest attempts to outline a cognitive theory of emotion in a paper he and his colleagues gave at a symposium held at Loyola University in 1968.[41] The symposium, organized by Arnold herself, gathered together an array of psychologists offering disparate, often competing views. Plutchik, Tomkins, and Schachter were there, among other well-known conference participants.

In their presentation to the meeting, published in the proceedings two years later, Lazarus et al. commented on the difficulty of defining emotions by observing that theorists had often assumed there must be some one "characteristic" unique to emotions. But that research had failed to reveal the one "thing" to which the noun "emotion" referred. This was not because of experimental or introspective failures, they suggested, but be-

40. Phil Hutchinson, *Shame and Philosophy*, 88.

41. Richard S. Lazarus, James R. Averill, and Edward M. Opton Jr., "Towards a Cognitive Theory of Emotion," in *Feelings and Emotions: The Loyola Symposium*, ed. Magda N. Arnold (New York and London, 1970), 207–32; hereafter abbreviated as "TCTE."

cause of the kind of category to which emotion words belonged. Emotion words do not refer to things, they proposed, but rather to "syndromes" in the sense that a disease is a syndrome. As syndromes, emotions are not characterized by a single symptom or set of symptoms, nor do they have one center or locus.

By comparing emotions to disease syndromes, Lazarus et al. appeared to be offering a natural-scientific approach to the affects. But then they seemed to change direction by mentioning Kenny's book. "In addition to referring to syndromes," they declared in citing Kenny's text, "emotion concepts are also relational; that is, they typically imply an object, just as the concept 'answer' implies a question" ("TCTE," 212). This statement is self-divided because in it the authors appear simultaneously to be making a "grammatical" claim in the spirit of Kenny's arguments—that emotions imply an object in the same way that the concept "answer" implies a question—and an empirical claim at odds with Kenny's position. For if, according to Lazarus et al., emotion concepts "typically" imply the existence of objects, then they may also atypically lack them, suggesting that the relationship between emotions and their objects is not a necessary and conceptual one, as Kenny had argued, but is only a contingent matter. Perhaps, though, it would be more accurate to say that the question-answer analogy proposed by Lazarus et al. was not "grammatical" in any Wittgensteinian sense at all, but was merely definitional. For if you want to know what an "answer" is, you can look the word up in a dictionary, which will tell you that an answer is a response to a question. But the meaning of "emotion" cannot be established by consulting a dictionary in that way, for there is no comparable definitional sense of the word, which is why Kenny thought it was necessary to undertake a conceptual analysis. In short, the authors' question-answer analogy was not adequate to the issue.

Indeed, what is notable about Lazarus et al.'s reference to Kenny's book is their failure to grasp the import of the latter's arguments. It is true that in the same paper Lazarus, Averill, and Opton made a few comments about the relevance of ordinary language usage and terms for psychology. They also cited a recently published paper on the value of ordinary language philosophy itself by his student and coauthor Averill, who had also read Kenny's work.[42] But not only did Lazarus et al. quickly pass on to other

42. In another paper, Averill in stated in words very close to Lazarus's that "The concept 'emotion' implies an object, just as the concept 'answer' implies a question. That is, emotional terms refer to relational as well as to intrinsic properties of the response." Averill added: "This should not, of course, be taken to mean that an emotion

topics, I think it is fair to say that the essential issues with which Kenny was dealing eluded them.[43] In particular, the authors did not appreciate the conceptual-Wittgensteinian considerations that drove Kenny's investigation and instead aligned his work with the kind of causal investigation of the emotions to which the latter was in fact opposed. Lazarus et al. proceeded as if what was required for an understanding of emotions defined as "syndromes" was to identify the inner cognitive appraisal processes causing emotional states and to link these states to their contingently attached external objects. We might put it that Lazarus et al. "psychologized" and operationalized the issue of intentionality in ways that pulled against Kenny's Wittgensteinian-inspired diagnosis. (I am reminded here

---

is more than its manifestations, any more than an answer is more than the statement which expresses it. The concept of object as it applies to emotional response is quite complex (Gosling, 1965; Kenny, 1963) and is not discussed here. At the risk of oversimplification, it may simply be said that each emotion is logically appropriate to a certain class of objects" (James R. Averill, "Grief: Its Nature and Significance," *Psychological Bulletin* 70 [1968]: 742). Gosling's 1965 paper, cited by Averill, was a critical review of Kenny's 1963 book in which the author discussed the complexity of the distinction between the concepts of "object" and "cause." Gosling focused especially on the familiar question of (so-called) "objectless emotions," such as states of "pointless depression," states that, he argued, posed an obstacle to Kenny's intentionalist position. See J. C. Gosling, "Emotion and Object," *Philosophical Review* 74 (4) (1965): 486–503.

43. In yet another paper, Averill referred to works by Ryle, Wittgenstein, Austin, and other authors, including papers by philosophers Bedford and George Pitcher that critiqued Cartesian-causal theories and emphasized instead the intentionality and object-directed nature of emotions in terms not unlike Kenny's. Averill's article was a brave attempt to offer criticisms of efforts to achieve scientific precision about language by operationalizing meaning, in the course of which he made some pertinent comments about Wittgenstein's and Ryle's critiques of the notion of private psychological states and their reification as mental events. See James R. Averill, "Operationism, Metaphysics, and the Philosophy of Ordinary Language," *Psychological Reports* 22 (1968): 861–87. But it is a sign of the difficulty of understanding Wittgenstein's contribution to ordinary language philosophy that in the course of his article Averill cited as having "much merit" (884) Jerry Fodor and J. J. Katz's famous (or infamous) attack on the philosopher Stanley Cavell's remarkable essays "Must We Mean What We Say?" and "The Availability of Wittgenstein's Later Philosophy," whose aim was precisely to defend Wittgenstein's procedures against Fodor and Katz's empirical-positivist misunderstandings and to say what he thought those procedures are, properly understood. In this connection, see Stanley Cavell, "Must We Mean What We Say?," *Inquiry* 1 (1958): 172–212; Stanley Cavell, "The Availability of Wittgenstein's Later Philosophy," *Philosophical Review* 71 (1962): 67–93; J. A. Fodor and J. J. Katz, "The Availability of What We Say," *Philosophical Review* 72 (1963): 57–71; Stanley Cavell, "Aesthetic Problems of Modern Philosophy," in *Must We Mean What We Say?* (New York, 1969), 73–96; Richard Henson, "What We Say," *American Philosophical Quarterly* 2 (1) (1965): 52–62; and Stanley Bates and Ted Cohen, "More on What We Say," *Metaphilosophy* 3 (1) (1972): 1–24.

of Stanley Cavell's brilliant thought that the achievement of Wittgenstein's *Philosophical Investigations* was to de-psychologize psychology.)[44] In sum, the authors adopted the very paradigms and modes of analysis that Kenny rejected and in doing so showed that they had not understood the stakes of the latter's book. They implicitly assumed a Cartesian view of the mind as a private psychic domain separated from a world of objects onto which the individual mind imposed its personal cognitions and evaluations. They thereby individualized and subjectivized affective meaning.

Lazarus, Averill, and Opton's inability to comprehend what Kenny was trying to do is not surprising. Philosophical work does not circulate in other disciplines the way empirical research does, and it usually takes several decades before it is actually understood and digested in ways that allow it to have an impact outside philosophy itself. It is only recently that philosophers have begun to focus on the question of intentionality in terms that throw light on what Kenny was trying to accomplish when he contended that, since mental states such as intentions are connected with their objects not by empirical laws but conceptually, mental states cannot be considered the cause of the emotions. As Stoutland has pointed out, at the time Kenny published his book the work of the "school" of the philosophy of intention and action to which he belonged, stemming from Wittgenstein's and Anscombe's thought, was neglected or even disdained by most mainstream philosophers, who argued that the philosophy of intention required lawlike, empirical generalizations that causally con-

44. Cavell writes: "We know of the efforts of such philosophers as Frege and Husserl to undo the 'psychologizing' of logic (like Kant's undoing Hume's psychologizing of knowledge): now, the shortest way I might describe such a book as [Wittgenstein's] *Philosophical Investigations* is to say that it attempts to undo the psychologizing of psychology, to show the necessity controlling our application of psychological and behavioral categories; even, one could say, show the necessities in human action and passion themselves" (Cavell, "Aesthetic Problems of Modern Philosophy," in *Must We Mean What We Say*, 91). In a footnote Cavell quotes from and then comments on a passage from Wittgenstein's text: "Consider, for example, the question: 'Could someone have a feeling or ardent love or hope for the space of one second—*no matter what* preceded or followed this second?' (*Investigations*, ¶583). We shall not wish to say that this is logically impossible, or that it can in no way be imagined. But we might say: given our world this cannot happen; it is not, in our language, what 'love' or 'hope' mean; necessary in our world that this is not what love and hope are. I take it that our most common philosophical understanding of such notions as necessity, contingency, synthetic and analytic statements, will not know what to make of our saying such things" (91). Everything challenging Lazarus's attempt to establish a science of emotions is contained in these words. The question posed by them is: *what* kind of scientific enterprise in psychology would make sense in the light of the considerations Cavell raises?

nected the phenomenon to be explained with the conditions that gave rise to it.[45] If the majority of philosophers rejected Wittgenstein's and Anscombe's undertaking, what chance did a busy experimental psychologist like Lazarus have of appreciating Kenny's arguments?

Besides, even if Lazarus had been able to master and accept them, their implications would have been unclear. Did it follow from Kenny's discussion that all experimental approaches to the affects would have to be given up as misguided? Or was the import of his conclusion rather that certain *kinds* of experimental strategies were ill advised—for example, those that imagined that the goal was to identify the putative internal psychic mechanisms thought to cause emotional reactions—but that more ecological or ethological approaches, involving the study of intact animals in their natural settings, might be warranted and valuable? Those questions do not seem to have even occurred to Lazarus at this time.[46] All in all, it is understandable that when Lazarus first read Kenny's book he was unable to grasp its mode of argumentation or its diagnosis of the problem of the intentionality of the emotions. It is interesting that he found his way to Kenny's work in those years, but a comprehension of its implications for psychology was beyond his reach.

It might seem that I have attributed more importance to Kenny's book than is warranted by Lazarus's reference to it in one paper. My rationale for paying attention to it is both historical and theoretical. In fact, Lazarus cited Kenny's text more than once during the 1970s precisely because it appeared to him relevant to certain fundamental questions about the nature of the emotions.[47] And the same issues at stake in Kenny's book would

45. Frederick Stoutland, "Introduction: Anscombe's *Intention* in Context," in *Essays on Anscombe's Intention*, eds. Anton Ford, Jennifer Hornsby, and Frederick Stoutland (Cambridge, MA, 2011), 6–8.

46. Moreover, I do not want to suggest that Kenny's analysis of the nature of emotions as intentional actions was so brilliantly and coherently carried out that its message ought to have been obvious. On the contrary, not only were the issues he was dealing with inherently difficult, it's not even clear how sound was his own grasp of Wittgenstein's ordinary language procedures and their implications for psychology. As the recent collection *Essays on Anscombe's Intention* makes evident, the interpretation of Anscombe's discussion of intention and action on which Kenny drew remains to this day a topic of debate and contestation.

47. See Richard Lazarus, Masatoshi Tomita, Edward Opton Jr., and Masahisa Kodama, "A Cross-Cultural Study of Stress-Reaction Patterns in Japan," *Journal of Personality and Social Psychology* 4 (1966): 622–33; James R. Averill, Edward H. Opton Jr., and Richard S. Lazarus, "Cross-Cultural Studies of Psychophysiological Responses During Stress and Emotion," *International Journal of Psychology* 4 (2) (1969): 91; and Richard S. Lazarus and James R. Averill, "Emotion and Cognition: With Special Reference to Anx-

continue to concern not just Lazarus but other emotion theorists as well, some of whom also read Kenny's text.[48] Indeed, as recently as 2013 the well-known philosopher of biology Paul Griffiths, whose views I mentioned in the introduction and will be discussing in chapter 6, cited Kenny's book in the course of reviewing current emotion research in philosophy.[49] That Griffiths favors a non-intentionalist account of the so-called basic emotions at odds with Kenny's intentionalism does not detract from the interest of the fact that, to this day, the issue of intentionality and indeed Kenny's views continue to haunt the field of the affective neurosciences.

It is worth noting that if Lazarus believed that affective meaning was to be considered the outcome of causal, psychic mechanisms or processes, then his position raised empirical expectations that he would soon find challenging to meet. For example, he needed to show experimentally that the appraisal process was separate from and prior to the resulting emotional response, a difficult requirement given that, as we shall see, according to Lazarus himself emotional reactions to appraisals are instantaneous, so there is virtually no time gap between the two "events." The

---

iety," in *Anxiety: Current Trends in Theory and Research*, ed. C. D. Spielberger (New York, 1972), 253–54. In these experiments, the general patterns of reaction were assessed on three dependent variables: skin conductance, distress rating, and a modified Nowlis Adjective Check List of Mood. The latter did not measure basic emotion categories but the following items: Concentration, Aggression, Pleasantness, Activation, Deactivation, Egotism, Social Affection, Depression, and Anxiety. For details concerning the Nowlis checklist, which originally contained 145 adjectives, see Vincent Nowlis, "The Development and Modification of Motivational Systems in Personality," in *Current Theory and Research in Motivation: A Symposium* (Lincoln: University of Nebraska Press, 1953), 114–38; and V. Nowlis and H. Nowlis, "The Description and Analysis of Mood," *Annals of the New York Academy of Science* 65 (1965): 345–55.

48. See for example W. G. Parrott, "The Role of Cognition in Emotional Experience," in *Recent Trends in Theoretical Psychology: Proceedings of the Second Biannual Conference of the International Society of Theoretical Psychology, April 20–25, 1987*, eds. W. J. Baker, L. P. Mos, H. V. Rappard, and H. J. Sam (New York, 1988), in which, in the process of critiquing Jamesian-feeling approaches to the emotions and defending a cognitivist position, the author states: "It is well known that this 'Cartesian' view of emotion succumbs to a variety of criticisms. Wittgensteinians have pointed out that this view falls prey to the problems of any 'private language' (Kenny, 1963)" (331).

49. Paul E. Griffiths, "Current Emotion Research in Philosophy," *Emotion Review* 5 (2) (2013): 215–22. See also Andrea Scarantino, *Explicating the Emotions*, unpublished doctoral dissertation, University of Pittsburgh, 2005, which cites and discusses aspects of Kenny's 1963 book in connection with the problem of intentionality and the latter's emphasis on the role of the formal object in emotion. Of course, Kenny would not have endorsed Scarantino's attempt to naturalize the intentionality of the emotions by appealing to Ruth Millikan's teleosemantic theory.

causal role of appraisal in emotion was also going to be hard to demonstrate empirically if, as he also believed, appraisal is frequently unconscious and automatic and hence is unavailable for conscious self-report.

Some of these issues did not surface as a challenge until the 1980s, when Zajonc launched the controversy over the role of cognition in emotion. In the meantime, Lazarus spent the 1970s completing his laboratory studies of stress and then turning to the clinic as offering a more natural setting for the study of affect. Unhappy with what he perceived as the limitations of laboratory research approaches, he decided to pursue his by-now-influential appraisal approach by pursuing in-depth studies of how individuals react to different life threats in more ordinary, everyday settings. The shift to the clinic yielded a series of reports in which Lazarus attempted to enlarge and make more precise his views about appraisal and coping processes and to address some of the methodological challenges involved in conducting appraisal research across time and in different clinical groups and situations.[50] But in the course of pursuing his agenda, Lazarus committed himself to certain views that further compromised his position on the intentionality of the emotions. It is to this aspect of his work that I now turn.

DIFFERENTIATING THE EMOTIONS

Among the views to which Lazarus was committed was that emotions are differentiated into natural classes or types. It is generally accepted by today's appraisal theorists that both Arnold and Lazarus endorsed the notion of discrete emotions as central to their respective enterprises. Not only do they associate the rise of Arnold's and Lazarus's appraisal concept with the claim that emotions are segregated into distinct, universal categories, they also link Arnold's and Lazarus's work to the idea that each basic emotion is tied to a specific type of appraisal. Thus, Roseman and

50. For Lazarus's reasons for transitioning to clinical research, see Richard S. Lazarus and Raymond Launier, "Stress-Related Transactions Between Person and Environment," in *Perspectives in Interactional Psychology*, eds. L. A. Pervin and M. Lewis (New York, 1978), 287–327; Richard S. Lazarus and Susan Folkman, "Transactional Theory and Research on Emotions and Coping," *European Journal of Personality* 1 (3) (1987): 141–69; Richard S. Lazarus, James C. Coyne, and Susan Folkman, "Cognition, Emotion and Motivation: The Doctoring of Humpty-Dumpty," in *Psychological Stress and Psychopathology*, ed. R. W. J. Neufeld (New York, 1982), 218–39; Richard S. Lazarus, "Vexing Research Problems Inherent in Cognitive-Mediational Theories of Emotion and Some Solutions," *Psychological Inquiry* 6 (1995): 183–96; and Richard S. Lazarus, *The Life and Work of An Eminent Psychologist*, chapter 9.

Smith, two prominent appraisal theorists, have recently suggested that among a number of specific problems the early appraisal theories were trying to solve, the first was how to account for the differentiated nature of emotional responses.[51] This is how Lazarus himself viewed the matter.

In her 1960 book, Arnold had already expressed her opinion that emotions formed distinct categories and that each distinct emotion was accompanied by a specific pattern of bodily activity.[52] Early on, Lazarus expressed a similar belief, even though he conceded that convincing evidence was still lacking. Thus, in 1966 he stated his assumption that the affects "are different from each other even though they appear to overlap and coexist" and claimed that there was "no good reason to assume the negative position that the names for different affects do not refer to distinguishable states and that the person is never capable of accurately and consistently reporting about them" (*PS*, 321). He repeated this belief on several occasions, stating in 1968, for instance, that the analysis of emotions conceived as distinct response syndromes made "mandatory, at some point, the classification of emotions." He remarked that "I have taken the position that anger, fear, grief, euphoria, love, etc., are distinctively different in their three major response components, the physiological, motor, and cognitive. A consequence . . . is that research must be

51. See Ira Roseman and Craig A. Smith, "Appraisal Theory: Overview, Assumptions, Varieties, Controversies," in *Appraisal Processes in Emotion: Theory, Methods, Research*, eds. Klaus R. Scherer, Angela Schorr, and Tom Johnstone (Oxford, 2001), 3–19. See also Angela Schorr, "Appraisal: The Evolution of an Idea," in *ibid.*, 20–34. Appraisal theorist Phoebe Ellsworth has recently observed that basic emotion theories were the dominant psychological theories at the time she and other authors in the 1980s tried to show that the emotions proposed by the basic emotion theorists could be associated with a distinctive combination of appraisals. However, the author points to two "regrettable consequences" of that step: first, it suggested that appraisal theorists were just like the basic emotion theorists in positing the emotions as natural kinds, a view Ellsworth now rejects in the light of criticisms by Barrett (2006) and others; and second, the commitment to the basic emotions distracted attention from certain fundamental assumptions of appraisal theory, namely (1) that appraisals are continuous, and not categorical; (2) that a person can feel emotional even if the combination of appraisals does not correspond to any of the basic emotions; and (3) that emotional experience is not a state, but a "process" in which changes in appraisals, bodily responses, and action tendencies feeding back into each other transforming the emotional experience. See Phoebe C. Ellsworth, "Appraisal Theory: Old and New Questions," *Emotion Review* 5 (2) (2013): 126–28.

52. Magda B. Arnold, *EP*, especially chapter 10. See also Reisenzein, "Arnold's Theory of Emotion in Historical Perspective," 920–51, on Arnold's debt to William McDougall's instinct theory for her ideas about the existence of "discrete" or "basic" emotions.

aimed at distinguishing descriptively the properties of different emotional reactions."[53] He often declared that one of the greatest hindrances to the study of emotions was the lack of an adequate scheme for their definition and classification and called for a nosology of emotional phenomena. He held on to this position for the rest of his career. In one of his last papers, he joined his appraisal theory to the discrete emotions position by stating that "The theoretical construct that eventually came to be emphasized by those adopting a discrete emotions outlook is *appraisal*, its central issue being what an individual must think and want in order to react with each of the emotions."[54]

As this last remark suggests, Lazarus linked the idea of discrete emotions to the existence of corresponding patterns of cognitive appraisal or meaning. Thus, as early as 1966 he stated that "different cognitive processes are associated with different affects. And these cognitive processes determine the form of coping with threat" (*PS*, 323). He reiterated this point, stating his view that it was likely that "one day *specific physiological patterns, specific cognitive patterns, and specific behavioral patterns, in association with given eliciting conditions, will be identified, and shown to be organized in such a way as to clearly distinguish one emotional state from another*" (his emphasis).[55] The notion that discrete emotions are attached to characteristic cognitive patterns or coping styles became a central motif of his work.[56] Lazarus thus called for a research program designed to identify objectively the cognitive, behavioral-expressive, and autonomic components of different emotional reactions.[57]

53. Richard S. Lazarus, "Emotions and Adaptation: Conceptual and Empirical Relations," 197, 256.

54. Richard Lazarus, "Relational Meaning and Discrete Emotions," in *Appraisal Processes in Emotion: Theory, Methods, Research*, eds. Scherer, Schorr, and Johnstone, 37.

55. Lazarus, "Emotions and Adaptation: Conceptual and Empirical Relations," 204.

56. "The point is that each emotion involves its own particular kind of appraisal," Lazarus wrote in 1970, "its own particular kind of action tendencies, and hence its own particular constellation of physiological changes which are part of the mobilization to action . . . [T]he motor-expressive features (facial expressions or bodily postures) are in many cases universal in the species, though modifiable by social custom." Lazarus, Averill, and Opton, "Towards a Cognitive Theory of Emotion," 218.

57. "[T]he cognitions determining different emotional responses (syndromes) must be identified, and in turn, this leads naturally to a search for the antecedent conditions, both in the stimulus configuration and within the psychological structure of the individual, which govern the expression of the component reactions . . . [T]his plan . . . must be filled in by substantive theoretical propositions about the intervening cognitions of anger, fear, grief, joy, etc., and the conditions under which they will emerge." Ibid., 229.

It is not difficult to motivate historically Lazarus's belief that emotions form discrete types or categories. For one thing, it seemed to him on phenomenological grounds that we experience different emotions. In this regard, he was reacting against the majority consensus of the time, which held that emotions could not be differentiated into discrete categories because the evidence for characteristic behavioral or physiological patterns of response associated with specific emotions was lacking. As Lazarus explained, for theoretical reasons the "emotion-as-motivation" theorists treated emotion as merely an intervening variable; the drive theorists likewise downplayed the idea of emotional differentiation; and the behaviorists tended to make a single emotion, such as anxiety, carry all the theoretical weight. The net result was that emotion was treated as an undifferentiated state of drive or arousal rather than as a differentiated array of experiences and somatic and action patterns. "The great importance attached to the drive or motivational properties of emotion led to the virtual disappearance of emotion as a substantive topic in psychology journals," Lazarus et al. noted in 1982. "Aspects of emotion that were not readily accommodated in these terms were neglected . . . Psychology suffered a severe form of tunnel vision."[58]

So strong were the doubts about the existence of discrete emotions that, as Lazarus et al. observed in the same paper, in 1941 and 1962 Duffy had argued that the emotion concept was worse than useless, and in 1951 Brown and Farber had likewise rejected emotion as a concept suitable for scientific study. Lazarus et al. even criticized Schachter's well-known cognitivist model on the grounds that although the latter emphasized the role of cognition in emotion, he treated cognition as involving a labeling of diffuse states of arousal in terms too close to that of the drive activation theorists to satisfy Lazarus's convictions about affective differentiation.

Against this background, it is comprehensible that in 1966, at the very moment he began to pay attention to the emotions, Lazarus welcomed Silvan Tomkins's "somewhat unsystematic" attempt to define the affects as discrete states (*PS*, 321). It is also intelligible that early on he was receptive to Ekman's views about the existence of discrete emotions and the leakage of nonverbal behaviors as sources of information about specific emotions.[59] Repeating the opinion that one of the greatest barriers

58. Lazarus, Coyne, and Folkman, "Cognition, Emotion, and Motivation: The Doctoring of Humpty-Dumpty," 224.

59. See for example Lazarus, "Emotions and Adaptation: Conceptual and Empirical Relations," 208, citing Paul Ekman, "Communication Through Non-Verbal Behavior: A

to the investigation of emotion had been the lack of an adequate classification, Averill, Opton, and Lazarus in a paper of 1969 offered what they called a "first approximation" of this kind. First, they identified three major components of any emotional response system: "input variables (stimulus properties)," the "appraiser subsystem," and "output variables (response categories)." *Input variables* or stimulus properties included "information" specific to objects with emotional meaning, nonspecific information such as novelty, intensity, ambiguity, as well as input due to the response feedback from the individual. The *appraiser subsystem* included the individual's primary and secondary appraisals of the stimulus situation, a subsystem that Averill, Opton, and Lazarus compared to the "decider system" in J. Miller's general systems theory, thereby implying a problematic equivalence between their appraisal theories and the latter's information-processing approach—a topic to which I will return in the next chapter. The authors then defined the *output* variables as involving three different emotional response categories: the *cognitive*, the *expressive*, and the *instrumental.* The cognitive response reactions included reappraisals or other modes of defense, and the authors here made a pitch for the importance of such cognitive modes of response.[60]

Especially interesting is what the authors had to say about the "expressive" and "instrumental" categories of response. They described expressive reactions as "stereotyped and rather discrete, e.g., weeping, laughing, and certain facial displays," and said of them that such expressive reactions "typically are not intentionally directed toward any goal." They included in this group "*biologic expressors*" such as animal displays and the fixed action patterns described by ethologists, as well as "acquired expressors" of the kind learned in childhood, as described by Ekman and Friesen and others. They contrasted those biological expressive responses with *instrumental* responses, consisting of "coordinated behavior sequences directed toward some goal, e.g., avoidance of harm or injury to another." The distinction between expressive and instrumental reactions was not considered hard and fast, and the authors' further subdivision of

Source of Information About an Interpersonal Relationship," in *Affect, Cognition and Personality: Empirical Studies*, eds. S. S. Tomkins and C. E. Izard (New York, 1965), 390–442. See also James R. Averill, Edward M. Opton Jr., and Richard S. Lazarus, "Cross-Cultural Studies of Psychophysiological Responses During Stress and Emotion," 88, for references to Paul Ekman and Wallace V. Friesen, "The Repertoire of Nonverbal Behavior-Categories: Origins, Usage, and Coding," *Semiotica* 1 (1969): 49–98.

60. James R. Averill, Edward M. Opton Jr., and Richard S. Lazarus, "Cross-Cultural Studies of Psychophysiological Responses During Stress and Emotion," 83–102.

instrumental responses into "symbols," "operators," and "conventions" was not especially clarifying.[61]

What should be remarked on is that in this paper we see the extent to which Lazarus's position was beginning to resemble Ekman's developing neurocultural theory of the emotions. From one point of view this is understandable because, as the authors report, Ekman was collaborating with Lazarus at just this time on experiments designed to determine the influence of cultural differences on emotional responses. In fact, one of the studies described by Averill, Opton, and Lazarus appears to have been the prototype for Ekman's soon-to-be-famous study of the (so-called) "spontaneous" reactions of Japanese and American students to stressful films. In 1963–1964, Lazarus had spent a year in Japan carrying out a series of experiments designed to elucidate the cultural differences, if any, between Japanese and American coping responses, by comparing their respective reactions to the *Subincision* film. In their first report of this cross-cultural study, Lazarus and his colleagues reported that although the Japanese subjects responded to the film in a manner quite similar to that of their American counterparts, there was one major cross-cultural difference. Specifically, unlike their American counterparts, whose levels of physiological arousal as measured by skin conductance fluctuated throughout the film viewing process in response to the sequence of threatening and less threatening or "benign" scenes, Japanese subjects not only demonstrated consistently high levels of skin conductance during both the control and stress films, but exhibited very little variation corresponding to the threatening or benign scenes of the film. In other words, the Japanese subjects appeared to be in a continuous state of physiological arousal during the film viewing, an arousal not reflected in their self-reports. After ruling out various possible explanations for these results, Averill, Opton, and Lazarus proposed that the psychological experiment itself was threatening to the Japanese because, unlike Americans, they were not used to being observed or evaluated under such experimental conditions and therefore found the experience highly stressful. The Japanese were therefore reacting to the entire situation and not to variations in the stimulus content.[62]

61. Averill, Opton, and Lazarus, "Cross-Cultural Studies of Psychophysiological Responses During Stress and Emotion," 88.

62. R. S. Lazarus, E. Opton, M. Tomita, and M. Kodoma, "A Cross-Cultural Study of Stress-Reaction Patterns in Japan," *Journal of Personality and Social Psychology* 4 (1966): 633–23. Lazarus made the same point in Averill, Opton, and Lazarus, "Cross-Cultural Studies of Physiological Responses During Stress and Emotion," 97.

That finding prompted further experiments back in the United States to determine whether the results changed if the Japanese were allowed to adapt to the experimental situation. Preliminary findings, also reported by Averill, Opton, and Lazarus, indicated that the Japanese students continued to show higher levels of skin conductance throughout the experiment, even as their self-reports failed to register their apprehensions. At this juncture, Ekman entered the scene as a co-partner with Lazarus in carrying out yet another experiment, this time focused on the facial expressions of the experimental subjects during the film viewings.[63] In other words, Lazarus and his group now incorporated Ekman's interest in facial expression into their experimental protocols. In a pilot study described in the same paper by Averill, Opton, and Lazarus, hidden observers rated the expressiveness of the Japanese and American subjects as they watched control and stress films. Motion pictures were also taken of the faces and, in a further variant, the authors reported videotaping the faces of both Japanese and American subjects as they viewed a specially edited film containing scenes from four stress films, while physiological changes were continuously recorded.

This last experiment appears to have been the prototype of the famous experiment on the difference between "spontaneous" reactions of Japanese and American students to stress films carried out by Ekman and Friesen. As discussed in chapter 2, in that experiment, first reported in

63. Ibid., 98, n. 4. For Ekman and Friesen's report on the origin of these experiments, see their "The Repertoire of Nonverbal Behavior: Categories, Origins, Usage, and Coding," 81: "This research is being conducted in collaboration with Lazarus, Averill, and Opton, utilizing their stress-inducing procedure . . . In our study we take motion pictures of the subject's facial expressions and hand movements without his knowledge, while he watches a stress film and a neutral film. We have collected pilot data, utilizing very brief samples of nonverbal behavior both in Japan and in the U.S. Our analysis of these records involves both the application of FAST and the collection of Japanese and U.S. observers' interpretation of the affect shown in the Japanese and U.S. films. The FAST analysis is designed to determine whether the same muscles move in response to stress in both cultures, and whether the same display rules are exhibited in the two cultural groups. The observers' interpretations of the stimuli will reveal whether both cultures interpret similarly the stress reactions of members of their own and of another culture. It is too early to report results from this study, other than the impressions that the procedures have worked and information is obtainable from the experiments." In Paul Ekman, Wallace V. Friesen, and Silvan S. Tomkins, "Facial Affect Scoring Technique: A First Validity Study," *Semiotica* 3 (1971): 57, the authors cite an unpublished manuscript, P. Ekman, R. S. Lazarus, W. V. Friesen, E. T. Opton, J. R. Averill, and E. J. Malmstrom, "Facial Behavior and Stress in Two Cultures." As far as I have been able to determine, that paper was never published in this form.

1972, Ekman claimed to have demonstrated the way in which cultural display rules controlling for polite behavior when in the presence of an authority figure caused the Japanese students, unlike the Americans, to cover over with polite smiles their natural-universal disgust response at seeing such stress films.[64]

On the basis of that experiment, which went on to acquire iconic status, and with reference to his cross-cultural judgment studies, Ekman argued that Tomkins was right to suggest the existence of a limited number of discrete, basic emotions or "affect programs," programs or patterns of emotional response that could be disguised by learned, culturally determined "display rules" but that under the right conditions would "leak out." In the same way, Lazarus began to suggest that there exist phylogenetically evolved, wired-in mechanisms of response, such as facial expressions, which could be camouflaged or inhibited, especially in humans, by the superimposition of culturally determined norms and behaviors, but which were a biological given or predisposition. It is as if Lazarus saw Ekman as providing a particular piece of the emotion puzzle (the facial expression piece), without believing that Ekman's views compromised his own commitment to the "cognitive" piece (the cognitive appraisal piece) of the same puzzle.

Yet for Lazarus to proceed in this way was to risk undermining the coherence of his position. By declaring, as Averill, Opton, and Lazarus did in their 1969 paper, that facial expressions are not directed toward any goal, Lazarus risked severing emotion from the intentional object in ways that conformed to Tomkins's and Ekman's views about the non-intentional status of the basic emotions but were not coherent with his own professed intentionalism. As I have shown in previous chapters, for Tomkins and Ekman the way to understand the emotions is that they are elicited or "triggered" by what we might call the emotional object, but the object is nothing more than a stimulus or trip wire for an inbuilt behavioral response. We might put that in Tomkins's and Ekman's account the intentional object of the emotion is turned into the causal trigger or "releaser" of the emotional "program" or response. The issue for them was to determine the causal conditions activating the automatic discharge of the affective response. On this model, emotional responses could and did interact with the purposive-cognitive systems in the brain, but they were in principle independent of them.

The trouble with such formulations for the coherence of Lazarus's proj-

64. Paul Ekman, "Universals and Cultural Differences in Facial Expressions of Emotion," in *Nebraska Symposium of Motivation 1971*, ed. J. Cole (Lincoln, NE, 1972), 207–83.

ect is that the intentional object as such tended to disappear from view. Indeed, in their 1969 paper Averill, Opton, and Lazarus mentioned the intentional object almost as an afterthought and with no clear connection to the theoretical framework that preceded it. Thus, having offered a provisional classification of the emotions, the authors added with yet another reference to Kenny's 1963 book that "the mere listing of reactions, no matter how complete, is insufficient to specify a response as emotional. Emotional concepts are *relational*, that is, they refer to responses in relation to objects, which may include complex situational and social factors (Kenny, 1963). In this respect, an emotion, which implies an object, is like an answer, which implies a question" (his emphasis).[65] But this by-now-familiar statement hangs, and when the authors again compared emotions to disease syndromes, the reader is left wondering whether, in referring to Kenny's work once again, Lazarus and his coauthors believed they were offering a conceptual analysis of emotions as intentional states or making an empirical claim about the intentional object—or both.

Moreover, as his ideas about the various components of emotions and their differentiation and classification evolved, Lazarus began to further develop the notion that appraisals fall into characteristic patterns according to whether there is a particular environmental harm or benefit for the person involved. He called these appraisal patterns "core relational themes" and tied those themes to specific emotions: thus, the appraisal of loss was linked to the emotion of sadness, and so on. His formulations came so close to Tomkins's and Ekman's "affect program" views as to be almost indistinguishable from them. In his much later book *Emotion and Adaptation* (1991), summarizing his life's work (Lazarus retired from teaching that year), he commented:

> *If* a person appraises his or her relationship to the environment in a particular way, *then* a specific emotion, which is tied to the appraisal pattern, always follows. A corollary is that *if* two individuals make the same appraisal, *then* they will experience the same emotion, regardless of the actual circumstances. I think of this as a *psychobiological principle*, which provides for universals in the emotion process of the human species and probably applies to other animals, too. In other words, we are constructed in such a way that certain appraisal patterns and their core relational themes will lead to certain emotional reactions. This biological principle is similar in function to

65. Averill, Opton, and Lazarus, "Cross-Cultural Studies of Psychophysiological Responses During Stress and Emotion," 91.

> the concept of *affect programs* . . . though my version of it is highly flexible, especially with respect to the input; therefore, there is considerable variation in the exact details of the emotional response.
>
> The psychobiological principle is essential for a cognitive-motivational-relational theory of emotion, because it implies a degree of universality—or commonality—in the emotion process, which seems evident observationally. (his emphasis)[66]

Here, Lazarus explicitly recognized the resemblance between his views and the Tomkins–Ekman "affect program" approach to the emotions, even as he claimed he attached greater importance than either Tomkins and Ekman did to the influence of individual experiences on the input side of the affective process.[67]

But, as the anthropologist Richard Shweder observed in a sympathetic yet shrewdly critical review of Lazarus's "great book," in which he focused on this passage, Lazarus's core relational themes were analyzed in such abstract terms, such as "irrevocable loss" rather than "irrevocable job loss" or "irrevocable loss of a child," that, despite his cognitivism, he ended up "short of fully defining the emotions by reference to their 'objects' (such as loss of a job in contrast to loss of a child)." The result was that Lazarus's appraisal theory made it possible to treat two mental appraisals as equivalent, regardless of differences in the events or objects they were about.[68] As Shweder put it: "It is assumed [by Lazarus] that mental processes do not function concretely and that, in the production of an emotional experience, there is no fundamental qualitative difference between, for example, 'I feel the way I feel *because my child has died*' and 'I feel sad *because I was fired from my job*'" ("EY," 323; his emphasis). I would supplement Shweder's remarks by pointing out that the differ-

66. Richard S. Lazarus, *Emotion and Adaptation* (New York, 1991), 191.

67. For the convergence between Lazarus's views and Ekman's position on discrete emotions, see Paul Ekman, *The Face of Man: Expressions of Universal Emotions in a New Guinea Village*, foreword by Richard S. Lazarus (New York, 1988); Paul Ekman, "An Argument for Basic Emotions," *Cognition and Emotion* 6 (3–4) (1992): 187; Richard Lazarus, "Universal Antecedents of the Emotions," and Paul Ekman, "Antecedent Events and Emotion Metaphors," in *The Nature of Emotion: Fundamental Questions*, eds. P. Ekman and R. J. Davidson (New York and Oxford, 1994), 146–49, 163–17; and Paul Ekman, "What We Become Emotional About," in *Feelings and Emotion: The Amsterdam Symposium*, eds. Antony S. R. Manstead, Nico Frijda, and Agneta Fischer (Cambridge, 2004), 119–35.

68. Richard A. Shweder, "Everything You Ever Wanted to Know About Cognitive Appraisal Theory Without Being Conscious of It," *Psychological Inquiry* 4 (4) (1993): 322–42; hereafter abbreviated as "EY."

ences between those two experiences of loss involve not just a difference in the two kinds of objects but also a difference in the very meaning of the concept of loss itself. On the one hand, the object in each case of loss is distinct, and that is important. On the other hand, the very grammar of loss is also different in each case. If we had to paraphrase the different kinds of losses involved in losing a child versus losing a job, we would soon find that our descriptions would diverge, because losing a child usually involves suffering and despair of a kind and depth that losing a book, or a game, or even a job does not.

Shweder went on to note in this regard the "dual ontological status" of Lazarus's abstract core relational themes, in that the themes had both a constitutive and a causal-empirical status. By 1991 this idea was something Lazarus himself had come to embrace by proposing that the appraisal-emotion relationship was both a synthetic (or causal) and a "logical" (or "analytic") one. But Shweder demurred. It was not just that he questioned whether appraisals were the kind of things that could be assessed empirically, especially if, as Lazarus himself believed, such appraisals were frequently automatic and unconscious. Shweder's fundamental objection was to the very idea that there were invariant causal connections between two types of events in the natural world, the making of an appraisal and a specific emotional response. He did not think the relationship between appraisals and emotions was a causal one at all, but rather was a matter of "conceptual necessity." Lazarus was right to think it was a "foregone conclusion" that, for example, sadness and loss are tied to each other, but "not as events in the empirical world." They are bound to one another, Shweder argued, because the appraisal condition (loss) is intrinsic to our concept of what it is like to be sad. As he put it:

> We are not biologically constructed in such a way that mental event A (the appraisal or loss) and mental event B (the experience of sadness) must go together. It is the idea of sadness that is so constructed. Internal to the idea of sadness is a connection to loss that we have no choice but to employ if we are to interpret others as sad. The link between the appraisal condition and the emotion is part of an a priori conceptual architecture, which is available for us to put to use in our attempts to comprehend others as persons and to arrive at a reading of their emotional life. ("EY," 324)

On this basis, Shweder suggested that many readers would query whether the postulated mental appraisals that Lazarus claimed were a causal condition of the emotion were anything other than a "reified redescription of the meaning of the emotion." "It left me wondering," Shweder wrote, "whether it is really necessary to reify meanings (describing them as if they were antecedent causal events) in order to acknowledge their central role in our mental life" ("EY," 323). Identifying himself as "in some sense or other" a cognitive appraisal theorist, Shweder nevertheless doubted that Lazarus's psychobiological principle was evident observationally. He concluded:

> I do not even believe that it is the kind of formulation (a "hypothesis") that might one day be supported or disproved by new evidence from cross-cultural research. In my view, cognitive appraisal theory is not so much a theory as a framework of concepts for generating interpretations about the mental life of others . . . I am skeptical . . . because I believe that the . . . way appraisals and emotions are "tied to" or "bound to" one another is a matter of conceptual necessity, not causal necessity . . . The connections and links built into cognitive appraisal theory (loss and sadness, transgression and guilt, etc.) are not there as a result of being observed. They arise out of the meanings inherent in our emotional state concepts (the "folk psychology," to use the contemporary philosophical parlance) that we use to "mind-read" the subjective states of others. ("EY," 324–25)[69]

It is not clear to me that Lazarus was ever able to reply adequately to these fundamental criticisms and concerns.[70] How he would try to do

69. For Shweder, the "ontological" status of appraisal is still a central topic. In 2014 he identified the same conflict between "causal" versus "constitutive" accounts of appraisal at work in four different theoretical approaches to the emotions, associated with the work of Agnes Moors, Jessica Tracy, Batja Mesquita and Michael Boiger, and Lisa Feldman Barrett, respectively. He argued with reference to Lazarus's earlier views that "the connection between an emotion concept (such as shame or pride) and its 'appraisal' condition is not falsifiable, but is rather a constitutive feature of its very meaning." See Richard A. Shweder, "Comment: The Tower of Appraisals: Trying to Make Sense of the One Big Thing," *Emotion Review* 6 (4) (2014): 324.

70. For a very different critique of the same 1991 book by Lazarus, undertaken from the perspective of a researcher committed to a neurobiological approach to the emotions, see Jan Panksepp, "Where, When, and How Does an Appraisal Become an Emotion? 'The Times They Are A'Changing,'" *Psychological Inquiry* 4 (4) (1993): 334–42. Here, Panksepp suggests that the "affect programs" mentioned by Lazarus as comparable to his psychobiological principle are isomorphic with the highly conserved, self-sufficient "emotional command systems" that Panksepp identifies as the fundamental affect systems of the mammalian brain. Panksepp thus suggests that the basic emotional systems function independently of the later-developing cognitive processes that may eventually be involved in the emotions. For Panksepp, appraisal thus comes late to the operation of the organism's fundamental emotional systems. For Panksepp's recent defense of his version of the basic emotions approach and Lisa Feldman

so is the story of the attempts he made, starting in the 1980s, to respond to the challenges posed by Zajonc and other critics on the basis of a position already marred by conceptual confusions and contradictions. These developments are the topic of the next chapter.

Barrett's post-Fridlund and post-Russell critique of the idea of emotions as discrete natural kinds, a critique in which Barrett implicates Lazarus, see Lisa Feldman Barrett, "Are Emotions Natural Kinds?," *Perspectives on Psychological Science* 1 (1) (2006): 28–58; Jan Panksepp, "Neurologizing the Psychology of Affects: How Appraisal-Based Constructions and Basic Emotion Theory Can Coexist," ibid., 2 (3) (2007): 281–96; and Lisa Feldman Barrett, Kristen A. Lindquist, Eliza Bliss-Moreau, Seth Duncan, Maria Gendron, Jennifer Mize, and Lauren Brennan, "Of Mice and Men: Natural Kinds of Emotions in the Mammalian Brain? A Reply to Panksepp and Izard," ibid., 2 (3) (2007): 297–312.

[CHAPTER FOUR]

RICHARD S. LAZARUS'S APPRAISAL THEORY II

The Battle Is Joined

> Zajonc had touched a nerve and uncovered an unresolved set of modern issues that apparently had lain dormant in the minds of many psychologists. . . . After the smoke had settled I could see that there were many mistakes in the way some of the issues had originally been cast.[1]

In 1980 Robert Zajonc published a paper, "Feeling and Thinking: Preferences Need No Inferences," which precipitated a long-lasting debate among researchers and clinicians as to the validity of his views. His arguments for the primacy of affect over cognition challenged all those who, like Lazarus, had endorsed a cognitive approach to the affects—an approach Zajonc treated as the dominant position among affect researchers in 1980.[2]

The essentials of Zajonc's arguments are easily summarized. According to Zajonc, emotions or affects are immediately given events, untouched or unmediated by cognition, judgment, or reflection. They are directly caused by a perception or sensory stimulus that does not need to be "recognized." In other words, our affective "preferences," tastes, and evaluations need no "inferences" because they are independent of cognition. Zajonc cited the work of Darwin and other researchers to suggest that affect systems have phylogenetic and ontogenetic primacy over cognition, that they are universal, and that they are activated in much the same way as reflexes. No cognitive or conceptual mediation is required. He also proposed on the basis of a variety of empirical and theoretical considerations that affect and cognition might be served by separate neuroanatomical systems.

Zajonc's arguments resonated with Tomkins's affect program theory.

1. Lazarus, "The Cognitive-Emotion Debate: A Bit of History," in *Handbook of Cognition and Emotion*, eds. Tim Dalgleish and Mick J. Power (London, 1999), 7–8.

2. Robert Zajonc, "Feeling and Thinking: Preferences Need No Inferences," *American Psychologist* 35 (1980): 151–75; hereafter abbreviated as "FT." Zajonc delivered this paper on the occasion of his receiving a Distinguished Scientific Contribution Award from the American Psychological Association.

As Zajonc observed, he himself focused more on the general quality of feelings documented in studies on taste preferences and aversions than on the specific emotions of surprise, anger, guilt, or shame discussed by Tomkins. But his emphasis on the independence of the affective system from the cognitive system confirmed what Tomkins had been arguing for years. It is therefore not surprising that Tomkins felt vindicated. "He is the first social psychologist to entertain a position I have argued for 20 some years," he observed of Zajonc, "that feeling and thinking are two independent mechanisms, that preferences need no inferences, that affective judgements may precede cognitive judgements in time, being often the very first and most important judgements. This article is a brilliant one that will, I think, be influential in loosening the unthinking hold of thought on social psychologists."[3] Other researchers were less impressed. "The underlying world view is that of mechanism," Sarbin responded critically, "and the job for the scientist is to find the connections between the elements of the machine. My reading of the arguments leads me to suggest that Zajonc's answer to the question, What is emotion? is to locate it in a category that would include knee-jerks, eye-blinks and the startle pattern."[4]

Zajonc's attack on cognitive theories of emotion precipitated something of a crisis for Lazarus. The challenge posed was not just that he was obliged to defend and clarify his views but that he was forced to confront certain "mistakes," as he saw it, in how both Zajonc and he had formulated their respective positions.[5] By contrast, Zajonc argued from strength, not because his standpoint was correct—it was flawed both conceptually and empirically—but because it resonated so widely with the zeitgeist. In another critical response, Parrott and Sabini observed of Zajonc's initial salvo:

> The continued influence of Zajonc's paper suggests that it expressed something about affect that many psychologists believe to be true and important, regardless of the specifics of logic and evidence with

3. Silvan S. Tomkins, "The Quest for Primary Motives: The Autobiography of an Idea," *Journal of Personality and Social Psychology* 41 (1981): 316–17. In his 1980 paper, Zajonc cites Tomkins's book *Affect, Imagery, Consciousness*, vol. 1 and 2; and Paul Ekman and Wallace V. Friesen, "The Repertoire of Nonverbal Behavior: Categories, Origins, Usage and Coding," *Semiotica* 1 (1969): 49–98.

4. Theodore R. Sarbin, "Emotion and Act: Roles and Rhetoric," in R. Harré, *The Social Construction of Emotions* (Oxford, 1986), 87.

5. Richard S. Lazarus, "The Cognition-Emotion Debate: A Bit of History," in *Handbook of Cognition and Emotion*, eds. Tim Dalgleish and Mick J. Power (London, 1999), 8.

which Zajonc supported his claims . . . Zajonc opposed a trend toward rationalism in modern psychology and instead directed attention to aspects of human experience championed by Romantic philosophers . . . It seems plausible that the popularity of Zajonc's proposals stem in part from our culture's continued acceptance of many of these romantic elements and from the neglect they had received in the research literature.[6]

I think this is correct, so far as it goes. Later, Fridlund likewise conjectured in a critical evaluation of Tomkins's and Ekman's two-factor emotion theory that the appeal and endurance of their position—and by implication Zajonc's—were owing to the exploitation of a "tacit but familiar Rousseauean romanticism," according to which emotional faces signify authenticity and rule-governed social faces denote the "inevitable loss of innocence forced by society."[7] And scholars before Fridlund, such as Theodore Sarbin and notably the philosopher of the emotions Robert Solomon, also traced such "unreasoning" or "hydraulic" views of the passions, which incorrectly separate them from rational thought, back to romanticism and even earlier.[8]

But I think it is important to recognize a more immediate reason for the appeal of Zajonc's thesis—a reason that has gone unremarked on by scholars. This is that, during the years leading up to Zajonc's claims, the theoretical framework for theorizing mental states, including the affects, had undergone a major shift, a shift anticipated by Tomkins in the 1960s and that during the 1970s had received significant reinforcement. Zajonc belonged to that moment in the history of American thought when the idea that the mind could be understood in information-processing terms had come to be accepted by many researchers as a breakthrough in psychology. As we have already seen in the previous chapter, thanks to the "New Look at the New Look" suggested by Erdelyi and others, many scientists had become convinced that phenomena such as subconscious de-

6. W. Gerrod Parrott and John Sabini, "On the 'Emotional' Qualities of Certain Types of Cognition: A Reply to Arguments for the Independence of Cognition and Affect," *Cognitive Therapy and Research* 13 (1) (1989): 58–59.

7. Alan J. Fridlund, *Human Facial Expression: An Evolutionary View* (San Diego, CA, 1994), 293.

8. Sarbin, "Emotion and Act: Roles and Rhetoric," 84–85; Robert C. Solomon, *The Passions* (New York, 1976), 139–49. The historian Barbara Rosenwein has suggested that the hydraulic model has a much longer history, going back well before the advent of Romanticism. See Jan Plamper, "The History of Emotions: An Interview with William Reddy, Barbara Rosenwein, and Peter Stearns," *History and Theory* 49 (May 2010): 251–52.

fense or subception could best be viewed, not as the workings of a suspect or mysterious Freudian "unconscious" held to be capable of defending against the perception of unwanted stimuli, but as the outcome of automatic, non-intentional, nonconscious information-processing systems. In short, Zajonc's paper testified to the success of the computer model of the mind. In the discussion that follows, I will consider the central importance of this development for the dispute over the nature of affect.

ZAJONC AND THE AUTOMATICITY OF AFFECT

Zajonc's reliance on computational models is evident at the very outset of his 1980 article when he explicitly laid out the cognitive position he was opposing in information-processing terms. On this model, affective reactions were seen to come at the end of a linear sequence of encoding steps leading from the stimulus input to cognitive representations and then to subsequent affective reactions and judgments. The series of posited stages suggested that considerable processing of information had to be accomplished before affects or feelings could occur. As Zajonc observed:

> An affective reaction, such as liking, disliking, preference, evaluation, or the experience of pleasure or displeasure, is based on a prior cognitive process in which a variety of content discriminations are made and features are identified, examined for their value, and weighted for their contributions. Once this analytic task has been completed, a computation of the components can generate an overall affective judgment. Before I can like something I must have some knowledge about it, and in the very least, I must have identified some of its discriminant features. Objects must be cognized before they can be evaluated. ("FT," 151)

This was not exactly a wrong description of the order of events postulated by Lazarus, except that he would not have put the issue in computational terms: he *did* think that cognition intervened or mediated between the originating stimulus and the emotional reaction. The value of posing the cognitive theory of the emotions in computational terms was that it enabled Zajonc to suggest that affective responses had to be rather slow, coming as they did at the end of a sequence of decision stages in the information-processing system. But according to Zajonc that was not the case: affective judgments often appeared to be so quick as to preclude the idea that there was time for any kind of cognitive processing to take place. Affect often comes first, according to his proposal, before we know precisely what it is we are reacting to.

Zajonc did not deny that in "nearly all cases" thought or cognition accompanied feelings, and feelings accompanied thought. Nevertheless, he argued that, while affect was always present as a companion to thought, the converse was not always true:

> In fact, it is entirely possible that the very first stage of an organism's reaction to stimuli and the first elements in retrieval are affective. It is further possible that we can like something or be afraid of it before we know precisely what it is and perhaps even *without* knowing what it is . . . [T]he fact that cognitions *can* produce feelings—as in listening to a joke, for example, where affect comes at the end with a punch line or as a result of post-decision dissonance—need not imply that cognitions are necessary components of affect. ("FT," 154)

On this basis Zajonc argued that feeling "arises early in the process of registration and retrieval, albeit weakly and vaguely, and . . . derives from a parallel, separate, and partly independent system of the organism" ("FT," 154).

In the course of making his case, Zajonc drew attention to earlier figures, such as Wundt, Bartlett, Osgood, and others, whose views in favor of the primacy of affect could be seen to lend authority to his position. But, in a paper that mixed argument with an appeal to empirical evidence, Zajonc also presented the results of an array of somewhat haphazardly collected more recent experimental studies that appeared to support his claims. These included studies suggesting that nonverbal cues influenced many human interactions independently of the content of the utterances involved; research by Ekman and Friesen implying the independence of nonverbal communication from semantic content or meaning; experiments by Kahneman and Tversky and others indicating that consumer choices are often influenced by preferences and feelings rather than rational decisions; investigations suggesting that affect and preferences influence face recognition and discrimination independently of cognitive factors; studies of the behavior of nonhuman animals in response to threatening stimuli, such as snakes, connoting that such responses happen so quickly as to preclude the intervention of cognition; research on preferences and attitudes that failed to identify the cognitive mediators involved; experiments indicating the difficulty of changing personal attitudes by various forms of communication, suggesting that affect is "fairly independent and often impervious to cognition"; and related inquiries.

Particularly important to Zajonc's thesis were investigations of the "exposure effect." That was the psychological phenomenon according to which people tend to develop a preference or liking for previously unfamiliar objects merely because they have been repeatedly exposed to them.

Starting in the late 1960s, Zajonc had carried out a series of laboratory experiments on the exposure effect using a variety of test stimuli such as drawings, facial expressions, nonsense words, Japanese ideographs, and "meaningless" Chinese characters, and in each case had appeared to demonstrate the positive effects of mere exposure. Even the subliminal exposure of images on a tachistoscope had produced the same positive results. He had extended his inquiries to nonhuman animals, showing, for instance, that tones played to unhatched chicks were subsequently preferred.[9]

Zajonc's experiments on the exposure effect were the start of his career-long interest in the operation of affective processes occurring outside the person's knowledge or awareness. Although he favored an interpretation of the findings that stressed the role of emotional influences independent of cognition, in an ongoing debate other scientists claimed that some sort of cognitive recognition must be involved. Indeed, as Zajonc acknowledged, the explanation of the exposure effect was very elusive. More generally, he admitted that his 1980 paper on affect was more of a speculative think piece than a solid argument based on incontrovertible evidence. "[B]uilding on the scanty evidence we now have," he conceded, "I have tried to develop some notions about the possible ways in which affect is processed as part of experience and have attempted to distinguish affect from processing of information that does not have affective qualities" ("FT," 151). Or, as he also stated: "Because the language of my paper has been stronger than can be justified by the logic of the argument or the weight of the evidence, I hasten to affirm that one of my purposes was to convince you that affect should not be treated as unalterably last and invariably post-cognitive" ("FT," 172). And in a subsequent article, he admitted that his hypothesized conceptual separation between affect and cognition was "in part, quite arbitrary" and was posited "in order to satisfy certain theoretical purposes."[10]

In short, Zajonc's paper had an explicitly polemical intent. Although

9. For a complete listing of Zajonc's publications on the exposure effect and related topics, see *Unraveling the Complexities of Social Life, A Festschrift in Honor of Robert B. Zajonc*, eds. John A. Bargh and Deborah K. Apsley (Washington, DC, 2001), 187–96. The essays in this volume testify to the impact of Zajonc's claims concerning the independence of affect from cognition on one influential segment of the psychology community.

10. R. B. Zajonc, Paula Pietromonaco, and John Bargh, "Independence and Interaction of Affect and Cognition," in *Affect and Cognition: The Seventeenth Annual Carnegie Symposium on Cognition*, eds. M. S. Clark and S. T. Fiske (Hillsdale, NJ, 1982), 212; hereafter "IAC."

he deplored the tendency to place affect at the end of a linear sequence of information processing, this did not mean that he himself eschewed information-processing theories as such. On the contrary, his arguments for affective primacy were fundamentally indebted to the theoretical reframing of certain issues in emotion research made possible by the computational ideas of the cognitive revolution.[11] Especially relevant to that reframing was Erdelyi's 1974 effort, already mentioned in the previous chapter, to resolve the apparent paradox of subception, or the unconscious response to subliminal stimuli, by reinterpreting the findings in information-processing terms. He had written:

> It may be noted . . . that the various encoding processes discussed in this section suggest one concrete resolution of the logical paradox issue raised earlier with regard to perceptual defense. If, because of its emotionality, some information in raw storage fails to be encoded into more permanent memory (short-term storage), the critical information becomes irretrievably lost to the perceiver after the rapid fading of the icon (or echo). The perceiver can be said to have defended against the input since it is quite unavailable to him beyond a fleeting moment. Yet, at a different level, he did "perceive" it in that it *was* available, if very briefly, in an iconic or echoic storage during partial or full analysis by long-term memory. Thus, not quite so paradoxically, he both perceived and defended himself against perceiving (further) the very same input.[12]

Or, as Richard Nisbett and Timothy Wilson had put Erdelyi's point: "We cannot perceive without perceiving, but we can perceive without remembering. The subliminal perception hypothesis then becomes theoretically quite innocuous: Some stimuli may affect ongoing mental processes, including higher-order processes of evaluation, judgment, and the initiation of behavior, without being registered in short-term memory, or at any rate without being translated into long-term memory."[13] The proposal carried within it the seeds of a theory according to which affective processing might occur independently of conscious perception or cognition because

11. In his 1980 paper, Zajonc cited both Ulric Neisser's text *Cognitive Psychology* (1967), a major text in the cognitive revolution, as well as the latter's *Cognition and Reality* (1976) ("FT," 174).

12. Matthew Hugh Erdelyi, "A New Look at the New Look: Perceptual Defense and Vigilance," *Psychological Review* 81 (1974): 20.

13. Richard E. Nisbett and Timothy Decamp Wilson, "Telling More Than We Can Know: Verbal Reports on Mental Processes," *Psychological Review* 84 (1) (1977): 240. For a further discussion of this paper, see below.

emotional stimuli use separate channels of information from those employed in conscious cognition.

This is precisely what Zajonc suggested in 1980. Indeed, he explicitly recognized the connection between the affective primacy debate and the earlier subception controversy, crediting Erdelyi as one of the few cognitive psychologists who had attempted to understand the role of motivational and emotional factors in perception and cognition ("FT," 153, 172). Equally important for the persuasiveness of Zajonc's thesis concerning the primacy of affect were two reviews published by Posner and Synder in 1975, cited by Zajonc, which similarly suggested that automatic processes might take place in information-processing channels separate from the channels for conscious processing. By automatic processes, Posner and Synder meant those that were said to occur without intention, conscious awareness, or interference from other mental activities, and their claim was that many habitual mental events, including cognition, the formation of impressions, and other emotionally laden reactions, operate automatically in this way. The authors noted that automatic pathway activation was evident in studies of the so-called "Stroop" effect, in which experimental subjects cannot avoid processing an input item or stimulus that they wish to ignore: thus, when given the task of naming the *ink color* of words in a Stroop effect test, they intend to avoid reading the *actual words* but find it impossible to do so completely. Other investigations similarly suggested that subjects were unaware of the activation pattern created by input words.

The explanation proposed by Posner and Synder for these and related findings depended on the cybernetic assumption that automatic activation and conscious decisions occurred in different information-processing channels or nervous pathways. The authors remarked in this connection that in no place was the distinction between automatic and conscious processes more needed than in the study of the role of emotions and affects in perception. Following up on Erdelyi's earlier discussion of subception, they proposed that the existence of two independent memory channels, one automatic and the other conscious, could account for the apparent paradox of perceptual defense, since one pathway could automatically transmit the representation of interest—the word and its associations—while another pathway could process the conscious awareness of the representation.[14]

14. M. I. Posner and C. R. R. Synder, "Attention and Cognitive Control," in *Information Processing and Cognition: The Loyola Symposium,* ed. R. L. Solso (Hillsdale, NJ, 1975), 55–85.

An important book, *Unintended Thought* (1989), edited by James S. Uleman with Zajonc's former student, John A. Bargh, testifies to the influence of Posner and Synder's work during these years.[15] The volume clearly demonstrates that issues of automaticity and "non-intentionality" had begun to occupy center stage just as Zajonc was taking up the cause of the primacy of affect. It also confirms that those issues were directly linked to the success of information-processing theories of the mind. As Uleman and Bargh reported, by the mid-1970s cognitive models of mental functioning were being developed in which information processing in reading or in answering questions about category judgments was seen as largely "uncontrolled, automatic 'spreading activation' phenomena." Nevertheless, as the editors also remarked, it was not until Posner and Synder's 1975 publications that the role of intention in controlling such processes was called into question:

> [T]he assumption of control was still alive and well in the research that appeared to challenge it. Models of semantic memory that posited a mechanistic spreading activation process assumed that such processing was instigated by an intention to search memory or to answer the experimentally given question (e.g., Is a penguin a bird?). The uncontrolled activation of a word's meaning was considered a consequence of the larger intentional act of reading. And the demonstrations of irrational decision making were taken as evidence of short-cut, heuristic decision *strategies*. None of these areas of research ever explicitly challenged the assumption of intentional control over the way information is processed in one's mind. (*UT*, xiv)

According to Bargh and Uleman, all that changed when in 1975 Posner and Synder suggested on the basis of the existing evidence that intentional control of such processes might not exist. Bargh and Uleman attributed the impact of Posner and Synder's intervention to the fact that those investigators explicitly formulated the issue as an empirical problem rather than as a theoretical one, thereby opening the way for further research on the role of intention. Bargh and Uleman stated:

> By framing [the question of intentional control] up front as an empirical question, and not as an assumption a model could make or could not make, the long-standing implicit assumption of intentionality was transformed into an explicit empirical question. More

15. James S. Uleman and John A. Bargh, *Unintended Thought* (New York and London, 1989); hereafter abbreviated as *UT*.

> than this, Posner and Synder specifically discussed the importance of studying the role of intentionality in the domain of emotional experience and impression formation, and in the areas of perception and memory retrieval processes more generally.
>
> As most social psychological models implicitly assumed the role of deliberate, calculated, conscious, and intentional thought, the degree to which such thought did occur in naturalistic social settings became of critical importance. (*UT*, xiv)

But by emphasizing the idea that the impact of Posner and Synder's intervention was largely the result of their treating the issue of "intentionality" as an empirical question, Bargh and Uleman risked underestimating the theoretical stakes involved. What Bargh and Uleman took for granted were assumptions about the information-processing capabilities of the mind that also informed Posner and Synder's work, assumptions that yielded an account of certain cognitive and other mental processes as completely automated-mechanical actions. Thus, thought and cognition were themselves conceptualized as entailing modes of information processing that could be unconscious, involuntary and "non-intentional." The picture offered was one of diverse mental activities capable of operating automatically in independent but parallel information-processing channels. One social psychologist, Ellen Langer, went so far as to declare on the basis of information-processing theories and experimental considerations that much of our apparently mindful-cognitive social interactions and behavior goes on in the complete absence of mindedness, emphasizing instead the "mindlessness of ostensibly thoughtful action."[16] As one contributor to *Unintended Thought* put it:

> Langer (1978) quite colorfully articulated a widespread discontent when she urged attribution theorists to "rethink the role of thought in social interaction." She pointed out that social inference is not always a conscious and deliberate act; rather, it is often the province of mindless automata. Attributional theories (according to Langer) have been too quick to assume that perceivers engage conscious processes in order to achieve their causal conclusions. The clarion call was widely appreciated, and if Langer did not quite set the stage for a psychology of unconscious social inference, she at least rented the theater . . .

16. E. J. Langer, "Rethinking the Role of Thought in Social Interaction," in *New Directions in Attribution Research*, vol. 2, eds. J. H. Harvey, W. J. Wicks, and R. F. Kidd (Hillsdale, NJ, 1978), 35–58.

> While all of this was happening, experimental psychology was itself becoming unaware. Almost a century earlier, William James . . . had dismissed the unconscious . . . and, indeed, the supernatural flavor of Freud's *Umbewusst* made the concept generally unpalatable to the emerging scientific discipline . . . However, when theorists such as . . . Posner & Synder (1975) . . . began to explicate the parameters of automatic "off-line" processes in a computer-based vernacular, the ghost somehow fled the machine. All this is to say that there has, in recent years, been a discernible shift in the wind: The role of awareness and control in the social inference process has been appropriately questioned, and the role of unconscious and unintentional processes has been approved for exploration.[17]

This was a view of mental functions that could or should have raised many questions about the nature of intention itself. If a thought process occurs below the threshold of consciousness or awareness, does this necessarily mean that it is non-intentional in the sense of lacking all semantic or conceptual or cognitive content? If a process occurs very rapidly, does this exclude the intervention of conceptuality? To their credit, some authors in Uleman and Bargh's volume did attempt to qualify some of the more extreme assertions being made about the automaticity and non-intentionality of thinking, cognizing, and affective judgments, and in doing so tried to probe the definition and meaning of terms such as "intention" and "automaticity." They questioned the idea that just because various mental processes appeared to occur automatically and below the threshold of awareness, that meant that such processes take place independently of any thinking or cognition.

Thus, in a thoughtful discussion focusing on affect and automaticity that made several references to Zajonc's work, among that of others, Isen and Diamond criticized recent authors who had eliminated the role of the person and meaning when analyzing the results of their experiments. On theoretical and methodological grounds, Isen and Diamond suggested that responses that had appeared to such researchers to be irresistible and obligatory reactions to stimuli were often the outcome of cognitive and interpretive processing. "[E]ven if affect (or some other stimulus) were processed automatically," they argued, "it might not mean that such

17. Daniel T. E. Gilbert, "Thinking Lightly About Others: Automatic Components of the Social Inference Process" (*UT*, 192). In their introduction to *Unintended Thought*, Bargh and Uleman state that Langer's chapter was "perhaps as influential as Posner and Synder's, and maybe more so within social psychology" (*UT*, xiv).

stimuli were truly irresistible because of their nature and would always *have to be* processed in that way. Rather, when stimuli are processed automatically, it may be the result of their nature in combination with other demands of the situation and the person's goals, plans, strategies, and perhaps even prior decisions to allow that kind of processing to occur, or 'set.'"[18] In the same vein, they observed that

> [C]ognitive functions that are "automatic" are sometimes depicted as "data-driven" or "bottom-up" (i.e., based entirely on the stimuli presented, with little or no cognitive mediation required or possible), and consequent behavior is seen as largely under the control of environmental factors and as involuntary. This aspect of the definition appears to be open to revision, however, as experiments increasingly are showing that meaning and interpretation do play a role in cognitive processing that otherwise fits the definition of "automatic," and that automatic processes can be interrupted or modified. (*UT*, 127–28)

Critically examining certain experiments and claims, the authors added: "Thus, the meaning of the unattended (automatically processed) message had been identified by subjects, and it would be difficult to argue that interpretation or inference of meaning did not occur in this 'automatic' process" (*UT*, 128). Isen and Diamond noted that some automaticity theorists might disagree with them, commenting in this regard that

> Bargh . . . for example, argues that the semantic meaning of words can be processed automatically, without awareness or intention. We are not convinced, however, that it has been demonstrated that such processing does not require cognitive capacity, a crucial element in the definition of automaticity. To establish that cognitive capacity is not required, it should be demonstrated that semantic meaning can be processed without being subject to the effects of set size, for example. (*UT*, 149)

On similar grounds, Isen and Diamond criticized interpretations of the Stroop effect as a purely "automatic" phenomenon and suggested instead that "the subject's goal or intent played a role in the process" (*UT*, 128).

Of particular relevance to the Zajonc–Lazarus debate was an experiment on the automaticity of affect by Hansen and Hansen, published in 1988. Hansen and Hansen drew on the Basic Emotion Theory of facial expressions associated with the work of Tomkins, Ekman, and Izard, as

18. Alice M. Isen and Gregory Andrade Diamond, "Affect and Automaticity" (*UT*, 127).

well as on Zajonc's papers, to suggest that the automatic processing of threatening or angry facial expressions has special adaptive or survival value. Using photographs of posed expressions selected from Ekman and Friesen's standard set, they purported to demonstrate experimentally that angry faces were processed so efficiently and effortlessly that when subjects were presented with one angry face in a crowd of happy ones, the angry face quickly "popped out" from the crowd in a way a happy face in a crowd of angry faces did not. The investigation therefore appeared to confirm Zajonc's and others' claim concerning the automaticity of affect processing in the case of evolutionary significant stimuli, such as threatening faces. But Isen and Diamond suggested that Hansen and Hansen's results fell short of establishing unambiguously that affect is processed automatically: because perceptual features could distinguish the target elements from the distracting or crowd faces, it was possible that the subjects were responding to these perceptual cues, rather than to affective valences as such (*UT*, 135).[19]

Isen and Diamond's general conclusion was that the distinction between automatic or non-intentional processes on the one hand and cognitive processes on the other was not as absolute as some investigators had assumed. They criticized the tendency of many researchers to treat processes that appeared "automatic" as machine-like phenomena occurring separately from the person, suggesting instead that automaticity might be better understood as a form of processing without attention. They therefore denied the existence of any simple opposition between habitual, skilled, or apparently automatic reactions and actions performed with conscious intention and control. On this basis, they suggested that

19. C. H. Hansen and R. D. Hansen, "Finding the Face in the Crowd: An Anger Superiority Effect," *Journal of Personality and Social Psychology* 54 (1988): 917–24. A few years later, Purcell et al. demonstrated that Hansen and Hansen's finding was the effect of a confound that occurred during the digital image processing (gray-scale pictures were converted into black-and-white pictures), resulting in conspicuous black spots that pertained only to the angry faces and not to the happy ones. Apparently, subjects in the experiment detected this confound and used it to discriminate between stimuli. When Purcell et al. corrected for this confound by using the original gray-scale pictures, the search asymmetry between angry and happy faces was not confirmed. See D. G. Purcell, A. I. Stewart, and R. B. Skov, "It takes a Confounded Face to Pop Out of a Crowd," *Perception* 25 (1996): 1091–108. For a recent review of experiments on this topic, see also Gernot Hortsmann and Andrea Bauland, "Search Asymmetries with Real Faces: Testing the Anger-Superiority Effect," *Emotion* 6 (2) (2006): 193–207. In their paper, Hortsmann and Bauland make clear that Hansen and Hansen were using Ekman and Friesen's standard photographs of posed facial expressions as stimuli in their experiment.

the automaticity of affective reactions was the result of "overlearning" or repeated exposure, producing effects of apparent irresistibility (*UT*, 138–39). In the case of habitual skills such as driving, for example, Isen and Diamond proposed that "chunking," or the reduction of the cognitive capacity required for skilled coping, only made it seem as if these processes were being processed automatically, without cognitive involvement (*UT*, 140).

In a somewhat similar spirit, in the same volume Gordon Logan observed that on the one hand, skills such as playing the piano or a sport seem to depend on habitual actions occurring automatically, without effort or conscious thought, but on the other hand, when necessary, such skills also require the performer to make very rapid, finely tuned, controlled adjustments. He suggested on the basis of these and related considerations that "automatic processing is not beyond control. Stroop and priming experiments, which provide most of the evidence for uncontrollable automatic processing, show that automatic processing is strongly influenced by intention and attention."[20] Isen and Diamond's general conclusion was similarly that "the influence of automatic processes increasingly can be viewed as subject to modification and intervention rather than as inevitable and irresistible. Moreover, the occurrence of seemingly automatic processes may sometimes be the result of decisions to employ attention elsewhere" (*UT,* 147). The issues at stake in this debate anticipated by nearly twenty years the recent important controversy between the philosophers Hubert Dreyfus and John McDowell over the "mindlessness" or "mindfulness" of habitual, skilled coping.[21]

Important as these and related criticisms of the problem of automaticity were, I think it is fair to say that in the extensive literature on automaticity at this time the larger philosophical questions concerning the nature of intentionality tended to be sidestepped. Indeed, the computational terms of most of the analyses almost guaranteed the absence of serious

20. Gordon D. Logan, "Automaticity and Cognitive Control" (*UT*, 64). Thus, according to Logan, experimental subjects responding to unconscious "priming" were sensitive to the actual meaning of the words used as priming stimuli, with the result that their responses were not the obligatory result of stimulus presentation, as theorists of automaticity proposed, but were under the subject's strategic control.

21. In his 2005 presidential address to the American Philosophical Association, Dreyfus attacked McDowell's views about mind-world relations by accusing him of exaggerating the role of mindedness in skilled coping. McDowell's response sparked a major exchange of views. It seems to me that McDowell got the better of Dreyfus in the dispute. The discussion can be followed in *Mind, Reason, and Being-in-the World: The McDowell-Dreyfus Debate*, ed. Joseph K. Schear (London and New York, 2013).

discussion of the topic: it was simply assumed that somehow a computational analysis of cognition and affect could be counted on to produce the necessary meaning at the end of the information-processing sequences. For our purposes, what is interesting is how Zajonc found support in contemporaneous ideas about automaticity and non-intentionality to pursue the idea that affect itself could be theorized in information-processing terms as largely automatic and non-intentional or noncognitive. Although in his 1980 paper he did not deny that we are often able to control our emotional expressions, he suggested that affective judgments are "inescapable" ("FT," 156).[22] This in turn suggested to him that feelings might be processed differently from cognitive content and that affects might therefore depend on an information-processing system distinct from that dedicated to cognition—a proposal he claimed went beyond that of Posner and Synder, who had restricted their suggestion to the idea that there exist separate cognitive systems. However, it was the very idea of discrete channels capable of performing independent, automatic information-processing functions, including affective information-processing ones, that was central to Zajonc's position.

A further significant consequence of Zajonc's arguments concerned the interpretation of unconscious mental activity. Prior to the information-processing revolution, unconscious processes were treated on Freudian principles as mental events that were not present to consciousness because they had been actively repressed. The concept of the unconscious was a dynamic-conflictual one involving the role of an ego capable of banishing the subject's unacceptable desires and wishes from conscious awareness. As I indicated in the previous chapter, phenomena such as subception or subliminal perception were thus viewed as the result of the individual person's unconscious ego defenses. But with the rise of information-processing theories of mental function, the dynamic unconscious of Freud was transformed. Unconscious activities were now viewed as forms of automatic, nonconscious information processing occurring in computer-style subsystems capable of acting independently of the mind's conscious control. The result was that the dynamic-conflictual dimension of the psychoanalytic unconscious was lost as mental actions were transformed into nonconscious, mechanically filtered processes and events.

That change in the understanding of unconscious activity was well un-

22. However, in a note Zajonc observed that the existentialists, such as Sartre, ascribed a substantial voluntary component to emotion, and conceded that because of the participation of sensory, cognitive, and motor processes, "the argument that emotions have some voluntary component is not without basis" ("FT," 156).

derway by the mid-1970s, as when in 1974 Erdelyi reformulated subception as a phenomenon due to information processing occurring outside the control of the conscious information-processing system. Commenting on this development a few years later, Erdelyi remarked that the experimental program associated with the "New Look" in perception and the problem of subliminal perception was

> riddled with controversy. By the late 1950s the general consensus emerged that the venture had been a failure and that no methodologically sound demonstration of unconscious perception—and more generally of unconscious processes—had been achieved . . . The failure of experimental methodology to corroborate the existence of unconscious processes was taken, as a matter of course, to reflect a failure of the concept rather than a failure of the extant methodology . . . The unconscious, having been given its chance in the laboratory, and having fumbled, seemed doomed as a scientific concept.[23]

But Erdelyi went on to suggest that at that very same moment the information-processing revolution in psychology offered a new departure. Although, as he wrote, that revolution had essentially no contact with psychoanalytic theory—"and perhaps all the more impressive on that account"—it generated a host of theoretical constructs that came close to constituting "rediscoveries of basic Freudian notions." "Censorship" thus became "filtering" or "selectivity"; the "ego" became an "executive control processes" or a "central processor"; "conflict" became "decision nodes"; "force," "cathexes," and "energies" became "weights" and "attention"; "depth" became "depth of processing" or "up-down processing"; the "conscious" became "working memory"; "unconscious" versus "preconscious" processes became "unavailable" versus "available" memories; and so on. "Since these new concepts had direct counterparts in the computer," Erdelyi added, "there was nothing remotely mystical about them. The concept of unconscious processes—if not the term itself—was not only *not* controversial but an obvious and fundamental feature of human information processing" (*P*, 59).[24]

23. Matthew Hugh Erdelyi, *Psychoanalysis: Freud's Cognitive Psychology* (New York, 1985), 59, 64; hereafter abbreviated as *P*.

24. Erdelyi noted that the introduction of new terms was "probably symptomatic of the continued nervousness on the part of experimental psychology about making its peace with the unconscious. The modern literature is replete with synonyms such as *automatic*, *inaccessible*, and *preattentive*, but only rarely does the unconscious itself

It is a sign of this new state of affairs that Erdelyi himself adopted "inaccessibility" as the criterion of the unconscious, discarding Freud's theory that it is dynamic repression which renders mental contents "inaccessible" or unconscious (*P*, 64). Indeed, he viewed the computer model (or "metaphor") of the mind as the solution that was needed to explain unconscious processes. "It is worth noting," he stated in yet another essay that clearly exhibits the theoretical underpinnings of these developments,

> that these philosophical difficulties of perceptual defense were resolved neither by experimental psychology (i.e., laboratory research) nor by philosophy but, instead, by an analogy. As soon as the rat gave way to the computer as cognitive psychology's central metaphor, the problems dissolved of their own weight, since it now became clear, indeed trivially obvious, that computers could selectively regulate their own input (and thus perceive at one level without perceiving at another level) through the operation of "control processes" that were neither mystical nor eternally regressive.[25]

Erdelyi's remarks make clear the extent to which what was at stake in these reformulations and in Zajonc's clams about the primacy of affect was indeed a theoretical change of perspective rather than the empirical development Uleman and Bargh claimed it was.

That this reformulation of the dynamic unconscious into an information-processing "cognitive unconscious" was motivated by a lack of comfort with the very notion of intention is made clear in another of Erdelyi's statements that deserves to be quoted at length:

> The ultimate problem with defense, as with so many other constructs in psychology, is that it is anchored to the notion of intention and purpose and is thus problematic on both philosophical and methodological grounds . . . [W]e have yet no explicit methodology of purpose. Consider the controversial phenomenon of *perceptual defense*, the purported elevation of perceptual thresholds to threatening stim-

---

make an appearance. There is no explicit discussion of the reason for this avoidance, though informal querying usually turns up a disinclination on the part of authors to be associated with *Freud's* unconscious, not any misgivings about nonconscious mentation. The psychoanalytic unconscious, it is usually claimed, carries excess theoretical baggage to which the typical modern investigator does not wish to give the appearance of subscribing" (*P*, 59–60).

25. M. H. Erdelyi and B. Goldberg, "Let's Not Sweep Repression Under the Rug: Toward a Cognitive Psychology of Repression," in *Functional Disorders of Memory*, eds. J. F. Kihlstrom and F. J. Evans (Hillsdale, NJ, 1979), 379.

> uli . . . Is such a phenomenon—lowered sensitivity to threatening stimuli—necessarily a *defense* phenomenon, as its name suggests? Not necessarily. Such a phenomenon could also be accounted for by the nondefensive disruption of perception and memory by emotion or by any other attention-disrupting event . . . Thus memory failures associated with anxiety or any other cognitive disturbance . . . need not be construed as evidence for repression; somehow—and here is where current methodological (or epistemological) techniques leave off—it must be demonstrated that the memory loss is intentional. A methodology needs to be developed capable of corroborating—or falsifying—clinical conceptualizations such as "intrusions were breakthroughs of themes defensively warded off . . . from conscious representation." (his emphasis)[26]

Erdelyi thus attempted to reframe the notion of unconscious intentionality and meaning in quantifiable, mechanical-computational terms. Kihlstrom in 1987 gave the name "cognitive unconscious" to the unconscious formulated in information-processing terms, and for the majority of cognitive psychologists the triumph of the computer model of the mind seemed virtually complete.[27]

26. M. H. Erdelyi, "Issues in the Study of Unconscious Defense Processes: Discussion of Horowitz's Comments, with Some Elaborations," in *Psychodynamics and Cognition*, ed. M. H. Horowitz (Chicago, 1990), 85–86. Isen and Diamond note an important difference between the psychoanalytic and the cognitive understanding of the unconscious: "[T]he phenomena addressed in the 'New Look' research differ in an important way from automatic processing as it is conceived in the cognitive literature. The 'New Look' literature deals with 'unconscious processing,' but primarily from the perspective of psychoanalytic theory, where unconscious, irresistible processes such as 'perceptual defense' and related phenomena are assumed to operate at substantial cost to the capacity of the conscious cognitive and other life systems. In particular, in the psychoanalytic tradition it is thought that keeping unconscious material from awareness requires expenditure of capacity. This is in contradistinction with the modern cognitive concept of automaticity, which is thought to free up capacity" (*UT*, 148).

27. John F. Kihlstrom, "The Cognitive Unconscious," *Science*, New Series, 237 (1987): 1445–52. For developments in the theory of unconscious information processing since Kihlstrom's influential formulation, see *The New Unconscious*, eds. Ran R. Hassin, James S. Uleman, and John A. Bargh (Oxford, 2005). The genealogy of the notion of a "cognitive unconscious" (as opposed to the Freudian unconscious) traces back to Helmholtz's inferential theory of perception, according to which there is a "gap" between our retinal stimulations and the objects we see, a gap that must be filled by unconscious inferences. Chomsky's linguistic theories and Fodor's computational psychology bolstered such ideas. For a helpful discussion of the origins of the notion of the cognitive unconscious and the confusion that arises when cognitive psychologists such as Kihlstrom and others attribute the human being's capacity for understanding

There are two further aspects of Zajonc's ideas about affective primacy that require comment at this juncture. The first is the emphasis he placed on the role of the somatic and motor aspect of emotional reactions, especially the role of the face. Indeed, in response to criticisms of his 1980 paper, his emphasis appeared to shift subtly. In subsequent publications, he did not deny that emotional "experience" itself, involving, as he stated, some form of "self-perception" as well as reported emotional experiences and judgments, must have a cognitive component. As he conceded in a 1982 paper: "When Mandler (1982) argues . . . that value judgments require cognitive participation, he cannot be contradicted. We are not in disagreement with Mandler on that point when he asserts that 'to say that one likes something requires access to stored knowledge.'" Rather, Zajonc's proposal now was that the "*expression* of emotion often requires no cognitive processes" (his emphasis). By "expression" of emotion he meant "those aspects of bodily states, changes, and acts that allow one individual to detect the emotional state of another, and often to identify the nature of that state." He compared such bodily expressions, such as facial expressions and somatic states, to those produced by the administration of alcohol, barbiturates, depressants, and other drugs.

On the one hand, by emphasizing bodily reactions, Zajonc appeared to isolate the motor or output side of the affective reaction from the input side, leaving intact the claims of cognitivists such as (the now named) Lazarus about the necessary role of cognition. On the other hand, by comparing emotional reactions to the behaviors and expressions produced in individuals by the administration of drugs or alcohol, Zajonc also included the input side of the emotional response in his analysis, implying the existence of some sort of sensory-physiological triggering mechanism to activate the affective process. "It is also the case that the administration of a variety of drugs generates a variety of emotional expressions," Zajonc wrote in this connection. "Alcohol, barbiturates, amphetamines, depressants, and hashish all have clear somatic and expressive consequences. While these expressions and somatic effects may be *accompanied* by cognitive processes, that is, by the experience of emotion, appraisal, or evaluation, they are not necessary for the generation of a variety of emotional reactions that have distinct forms of expression. One needs no labeling or

---

to unconscious information-processing events in the mind-brain, see Wes Sharrock and Jeff Coulter, "Revisiting 'The Cognitive Unconscious,'" in *Perspicuous Presentations: Essays on Wittgenstein's Philosophy of Psychology*, ed. Daniele Moyal-Sharrock (Basingstoke, 2007), 95–113.

evaluation or appraisal to suffer (or enjoy) the affective consequences of ten ounces of alcohol given unobtrusively" ("IAC," 213; his emphasis). This statement suggests that Zajonc was actually theorizing emotions in purely sensory-motor terms. Thus, he observed in opposition to Schachter and Singer's cognitive views:

> Because we are seldom without *some* thoughts, the question of whether cognition is or is not necessary to emotion may seem quite pointless. It is not . . . Thus, if we imagine a state where an individual had no cognitions whatever, and either was or was not administered alcohol (without knowing, of course), then this individual would express (i.e., show) a different emotional state in one case than in the other. After all, the biological consequences of alcohol do have autonomic, visceral, glandular, and muscular effects that have all the earmarks of an emotion or mood, and they would have these effects regardless of the thoughts that the individual might be entertaining at the moment. Alcohol effects are not "all in the mind." ("IAC," 213–14)

Zajonc added as a further argument that the original cognitive bases of certain emotions can be forgotten or dissociated, as when we feel joy on meeting a friend whom we have not seen for a long time, even though we have entirely forgotten why we came to like that person in the first place. He also noted that, in a variety of pathological states and in hypnosis, affect can be aroused by the introduction of a specific stimulus without the individual connecting the one with the other.

These are hardly decisive arguments in favor of the idea that emotional reactions can be understood as purely sensory-motor phenomena. Too many questions are begged. For instance, is it really plausible or coherent to suggest that we might "recognize" an old friend sufficiently well to feel joy at seeing him, but that our response lacks cognitive content because in the moment we cannot remember why we ever liked him? Zajonc's assumption seems to be that although we recognize our friend (surely a cognitive achievement?), our response must be noncognitive because it is immediate—a dubious proposition, although one routinely made at the time, and not just by Zajonc.[28] But his statements have the merit of revealing the extent to which he increasingly tended to conceptualize pref-

28. See for instance the comments on this topic in the discussion generated by John Haugeland's paper "The Nature and Plausibility of Cognitivism," *Behavioral and Brain Sciences* 2 (1978): 215–60.

erences as "*primarily affectively based behavioral phenomena*" (his emphasis).[29] The result was an accent on the behavioral-expressive-motor aspects of emotional reactions, conceived as obligatory, quasi-reflex sensory-motor phenomena that at the limit occur independently of any cognitive content. In this way Zajonc aligned his position with that of Tomkins, Izard, and Ekman, whose work he began to cite with some regularity.[30] In addition, Zajonc began to develop a "Vascular Theory of Facial Efference," which assumed a relationship between facial expression and subjective feeling states in terms close to Tomkins's ideas about the role of facial feedback in generating specific emotional reactions.[31]

Another aspect of Zajonc's claims about affective primacy that deserves mention is his often-repeated observation that, if the cognitive processes held by Lazarus and others to be a necessary element in emotion are *unconscious*, then the case could not be proved, since self-reports or introspection, which depended on the subject's awareness, would be powerless to provide evidence of such unconscious mental states. As Zajonc et al. put the issue in 1982:

> The final proposition to be examined in this context is that cognition is always a necessary element of emotion, but that sometimes it is unconscious and thus it cannot be detected. This argument unfortunately cannot be disproved. If we find an emotion for which we cannot detect an observable cognitive correlate, then we would simply conclude that some cognitive processes did participate as necessary components, but they were unconscious and, therefore, accessible neither to the actor nor to the observer. Such a theory would have no constraints at all and it would be quite useless. ("IAC," 214)

29. Robert B. Zajonc and Hazel Markus, "Affective and Cognitive Factors in Preferences," *Journal of Consumer Research* 9 (1982): 124; emphasis in the original. Cf. Robert B. Zajonc and Hazel Markus, "Affect and Cognition: The Hard Interface," in *Emotions, Cognition, and Behavior*, eds. C. E. Izard, J. Kagan, and R. B. Zajonc (New York, 1984), 73–102.

30. See for example R. B. Zajonc, "On the Primacy of Affect," *American Psychologist* 39 (2) (1984): 118. In 1994 Ekman cited Zajonc's views in support of the notion of the automaticity of appraisal in basic emotion responses. See Paul Ekman, "All Emotions Are Basic," in *The Nature of Emotion: Fundamental Questions*, eds. Paul Ekman and R. Davidson (New York, 1994), 16.

31. For recent positive assessments of Zajonc's research, see *Unraveling the Complexities of Social Life: A Festschrift in Honor of Robert B. Zajonc*, eds. John A. Bargh and Deborah K. Apsley (Washington, DC, 2001); and "The Contributions of Robert Zajonc," *Emotion Review* 2 (4) (October 2010): 315–62, with contributions by former Zajonc students and others.

Since Lazarus did believe that many of the cognitive processes at work in emotion were unconscious, Zajonc's argument put him in an awkward position. Was his claim that emotions are intentional-cognitive states an empirical thesis demanding experimental or clinical proof? How could such proof be delivered if the mental states involved were unavailable to consciousness? Or was his claim a constitutive one about the inherently cognitive nature of the affects? If the claim was a constitutive one, then it might seem that debates about operationalizing appraisal in order to provide empirical evidence for the causal role of cognition were beside the point. But since at times Zajonc himself said there were no empirical means to decide his own position on affective primacy, it could easily come to seem that the two men were simply talking past each other.

LAZARUS VERSUS ZAJONC: MEANING AND THE COMPUTER MODEL OF THE MIND

It is in this context and against this background that Lazarus's responses to Zajonc's challenge must be assessed. Lazarus was not the only participant in the ensuing controversy. Some researchers, such as the social constructionists Sarbin and Averill, offered their own defenses of the intentionalist-cognitivist position. Others, as we have seen in the case of Tomkins, welcomed Zajonc's views. But Lazarus was a major player in the debate because of his long-standing advocacy of the cognitivist-appraisal approach and because he understood himself to be a principal target of Zajonc's attack. In a series of publications, Lazarus returned again and again to the issues, as he saw them, at stake in the controversy. I think it is fair to say that, more than many of his peers in the emotion research field, he was sensitive to some of the larger theoretical issues involved. But throughout the 1980s and 1990s, and indeed for the rest of his career, he continually scrambled to rearticulate his position in ways that would not leave him vulnerable to criticisms of one kind or another. The fact that he was unable to do so undoubtedly reflects his limitations on a theoretical and philosophical level. But it also testifies to the sheer difficulty of properly framing those issues, a difficulty that persists to the present day.

It is striking that the very first objection Lazarus raised against Zajonc concerned the dependence of the latter's claims about the primacy of affect on the computer model of the mind. "Here and elsewhere," Lazarus observed in his first response in 1981,

> Zajonc relies exclusively on the currently dominant approach to information processing, which is particularly compatible with the

> position that emotion cannot be viewed as post-cognitive. In this approach, information and meaning are tortuously built up from the initial receipt of meaningless bits of sensory input to their registration, followed later by encoding, storage, retrieval, or whatever. The origins of this approach lie in information theory . . . which uses the computer as an analogue of the mind. The mind scans a stimulus array to extract information or meaning from the environment in which it resides.
>
> If the meanings underlying emotion were indeed derived in this fashion, emotion would necessarily arise at the end of several stages of information processing, a premise which embarrasses the concept that personal meaning leads to emotion because emotion occurs so early in an encounter, often right at the start to facilitate adaptation.[32]

It was the idea that, on that model, cognition must intervene late in the process and that affect must also arrive late if it was post-cognitive that concerned Lazarus and that he rejected. He pointed out that there were other ways to think about cognition. As he noted, Erdelyi and other information theorists had already suggested, before Zajonc, that emotion could arise very early in any encounter, while the philosophers Michael Polanyi and Hubert Dreyfus had both completely rejected the computer model of the mind ("CR," 222).

The important point for Lazarus in these alternative ways of thinking about cognition was their implication that, far from coming late in an information-processing sequence, "meaning is an immediate feature of every transaction with the environment, whose features are in some degree attended and responded to selectively on the basis of needs, commitments, beliefs, and cognitive styles, in other words, properties that the person brings to the encounter which define what is meaningful" ("CR," 222). Moreover, he argued that the stimulus did not need to be known in detail for "instant recognition of its relevance or potential relevance for one's well-being—in effect, for a person or animal to appraise it as positive, harmful, threatening, or challenging" ("CR," 222). He observed in this regard that many psychologists, including Freud, had assumed that persons and organisms can respond to the world as meaningful even if perceptions are primitive and the precise details of objects or events are not recognized. He commented that this was what the earlier New Look

32. Richard S. Lazarus, "A Cognitivist's Reply to Zajonc on Emotion and Cognition," *American Psychologist* 36 (2) (1981): 222; hereafter abbreviated as "CR."

debates about subception were all about: as he and McCleary had discovered, although subjects in their experiment perceived tachistoscopically presented stimuli incorrectly, they nevertheless correctly discriminated threatening from nonthreatening ones. In effect, people could "instantly 'recognize'" that a situation was dangerous or benign and act accordingly, without having much information about it "and/or without consciously having examined the cognitive premises on which the appraisal is based" ("CR," 223).

Although Lazarus generously characterized Zajonc's view that emotion and cognition are best regarded as separate psychological systems as a "legitimate" theoretical proposition, he contended that it was wrong to reify emotion and cognition as independent and merely interactive because, as Zajonc himself acknowledged, in nature they are "normally fused." Their separation was not required by the fact that emotion comes at the outset of an encounter, because meaning is "commonly present" at the start, even if it is ill defined, unconscious, or unverbalizable and the environmental event is barely discernible or ambiguous. "The issue is not whether the cognitive activity underlying emotion is rational (objective) or irrational, consciously or deliberately arrived at, or primitive, but that cognitive activity concerning the significance of the event *always* underlies feelings and is an essential aspect of them," Lazarus concluded. "I would argue that there are no exceptions to the principle that emotion is a meaning-centered reaction and hence depends on cognitive mediation" ("CR," 223; his emphasis).

Lazarus's initial statement set out many of the terms of the ensuing debate with Zajonc. His critique of the computer model of the mind and his situating of the issues at stake in terms of the prior controversy over subception remained leitmotifs of many of his subsequent publications. His references to the work of Polanyi and Dreyfus were not his first. Already two years earlier, Folkman, Schaefer, and Lazarus had offered some shrewd criticisms of information-processing approaches to cognition and emotion. The authors' objections were sufficiently close to criticisms made especially by the philosopher Hubert Dreyfus, work they had just come across, that it is not surprising to find an addendum at the end of their article recommending Dreyfus's "sophisticated and carefully reasoned" discussion, in his 1972 book *What Computers Can't Do: A Critique of Artificial Reason*, of the limitations of the computer as a model for human reasoning.

One of the issues addressed by Folkman, Schaefer, and Lazarus was that emotions did not conform to the computer's simple binaries of either/or. Many years earlier the psychologist Robert Abelson—who had completed his PhD at Princeton under Silvan Tomkins—had expressed this problem rather vividly when, in a paper of 1963 cited by Zajonc on the computer simulation of "hot" cognition (or "cognition dealing with affect-laden objects"), Abelson had commented on the "uncongeniality of Gestaltist dynamic terms" for the construction of a "tight process model." "It is hard to express this difficulty," Abelson had written, "but I can perhaps indicate the problem by saying that, for Heider [a Gestaltist], things flow instead of click." That difficulty had not prevented Abelson from trying to get emotional evaluations to "click" by making them conform to a logical, either/or, decision-tree model of information processing.[33] In their 1979 article, Folkman, Schaefer, and Lazarus had made a similar point as Abelson's, though from a more critical perspective, when they had observed that the nature of emotion as "information" was "unclear." As a result, they had observed,

> emotions seem to be conceived of as operating by proxy on cognitions through their influence on programme decision points. Information processing is thereby reduced to a series of binary decisions—will the information be further processed or not? This does not seem faithful to the complex processes that are involved in the creation of meaning which characterize so many human adaptive activities, and to the possibilities inherent in tacit knowing (Polanyi, 1958).[34]

The reference was to Polanyi's book *Personal Knowledge*, in which the author argued that our perceptions of the world presuppose a great deal of background or "tacit knowledge," and that it was precisely that kind of unformalized and unformalizable tacit knowledge, necessary for understanding even the simplest situations, that computers could not simulate.[35]

In a related argument, Folkman, Schaefer, and Lazarus had suggested that information-processing theories were especially unsuited to modeling the remarkable ability of human beings to tolerate ambiguity and uncertainty in the environmental display ("CPM," 276–82). This was a topic that Dreyfus had addressed in an especially brilliant manner in his book

33. Robert P. Abelson, "Computer Simulation of 'Hot' Cognition," in *Computer Simulation of Personality: Frontier of Psychological Theory*, eds. Silvan S. Tomkins and S. Messick (New York and London, 1963), 278.

34. S. Folkman, C. Schaefer, and R. S. Lazarus, "Cognitive Processes as Mediators of Stress and Coping," in *Human Stress and Cognition: An Information-Processing Approach*, eds. V. Hamilton and D. M. Warburton (1979), 275; hereafter abbreviated as "CPM."

35. Michael Polanyi, *Personal Knowledge: Towards a Post-Critical Philosophy* (Chicago, 1958).

when discussing the difficulty researchers were already experiencing in developing programs that would allow a computer to recognize patterns and respond successfully to ambiguous stimuli and situations. On the basis of these and related considerations, Dreyfus had predicted that the seductive artificial intelligence research program that had come to dominate much of cognitive psychology would falter precisely because of the inadequacy of the researchers' epistemological, psychological, and ontological assumptions. He had traced those assumptions back to the Cartesian idea that human understanding consists in forming and using symbolic representations built up out of primitive, disembodied, context-free facts, elements, or features, and that explicit rules or instructions for manipulating those facts or elements or features could be determined.

Dreyfus had countered on Wittgensteinian, Heideggerian, and phenomenological grounds that human understanding did not work in that way. Rather, the human mind's sense of relevance was holistic or global: it depended on background "knowhow," ordinary coping skills, personal experience, and knowledge that could not be represented in formal rules of that kind—an impossibility, he thought.[36] In short, Dreyfus had claimed that ordinary human experience and everyday common sense resisted capture by the context-independent and digitalized formalizations of artificial intelligence researchers. The latest developments in computer research had seemed to him to confirm his sense of an emerging crisis in the field.

Because they thought so well of Dreyfus's arguments, Folkman, Schaefer, and Lazarus had ended their paper by citing the following passage from Dreyfus's book, a passage they felt captured the essence of the latter's critique of information-processing theories of the mind:

> "The psychological, epistemological, and ontological assumptions [of artificial intelligence and computer simulation] have this in common: they assume that man must be a *device* which calculates according to rules on data which take the form of atomic facts. Such a view is the tidal wave produced by the confluence of two powerful streams: first, the Platonic reduction of all reasoning to explicit rules and the world to atomic facts to which alone such rules could be applied without the risks of interpretation; second, the invention of the digital computer, a general-purpose information-processing device, which calculates according to explicit rules and takes in data in terms of atomic elements logically independent of one another.

36. Hubert L. Dreyfus, *What Computers* Still *Can't Do: A Critique of Artificial Reason* (Cambridge, MA, 1992), 3; hereafter abbreviated as *WCSCD*.

In some other culture, the digital computer would most likely have seemed an unpromising model for the creation of artificial reason, but in our tradition the computer seems to be the very paradigm of logical intelligence, merely awaiting the proper program to accede to man's essential attribute of rationality.

The impetus gained by the mutual reinforcement of two thousand years of tradition and its product, the most powerful device ever invented by man, is simply too great to be arrested, deflected, or even fully understood. The most that can be hoped is that we become aware that the direction this impetus has taken, while unavoidable, is not the only possible direction: that the assumptions underlying the conviction that artificial reason is possible are assumptions, not axioms—in short, that there may be an alternative way of understanding human reason which explains both why the computer paradigm is irresistible and why it must fail. (*WCSCD*, 231–32; his emphasis)"[37]

Especially in the early years of his debate with Zajonc, Lazarus returned often to this critique. "The most serious mistake in Zajonc's analysis lies in his approach to cognition, which is characteristic of much present-day cognitive psychology," Lazarus repeated in 1982. "In this approach information and meaning stem from the conception of mind as an analogue to a computer . . . This conception has been rebutted by Dreyfus (1972) and Polanyi (1958, 1966), and others, although the rebuttal has not affected the mainstream of cognitive psychology." Lazarus rejected the idea that meaning could be built up from essentially meaningless stimulus display elements or "bits," arguing once again that humans did not have to wait for revelation from information processing to "unravel the environmental code." "As was argued in the New Look movement in perception," he commented, "personal factors such as beliefs, expectations, and motives or commitments influence attention and appraisal at the very outset of any encounter. Concern with individual differences leads inevitably to concern with personal meanings and to the factors that shape such meanings . . . Information processing as an exclusive model of cognition is insufficiently concerned with the person as a source of meaning."[38]

37. Folkman, Schaeffer, and Lazarus cited the original 1972 edition of Dreyfus's book, 143–44.

38. Richard S. Lazarus, "Thoughts on the Relations Between Emotion and Cognition," *American Psychologist* 37 (9) (1982): 1020; hereafter abbreviated as "T."

Lazarus remarked that the immediacy of emotional responses accorded well with Zajonc's opinion that affective "meanings" come early before the person fully knows or recognizes what the object or event is. "However," he observed, "I reject the assumption that this early presence means that it is detached from or independent of cognitive appraisal." He argued that if one accepted, as Zajonc did, that meaning lies at the end of a sequence of cognitive processing, then the fact that we can emotionally react instantly, at the very start of a transaction, forces us to abandon the idea that emotion and cognition are necessarily connected and to adopt instead the position that emotion and cognition are separate psychological systems. But Lazarus repudiated the information-processing model on the grounds that "we do not have to have complete information to react emotionally to meaning . . . The meaning derived from incomplete information can, of course, be vague: we need to allow for this type of meaning as well as for clearly articulated and thoroughly processed meaning" ("T," 1021).

"THERE WERE MANY MISTAKES IN THE WAY SOME OF THE ISSUES HAD ORIGINALLY BEEN CAST"

In the ensuing years, Lazarus made similar statements and arguments.[39] Of course, as the debate unfolded in charges and countercharges, he raised other complaints against his antagonist. One of them, evident early on in the debate, was the charge that Zajonc's statements were often vague or ambiguous, as was shown in the many qualifying remarks the latter made in speaking of affect or feeling. For example, Lazarus quoted Zajonc as writing that "'affective judgments may be *fairly* independent . . . of perceptual and cognitive operations commonly assumed to be the basis of these affective judgments . . . Affective reactions can occur without *extensive* perceptual and cognitive coding'" and so on, suggesting that affect was not always independent of cognition and doing so in terms that made it difficult to pin down Zajonc's argument ("T," 1021; Lazarus's emphasis). This was a complaint also expressed by several other commentators.[40]

39. Thus, Lazarus more than once characterized Zajonc's embrace of the computer model of the mind as his "most serious mistake." See for example Richard S. Lazarus and Susan Folkman, *Stress, Appraisal, and Coping* (New York, 1984), 276.

40. See for example James A. Russell and Lisa Woudzia, "Affective Judgments, Common Sense, and Zajonc's Thesis of Independence," *Motivation and Emotion* 10 (2) (1986): 169–82; and W. Gerrod Parrott and John Sabini, "On the 'Emotional' Qualities of Certain Types of Cognition: A Reply to the Arguments for the Independence of Cognition and Affect," *Cognitive Therapy and Research* 13 (1) (1989): 49–65.

But Lazarus was himself vulnerable to the same accusation of vagueness or equivocation. He too qualified his claims in ways that muted the force of his position. This is evident, for example, in his suggestion that cognition and emotion were "usually fused," a formulation that could not help but suggest their dissociation or separation under unusual or abnormal conditions, making it unclear how cognition could also be the "necessary and sufficient condition of emotion" ("T," 1019). The problem is also evident in Lazarus's loose employment of the term "information." In his book Dreyfus had pointed to the ambiguity inherent in the expression "information processing." The expression in its ordinary, everyday usage simply meant that the mind takes account of meaningful data and transforms it into other meaningful data. But the cybernetic theory of information had nothing to do with meaning in that ordinary, everyday sense. As Dreyfus had explained, it was a "nonsemantic, mathematical theory of the capacity of communication channels to transmit data. A bit (binary digit) of information tells the receiver which of two equally probable alternatives has been chosen" (*WCSCD*, 165).

He had noted that Claude Shannon was perfectly clear that his theory of information, worked out for telephone engineering, carefully excluded as irrelevant the meaning of what was being transmitted. He had quoted Shannon as stating: "'The fundamental problem of communication is that of reproducing at one point either exactly or approximately a message selected at another point. Frequently messages have *meaning*; that is, they refer to or are correlated according to some system with certain physical or conceptual entities. These semantic aspects of communication are irrelevant to the engineering problem'" (*WCSCD*, 165; Shannon's emphasis). Dreyfus had also quoted Warren Weaver's "even more emphatic" comment on Shannon's work: "The word *information*, in this theory, is used in a special sense that must not be confused with its ordinary usage. In particular, *information* must not be confused with meaning. In fact, two messages, one of which is heavily loaded with meaning and the other of which is pure nonsense, can be exactly equivalent, from the present viewpoint, as regards information. It is this, undoubtedly, that Shannon means when he says that 'the semantic aspects of communication are irrelevant to the engineering aspects'" (*WCSCD*, 165; Weaver's emphasis).[41]

41. For a related critical discussion of the use of the concept of "information" in the behavioral and cognitive sciences, focused chiefly on Fred Dretske's attempt to provide a naturalist explanation of meaning based on the handling and coding of information, see Jeff Coulter, "The Informed Neuron: Issues in the Use of Information Theory in the Behavioral Sciences," *Mind and Machines* 5 (1995): 583–96. See also Rupert

Dreyfus had warned against shifting between the ordinary usage of the term "information" and the technical sense the term had acquired in cybernetic theory. He had suggested that much of the plausibility of cognitive simulation derived from the exploitation of the ambiguity between the different meanings of the term. The role of the programmer was to make the transition from statements that were meaningful (that contained information in the ordinary sense) to the strings of meaningless bits with which a computer operated (information in the technical sense). The ambition of artificial intelligence was to program the computer to do this translating job itself. But Dreyfus argued that it was by no means obvious that the human translator could be dispensed with. By quietly shifting back and forth between the ordinary and the technical meaning of the term "information," artificial intelligence researchers illegitimately gave the impression that the computer had already achieved this task of transformation. Precisely because Dreyfus thought it an open question whether human intelligence depends on rule-like operations performed on discrete elements, he suggested that authors should be careful to speak and think of "information processing" in quotation marks when referring to human beings.

This was a recommendation that was routinely ignored by a large majority of cognitive psychologists who exploited the ambiguities inherent in the term in order to gloss over the question of how the processing of information in the cybernetic sense could produce meaningful information in the ordinary, everyday sense of the term. Lazarus was one of those who ignored Dreyfus's warnings about terminology. At the start of his controversy with Zajonc, Lazarus had stated that "meaning exists not merely in the environmental display, but inheres in the [individual's] cognitive structures and commitments," which is to say that according to Lazarus the input from the environment is itself inherently meaningful—even if, according to the separation he imposed between perception and appraisal, it is not yet individually or personally meaningful ("T," 1022). But as the debate with Zajonc progressed, he began to theorize the input and the appraisal process in terms so close to that of his information-processing contemporaries that he risked losing a grip on his best insights.

Take for instance Lazarus's response to an influential paper by the neuroscientist Joseph LeDoux. In 1986 LeDoux proposed an exclusively neurophysiological or "neurobiological" theory of emotion based on his research on fear conditioning in rats. His most important (and now

---

Read, "The 'Hard' Problem of Consciousness Is Continually Reproduced and Made Harder by All Attempts to Solve It," *Theory, Culture & Society* 25 (2008): 52–86.

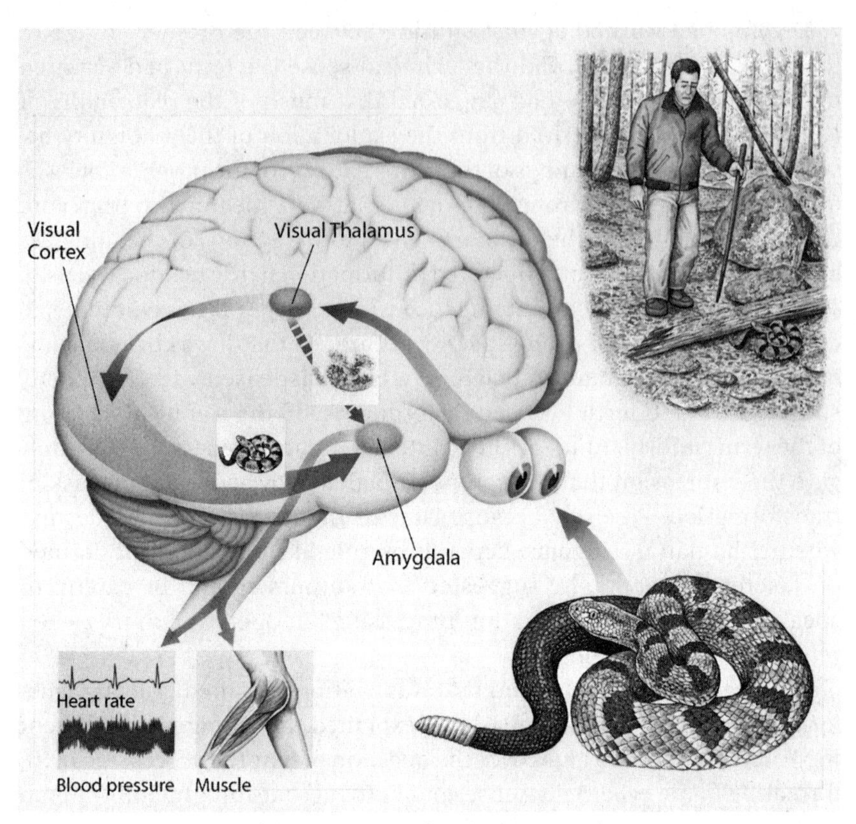

Figure 6 *Joseph LeDoux and the Two Sources of Fear*. Source: Joseph LeDoux, "Emotion, Memory and the Brain," *Scientific American* 270 (6) (1994): 50–56. Illustration by Roberto Osti, reproduced with permission.

well-known) claim was that there exists a distinct neuronal pathway capable of transmitting afferent signals from sensory processing areas directly to the so-called "limbic brain," especially to the subcortical amygdala, without involving the cortical regions, thereby permitting a form of automatic unconscious affective processing independent of cognitive functions. LeDoux described that pathway as a "quick and dirty" means by which the "crude" or "simple" "perceptual inputs" involved in rapid emotional responses could be transmitted and evaluated independently of consciousness and cognition (see Figure 6).

On this basis, he supported Zajonc's ideas about the primacy of affect. "[T[he stimulus features proposed by Zajonc to mediate unconscious preference formation are remarkably compatible with the crude processing capacities of the thalamo-amygdala pathway," LeDoux stated. "And his finding that preferences can be formed faster than objects can be recognized is consistent with the fact that the thalamo-amygdala pathway is a rapid processing channel."[42] According to LeDoux, the cortical-conscious system carried out the "understanding" or "cognitive interpretation" of the affective response only belatedly, and "after the fact."

In his discussion, LeDoux without embarrassment mixed a phenomenological level of analysis, in which it makes sense to talk of a person's capacity to perceive and interpret the world, with a neurophysiological level of analysis, in which it does not. "The purpose of this paper is to examine what is known about the neural pathways by which the brain interprets the emotional significance of environmental events," he announced ("SSE," 237). But the brain does not "interpret" the significance of events, it does not interpret anything: it is an energy-transforming organ that transmits neuronal impulses. LeDoux thus committed what, as I noted in the introduction, Bennett and Hacker in their critical examination of the philosophical foundations of neuroscience have called the "mereological fallacy" and what Sharrock and Coulter, in their analyses of the pitfalls of the computer model of the mind, have similarly described as the mistake of stripping from the mind or person the capacity to make meaning in order to hand that capacity over to the brain or to subpersonal levels of processing.[43]

What made LeDoux's mixing of the language of phenomenology with the language of neurophysiology appear so reasonable to so many researchers rather than the incoherent "gibberish" Dreyfus dubbed it (*WCSCD*, 179) was precisely that he employed the language of "information processing," with all its inherent ambiguities. "How does the brain transform patterns of physical energy in the environment into emotionally significant events?" LeDoux asked. "Emotion will be viewed here as an information processing function of the nervous system and will be examined accordingly—by tracing the fate of the initiating stimulus as it undergoes various transformations from sensory to meaning analysis" ("SSE," 237). In neurological terms, he added, "the task at hand is to follow the stimulus as it is received by the primary sensory system and transmitted to brain areas that evaluate the meaning of the stimulus and initiate appropriate changes in autonomic, endocrine, and behavioral ac-

42. Joseph E. LeDoux, "Sensory Systems and Emotion: A Model of Affective Processing," *Integrative Psychiatry* 4 (1986): 241; hereafter abbreviated as "SSE." The paper was followed by commentaries by various authors, including Lazarus.

43. In addition to the work of Bennett, Hacker, Sharrock, and Coulter, cited in the introduction, see also Anthony Kenny's discussion of the "Homunculus Fallacy" in *The Legacy of Wittgenstein* (Oxford, 1984), 125–36.

tivity" ("SSE," 237). So taken for granted was this information-processing approach that the ambiguity of LeDoux's terms, such as "initiating stimulus," in the posited sequence of neuronal events went unremarked.[44]

As Dreyfus observed of pioneering cognitive psychologist Ulric Neisser's and Fodor's influential use of similar terms, if "sensory input" was energy, then it was only necessary for it to be transformed into other energy in the brain. But if the input was a "simple perceptual event," then the author needed to explain what this percept was. Dreyfus commented that as philosophers had ceased to believe in sense data, theorists such as Neisser could not introduce the idea of a primitive percept "without a great deal of argument and evidence." Phenomenologically, Dreyfus argued, we directly perceive objects, not sense data. If Neisser wanted to shift the notion of input from the physical to the perceptual, it was up to him to explain what sort of perception he had in mind and what evidence he had that such a percept—neither a pattern of rays of light nor a perspectival view of an object—really existed. Of course, as Dreyfus further remarked, the notion of "information" was meant to resolve such confusions, since according to Neisser "sensory input" was conceptualized as a series of bits and it was those bits of information that were somehow transformed into meaning. But Dreyfus argued that it remained unclear what this information was, as long as the notion of "sensory input" remained ambiguous and as long as the question of how information was supposed to be related to the "stimulus" input, whether energy or direct perception, was not addressed (*WCSCD*, 179–87).

Dreyfus's strictures could be applied to LeDoux. The latter's declaration that he had demonstrated how the brain "interprets" the emotional "significance" or meaning of the "initiating stimulus" likewise begged many fundamental questions. Of course, it might also be argued that what was at stake for LeDoux was not exactly how the brain makes meaning, since his claim was that, in line with Zajonc's thesis, in the case of emotions the "initiating stimuli" were so crude and simple and the hypothesized information-processing circuits so minimal, that no cognition or meaning was involved at all. As he wrote:

44. LeDoux used various different words for the input signal: "afferent signals," "sensory messages," "sensory input," "the initiating stimulus," "emotional information," "sensory stimuli," "incoming signals," "sensory signals," "sensory messages," "natural environmental stimuli," "simple perceptual events," "undiscriminated stimuli," "simpler stimulus features," "sensory information," "peripheral stimuli," and "crude stimulus information."

> The studies that have implicated the cortico-amygdala pathways in emotional processing have generally involved complex sensory events, such as natural environmental stimuli or test items that must be distinguished on the basis of perceptual characteristics. By contrast, emotional responses coupled to simple perceptual events, such as the onset of undiscriminated stimuli or the presence or absence of a signal, survive decortication. These observations suggest that the relay of sensory input to the amygdala from the cortical sensory systems is required to process the significance of complex stimulus information, whereas simpler stimulus features can be evaluated by way of direct projections to the amygdala from sensory neurons in the thalamus. ("SSE," 240–41)

Likewise, Zajonc's thesis was that "preferences" (or affects) need no inferences because they occur without any cognitive processing at all, thus providing a "quick and dirty" response to certain rudimentary stimuli in ways that were of evolutionary advantage to the organism.

This is the terrain onto which Lazarus stepped. In his comments on LeDoux's paper, Lazarus started out promisingly enough by suggesting that usually neurophysiology had little to offer cognitive or "relational" theorists, such as himself, because neurophysiological concepts of emotion could not deal adequately with complex psychological processes involving meaning-centered activity. Stating that "Cognition is the *key* to emotion," Lazarus observed that the distinction LeDoux made between "cognitive" versus "value" transformations made no sense to him because, as long as it was not hardwired and automatic, "a value transformation must also be cognitive, even a relatively primitive valuing of good and bad without cognitive detail. The crucial issue is how the cognitive activity in emotion works and is reflected in pathways of neural activity in the brain" (his emphasis). Moreover, Lazarus also observed that cognitive theorists did not speak of the "stimulus" per se, but embedded it "in the person-environment relationship, since it takes both environmental and organismic (person) factors operating together to generate an appraisal or harm and benefit and, hence, emotion" ("SSE," 245–46).

So far, so good, one might think, although this last statement risked muddying the waters because it implied the existence of a separation between the person and the "stimulus" or "environment" requiring an intervening appraisal process to close that separation, when everything Lazarus had also said about an agent's direct apprehension of an already meaningful world might have seemed opposed to such a formulation.

Lazarus's declaration, however, was consistent with what he had always argued. As I have already emphasized in the previous chapter, from the start of his work on emotion Lazarus regarded the perception (or what he now called the "cognition") of an object and its appraisal as two separate processes. He thus posited a problematic distinction between the "cold" cognition of the object or environment, defined as a neutral cognition of no particular meaning to the individual, and a "hot" cognition of the same object or environment, a cognition that was meaningful to the subject in some personal way.[45]

What was new in Lazarus's commentary on LeDoux was that he now deployed the term "information" in his discussion. "Appraisal is sometimes confused with information," Lazarus stated. "*Information* is affectively neutral or cold knowledge about the environment and how it operates. *Appraisal* is a cognitive evaluation of the significance of a transaction for well-being. As such, it includes the evaluation of information. The appraisal becomes hot or emotional when the transaction is evaluated as involving a personal stake" ("SSE," 245; his emphasis). But what was the status of "cold knowledge" that Lazarus now identified as "information"? It seems from the context that by "cold" information Lazarus did not mean information in the cybernetic sense of meaningless bits of data, but information in the ordinary sense of meaningful (if slightly "impersonal") data about the world. Yet, in light of Dreyfus's observations about the ambiguity of the term "information," it was careless of Lazarus to use that word without the scare quotes Dreyfus had recommended. His employment of the term could not help but confuse an already confused situation.

Matters became potentially even more muddled when Lazarus disclosed what it was about LeDoux's proposal that really interested him. This had less to do with the distinction between "information" and "appraisal" itself than with the light LeDoux's ideas about neural pathways seemed to throw on the way cognitive activity in the emotions might work. In particular, Lazarus welcomed LeDoux's idea that there were two independent neural pathways by which "information from the environment having significance for well-being" could be processed, one a thalamo-amygdala pathway and the other a neo-cortical route. "Appraisals may be rational, deliberate, and conscious, on the one hand, or primitive, inde-

45. Lazarus subsequently acknowledged that his distinction between cold "knowledge" and hot "appraisal" was unstable: as knowledge involved a sense of self and the relation of the self to the world, it was unclear in what sense cold knowledge could be as purely "impersonal" as Lazarus had claimed it was. See Richard Lazarus, "Cognition and Motivation in Emotion," *American Psychologist* 46 (1991): 354.

liberate, and unconscious, on the other," Lazarus observed in this regard. "The difference implies a distinction in the way meaning is created that is relevant to LeDoux's analysis of neural circuits . . . It is tempting to think that primitive, indeliberate, and unconscious appraisals are represented in the thalamo-amygdala pathway, and rational, deliberate, and conscious appraisals on the other hand require neo-cortical involvement" ("SSE," 245–46). On this basis Lazarus praised LeDoux's "wonderful clarity" ("SSE," 246) for explaining the evolutionary advantage of such a primitive neural processing system in the thalamo-amygdala neural pathway, because such a pathway permitted quick defensive reactions to situations of threat. Those reactions could subsequently be aborted or attenuated after more detailed perceptual analysis by the cortical sensory systems. Lazarus therefore concluded by suggesting that LeDoux's thesis about the existence of two independent neural pathways could help resolve the debate taking place between himself and Zajonc, by suggesting a physiological rationale for distinguishing between two ways meaning could be achieved cognitively: one that was primitive, unconscious, and rapid, the other more complex, conscious, and deliberate.

Of course, Lazarus took the thesis of the existence of two separate neural pathways for processing affect in a direction different from that proposed by LeDoux. In line with Zajonc's claims, LeDoux thought that the thalamo-amygdala pathway route permitted "quick and dirty" emotional information processing in the absence of cognition. But now Lazarus was suggesting that the same thalamo-amygdala pathway provided the neural basis for emotional reactions subtended by cognitive appraisals of an automatic and unconscious kind. As Lazarus later observed, at the start of his thinking about appraisal he had felt that Arnold had "underemphasized the complexity of evaluative judgements," with the result that in his own treatment of the topic he had stressed the "complex, judgmental, and conscious process" involved. But, as he also observed, he had subsequently become more impressed with the "instantaneity of the process of appraising" and indeed its unconscious or automatic aspects.[46] LeDoux's findings impressed Lazarus, then, precisely because they suggested how these two different ways of generating meaning could be subserved by separate neural pathways, and from then on he repeatedly stressed this point.

The problem was that, once Lazarus endorsed LeDoux's idea of multiple levels of affective processing, he risked seeing his position collapse into that of information-processing theorists, for whom the idea of a hier-

46. Richard S. Lazarus, "Relational Meanings and Discrete Emotions," 51.

archy of processing systems operating largely independently of cognitive control was so attractive. The difference between Lazarus's concept of minimal, automatic, and unconscious cognitive processing and the concept of minimal, automatic, noncognitive processing thus threatened to become a distinction without a difference.[47] This development is well illustrated by cognitive psychologist Arvid Kappas's recent assessment of appraisal theory. He has suggested, in reference especially to Arnold's positions, that there is no discernible difference between her thesis of the intuitive and automatic nature of appraisals and Zajonc's ideas, and thus that "Arnold held basically the same views of those of Zajonc." Given Kappas's own identification of rapid unconscious processes with automatic subconscious information processing, his view of the matter is hardly surprising.

In fact, Kappas also considers Ekman's concept of "autoappraisers," defined as mechanisms capable of automatically detecting dangerous events and of triggering the relevant basic emotions, as virtually identical to Arnold's opinions. Similarly, he quotes Antonio Damasio as stating: "'Does the notion that emotions can be triggered nonconsciously and automatically deny the classical notion of an "appraisal" phase preceding emotions? Not at all. The process by which, at a given moment, an object or situation *becomes* an emotionally competent stimulus often includes a conscious, cognitive appraising of the circumstances. Besides, even when the process is nonconscious, the current context may play a role and enhance or reduce the competence of the stimulus.'" To which Kappas adds: "the *nonconscious and automatic process* described by Damasio, and contrasted by him to conscious, cognitive appraisal, is *appraisal* in her [Arnold's] usage. Using the word *appraisal* to refer to both processes, she described the same processing sequence as Damasio."[48]

But the collapse of Arnold's (and Lazarus's) notions of automatic, unconscious appraisal into information-processing views of the phenomenon has had as a consequence that the words *appraisal* and *cognition* have lost their connection to issues of meaning. Forms of cognition that are less complex than reflective thought are no longer distinguishable from

47. Thus, in his paper "Relational Meanings and Discrete Emotions," Lazarus suggested that it was precisely the possibility of thinking "in terms of stepwise information processing" (of the kind found in LeDoux's work) that enabled him to think about multiple levels of "processing" and to accept the idea of automatic, unconscious appraisals.

48. Arvid Kappas, "Appraisals are Direct, Immediate, Intuitive, and Unwitting . . . And Some Are Reflective," *Cognition and Emotion* 20 (7) (2006): 952–75.

purely mechanistic or neuronal forms of transmission. More generally, an important fallout from Lazarus's acceptance of LeDoux's thalamo-amygdala thesis was that the role of the person and the person's intentions in the making of meaning, so central to Lazarus's original ideas about the appraisal process in emotion, was lost sight of in favor of an emphasis on the role of various internal mental mechanisms or subpersonal processing systems. A related issue, of course, is whether LeDoux's theory, extended to humans, concerning two pathways to fear—a slower-acting cortical-mediated pathway, and a "quick and dirty" subcortical, thalamo-amygdala pathway—is indeed correct. As several studies have shown, and as we have already seen in chapter 2, among other criticisms of LeDoux's theory is the fact that the amygdala is not exclusively sensitive to fear situations, but is activated in response to both negative and positive affective stimuli.[49]

But Lazarus hardly noticed those dangers, so keen was he to find support for the thesis of two different ways of generating meaning, and especially for the idea of automatic, unconscious appraisals. Thus, in his 1986 response to LeDoux and subsequent writings, he appealed to proposals by Buck (1985), Baron and Boudreau (1987), and Leventhal and Scherer (1987), among others, all of whom offered versions of the idea that there are dual levels of cognitive processing, one deliberate and conscious, the other automatic and unconscious. Some of those researchers proposed an account of appraisal in terms derived from the perception theorist James J. Gibson's thesis of affordances; others couched their ideas in the very language of information processing that Lazarus had earlier criticized but that now went uncontested and unremarked by him. All such proposals

49. For discussions and critiques of theories concerning the role of the amygdala in the fear response, in a very large literature see Luiz Pessoa and Ralph Adolphs, "Emotion Processing and the Amygdala: From a 'Low Road' to 'Many Roads' of Evaluating Biological Significance," *Nature Reviews Neuroscience* 11 (11) (2010): 773–82; Luiz Pessoa, "Emotion and Cognition and the Amygdala: From 'What Is It?' to 'What's to Be Done?'," *Neuropsychologia* 48 (12) (2010): 3416–29; Lisa Feldman Barrett and Tor D. Wagner, "The Structure of Emotion: Evidence from Neuroimaging," *Current Directions in Psychological Science* 15 (2) (2006): 79–83; Kristen A. Lindquist, Tor D. Wager, Hedy Kober, Eliza Bliss-Moreau, and Lisa Feldman Barrett, "The Brain Basis of Emotion: A Meta-Analytic Review," *Behavioral and Brain Sciences* 35 (3) (2012): 121–202, with peer commentary by Ross Buck, Vittorio Gallese, Luiz Pessoa, Andrea Scarantino, Klaus Scherer, and many others. For a critical discussion of dual-process theories of emotional processing and their ties to basic emotion theory, see also Lisa Feldman Barrett, Kevin N. Oscher, and James J. Gross, "On the Automaticity of Emotion," in *Social Psychology and the Unconscious: The Automaticity of Higher Mental Processes*, ed. J. Bargh (New York, 2007), 173–217.

were grist for the latter's mill, so long as they distinguished between conscious and deliberate cognitions on the one hand and unconscious and automatic ones on the other.[50]

Yet, of all the issues that preoccupied Lazarus during these years, it was precisely the concept of appraisal itself, and especially unconscious appraisal, that gave him the most trouble, on both empirical and conceptual grounds. Lazarus was committed to the idea that appraisal, or "hot" cognition, played a causal role in emotion. One critical issue that arose for him, then, was how the claim for the causal role of appraisal was to be tested empirically, if so many appraisals were involuntary and unconscious and were therefore unavailable for conscious self-report. As we have seen, this was a question that Zajonc had raised from the start of his debate with Lazarus. The second critical issue concerned the very idea that the role of appraisal was a causal one. The two issues were linked: if the relation of appraisal or cognition to emotion was not a causal but a necessary or constitutive one, in the sense of the philosopher Anthony Kenny's earlier analysis of the intentionality of emotion, then certain kinds of empirical tests of the causal role of cognition in emotion would appear to be beside the point. On the other hand, if appraisal's role in affect was indeed causal, then the empirical question had to be addressed.

"TELLING MORE THAN WE CAN KNOW"

In the late 1980s and the 1990s, Lazarus found himself highly praised for his achievements, but also pressed even by those on the cognitive side of the debate to address certain criticisms of his position and formulations. In his efforts to answer his critics, Lazarus was forced to do some hard thinking about what kind of claim he was making when he insisted that emotions are cognitive states. He found himself obliged to focus critically on the experimental methods then being used to prove the cognitive nature of the emotions. And he also felt required to pay renewed attention to difficult questions about the nature of causality in psychology and by implication the nature of intentionality, questions that had concerned him at the start of his research and that now came back to haunt him.

50. Ross Buck, "Prime Theory: An Integrated View of Motivation and Emotion," *Psychological Review* 92 (3) (1985): 389–423; Reuben M. Baron and Louis A. Boudreau, "An Ecological Perspective on Integrating Personality and Social Psychology," *Journal of Personality and Social Psychology* 53 (6) (1987): 122–28; Howard Leventhal and Klaus Scherer, "The Relationship of Emotion to Cognition: A Functional Approach to a Semantic Controversy," *Cognition and Emotion* 1 (1987): 3–28.

In order to clarify his views, Lazarus called on the writings of a disparate group of philosophers, including Sartre, Heidegger, Merleau-Ponty, and Robert Solomon, among others. But I think it is fair to say that his modest engagement with their work did not do much to resolve the problems that troubled him. The result was a rather promiscuous combination of arguments, few of which were entirely satisfactory and some of which left him vulnerable to further criticisms.

We can track these developments in several important publications by Lazarus and his critics at this time. In a 1988 paper, Lazarus and Smith focused on the research methods being used to identify the specific cognitive dimensions shaping the emotional response. Basing themselves on Lazarus's by-now-familiar, although problematic, distinction between an individual's "objective" or "abstract" "knowledge" of the world (or what the authors also called "perceptual information" or "cold" cognition) on the one hand, and personal "appraisal" (or "hot" cognition) on the other, Lazarus and Smith rejected several of the techniques then being deployed to study cognition on the grounds that those techniques merely examined an experimental subject's "objective knowledge" rather than his or her individual appraisals. The details of those various techniques need not concern us here, except to note that, for Lazarus and Smith, properly designed imagery/role-playing methods, of the kind he and his colleagues had performed, involving retrospective reports by subjects of actually experienced emotional encounters, had the best potential for studying the appraisal process.[51]

Those methods depended on the subjects' self-reports of their experiences, and Lazarus and Smith were well aware of the problems said to be posed by that reliance. As they noted, in a highly influential paper with the title "Telling More Than We Can Know: Verbal Reports on Mental Processes," Richard Nisbett and Timothy D. Wilson (1977) had argued that subjects have little or no direct introspective access to their higher-order cognitive processes, because they are sometimes unaware of the existence of a stimulus that has influenced their responses, or are unaware of their actual responses, or that a stimulus has affected their responses.[52] Nisbett

51. Richard S. Lazarus and Craig A. Smith, "Knowledge and Appraisal in the Cognition-Emotional Relationship," *Cognition and Emotion* 2 (4) (1988): 281–300. See also Richard S. Lazarus and Susan Folkman, "Transactional Theory and Research," *European Journal of Personality* 1 (3) (1987): 141–69, in which the authors discuss the difficulty of separating the antecedent appraisal process from the emotional outcomes, given the circularity inherent in the concept of cognitive appraisal.

52. Richard E. Nisbett and Timothy D. Wilson, "Telling More Than We Can Know: Verbal Reports on Mental Processes," *Psychological Review* 84 (3) (1977): 231–59; here-

and Wilson had claimed that when people attempt to report on their cognitive processes, that is, on the processes mediating the effects of a stimulus to which they are exposed, they do not do so on the basis of a true introspection. Instead, people base their reports on a priori, implicit causal theories or judgments about the extent to which a particular stimulus is a plausible cause of a given response. In order to explain the negative findings on self-report accuracy, Nisbett and Wilson drew a distinction between the *contents* of the mind of which persons have private knowledge and the *processes* of the mind to which they have no access at all.

Nisbett and Wilson's paper and related publications on this topic appeared to constitute a potent critique of self-reports, the most widely used method in cognitive psychology, and for that reason they provoked much discussion. Since the employment of such self-reports was central to Lazarus's approach to the assessment of the appraisal process, it is not surprising that he saw the need to address Nisbett and Wilson's pessimistic conclusions. According to Nisbett and Wilson, individuals are far more likely to have introspective access to the products or "contents" of their cognitive activities than to the (unconscious) causal "processes" themselves. This implied that subjects are better able to report veridically *what* they are thinking and feeling than *why* they are, or indeed *how* their cognitions are influencing their feelings. In their comments, Lazarus and Smith accepted that conclusion, noting that they themselves were well aware that the subjects in their studies had been asked to report almost exclusively on the contents of their thoughts and feelings, and that it had been left to the researchers, rather than their subjects, to infer the underlying causal processes. They even justified their methods on this basis, arguing that there was a reduced risk of distortion when the investigator rather than the subject made the interpretive inferences. Lazarus and Smith's general position was that, when all was said and done, none of the methods used by investigators to examine cognitive activity in emotion was ideal or free from uncertainties. But some methods were better than

---

after abbreviated as "TMK." In their paper, Nisbett and Wilson observe that in his work on tacit knowledge Michael Polanyi had argued that "we can know more than we can tell," meaning that people can perform skilled activities without being able to describe what they are doing and can make fine discriminations without being able to articulate their basis. Nisbett and Wilson describe their own research as suggesting the opposite: that we sometimes tell more than we can know. This is the same Wilson who later published *Strangers to Ourselves: Discovering the Adaptive Unconscious* (Cambridge, MA, 2002), a work cited by Craig DeLancey (see the introduction) and the neurohistorian Daniel Lord Smail (see chapter 7).

others, especially those they advocated, methods that included paying close attention, in the spirit of psychoanalysis, to the discrepancies and contradictions expressed by experimental subjects.[53]

It is important to recognize that Nisbett and Wilson's critique of introspection was premised on the assumption, so central to cognitive psychology, that the contents of mental states were largely the product of hidden, unconscious information-processing events that were unavailable to the self or to consciousness. They had quoted approvingly several cognitive psychologists to this effect. For example, they had cited Ulrich Neisser as stating that "'The constructive processes [of encoding perceptual sensations] themselves never appear in consciousness, their products do'" ("TMK," 232). It was precisely in terms of the assumption of hidden levels of information processing that Nisbett and Wilson explained the inability of persons responding to subliminal stimuli to verbally report on their experiences ("TMK," 239.)

It will be apparent that Nisbett and Wilson's claims trenched on long-standing philosophical debates over the causes and explanation of human behavior. In particular, their arguments amounted to an indictment of intentionality. Their suggestion was that human behavior is determined by causes external to the person as an intentional subject, indeed that human beings are not advantaged over observers in their knowledge of their own intentions because the causes of their behaviors are subpersonal information-processing mechanisms over which they have no conscious control. Such claims had much in common with Zajonc's view that emotions are non-intentional, noncognitive states falling entirely outside the scope of the subject's intentions and volition. But what if those claims were mistaken? What if the idea that emotions depend on internal unconscious cognitive processes and mechanisms was wrong, so that methods aimed at providing objective empirical evidence for them were also incorrect? This would not necessarily rule out all empirical studies of the emotions. But it would call into question investigations premised on the idea that there exist hypothetical, unconscious, causal computations or cognitive appraisals "in the mind/brain" of the kind Nisbett and Wilson presupposed.

Dreyfus in his 1972 critique of artificial intelligence had rejected precisely such premises and ideas. Moreover, in various impressive articles

53. See Richard S. Lazarus, "Vexing Research Problems Inherent in Cognitive-Mediational Theories of Emotion, and Some Solutions," *Psychological Inquiry* 6 (3) (1995): 183–96.

certain participants in the debate generated by Nisbett and Wilson's 1977 paper likewise pointed out deficiencies in the latter's anti-intentionalist conceptual framework. For instance, in 1981 the psychologists Sabini and Silver in an especially brilliant paper focused on several ambiguities in Nisbett and Wilson's arguments. It's not enough to say in this connection, as Parkinson and Manstead have done with reference to Sabini and Silver's article, that there are difficulties with Nisbett and Wilson's process/content distinction because it has proved "nearly impossible to draw up a rigorous definition of what is, or is not, supposed to count as a 'process.'"[54] Sabini and Silver's brief yet elegant analysis was far more interesting and potentially devastating than that: they called into question the whole idea of introspectible *cognitive processes* being the kind of thing to which persons don't have access. In brief, the authors first discussed what it means for someone to give a description of a behavioral process, such as baking a cake, in order to demonstrate that the person requires much more than "access to his baking process." To describe the process accurately, the person has to understand what counts as irrelevant among all the things he did while generating the cake (e.g., answering the telephone, closing a door, scratching a knee, and so on) and what counts as relevant (turning on the oven, mixing the ingredients, placing the cake in the oven, removing it, and so on). In other words, according to Sabini and Silver, Nisbett and Wilson were wrong to think that having access alone ensures that someone is able to answer causal questions, because the individual has to be forearmed with an understanding of what it means to bake a cake in order to make use of the access available to him. Sabini and Silver applied the same analysis to cognitive processes, selecting as their example the case of perfect introspective access to a computer's program.

On this basis, the authors suggested that the entire picture of an individual introspecting his mind to find out the causes of behavior was a mistake, because to make sense, any report of one's cognitive processes cannot just be a report of putative internal processes but has to involve a reference to the objective circumstances in the world in which the performance has taken place. Knowing the world in all its rule-governed, everyday public aspects is part of what knowing a process is. In sum, and with a brief reference to Wittgenstein, the authors proposed that the rules governing our behavior are not individual mental contents inside the head/brain but are shared social products, a conclusion implicitly suggesting

54. Brian Parkinson and A. S. R. Manstead, "Appraisal as a Cause of Emotion," *Review of Personality and Social Psychology*, ed. M. S. Clark, 13 (1992): 131.

that the entire program of cognitive psychology, with its emphasis on introspecting internal causal processes or mechanisms, was misguided.[55]

But as we have seen, even as Lazarus tried with more or less success to resist succumbing to the information-processing presuppositions that informed the work of Nisbett and Wilson and that of so many of his contemporaries, he shared many of their basic assumptions: he too believed that human emotional life was determined by cognitive appraisals intervening in the mind as largely unconscious causal events in the affective process. In this context, and after more than thirty years of research on the emotions, Lazarus sought to defend his cognitive views and justify his approach. His defense took many forms, and it would take us too far afield to discuss in detail his numerous discussions of the familiar issues that concerned him. One move he now made, in his big book *Emotion and Adaptation* (1991) and elsewhere, was to recognize the existence of two different kinds of causal relationship between cognition and emotion. He described one causal relationship as "logical" or "analytic," by which he meant that certain appraisals "logically imply particular emotions without any necessary causal ascription." He described the other causal relationship as "synthetic," by which he meant that "certain patterns of appraisal *cause* particular emotions." Lazarus's claim was that if the description of each emotion was to include the cognitive appraisal that generated it as well as its behavioral and physiological correlates, as when anger includes an image of the offense and of the blameworthy person (or object), and the offense and blame are said also to cause the anger, then "*both* kinds of causal relationship between cognition and emotion, the synthetic and the logical, apply, which avoids an either/or position. This reasoning is consistent with the metatheoretical concept of *reciprocal causality* and with my reluctance to overemphasize material causes" (his emphasis).[56]

Lazarus's statements were hardly perspicuous, and whatever he meant by "logical" causality, he did not connect this concept to the Wittgensteinian-grammatical analysis of Anthony Kenny, whose ideas he had invoked earlier, or to Sabini and Silver's (1981) discussion. Nor did Lazarus's brief references to Sartre's views on the necessary inclusion of the intentional object in emotion or his allusions to Bandura's ideas

55. John Sabini and Maury Silver, "Introspection and Causal Accounts," *Journal of Personality and Social Psychology* 40 (1) (1981): 171–79. A valuable summary of objections to Nisbett and Wilson's claims was provided by Peter A. White, "Knowing More about What We Can Tell: 'Introspective Access' and Causal Report Accuracy 10 Years Later," *British Journal of Psychology* 79 (1988): 24–25.

56. Richard S. Lazarus, *Emotion and Adaptation* (New York and Oxford, 1991), 172–73.

about reciprocal determinism do much to clarify the difficult philosophical issues involved (*Emotion and Adaptation,* 131–32, 172–74).[57] Moreover, nothing followed from Lazarus's effort to distinguish between "logical" and "synthetic" causality: his views on the causal-mediating role of appraisals in emotion, especially the role of unconscious, automatic appraisals, and the methodological challenges of proving empirically the role of such unconscious appraisals, remained the same. Indeed, all the basic components of his position stayed intact, including his idea that certain patterns or styles of appraisal caused specific, discrete emotions, to the point that, as he acknowledged, his position became almost indistinguishable from the "affect program" theories of Tomkins and Ekman. The only difference between his views and theirs in this regard was that on the input side Lazarus wanted to make the affect program theory more responsive to cultural influences by emphasizing the importance of individual experience, history, and personality in the learning process. But for him the output side remained a matter of the leakage or "readout" of discrete emotions discharging in universal, signature patterns of facial expressions and autonomic-behavioral reactions, just as Tomkins, Ekman, and their followers had proposed.

As we saw in the previous chapter, in 1993 the anthropologist Richard Shweder forcefully criticized Lazarus's views on the grounds that the relation between appraisal or meaning and emotion was not a causal one but a matter of conceptual necessity, a position not unlike that of Kenny. But Lazarus rebuffed Shweder's criticisms: he rightly saw that Shweder was not offering a "minor quibble" about Lazarus's views but a major threat to his attempt at a scientific account of the emotions—even if, as Lazarus acknowledged, it was not easy to demonstrate empirically the role of unconscious appraisals, and even if, as he also confessed, he regarded logic as a "tricky business."[58]

Also challenging to Lazarus were criticisms made by the psychologists Parkinson and Manstead at this time. It is not easy to do justice to the lat-

57. See also Richard S. Lazarus, "Cognition and Motivation in Emotion," 352–67, for his further discussion of these issues. Lazarus's difficulty in mastering these concepts shows up in a subsequent comment in which he attributes to the "synthetic" form of causality, not the "logical" form, the notion that "appraisals do not produce emotions, but imply them"—a reversal of his use of the terms in his 1991 book. See Lazarus, "Emotions Express a Social Relationship, but It Is an Individual Mind that Creates Them, Author's Response," *Psychological Inquiry* 6 (3) (1995): 261.

58. Richard S. Lazarus, "Author's Response: Lazarus Rise," *Psychological Inquiry* 4 (4) (1993): 343.

ter's several commentaries on Lazarus's work in the 1990s, some of which strike me as confused or ambiguous. There are moments in their texts when, citing certain philosophical and related discussions, they suggest that cognition's role in emotion is a necessary or conceptual one because emotions always have a reference to some intentional object. At other moments they appear to reverse course in order to adopt an empirical-causal approach to the affects. On this latter basis, they claimed that the empirical-introspective evidence cited by Lazarus and others in favor of the causal role of appraisals was inadequate. This raised for them the possibility that, far from cognition being a necessary component of the emotional response, certain emotional states were not tied to cognitions or objects at all but were caused by a range of possible noncognitive processes.[59] As Parkinson wrote in 2001, "current available data do not support an exclusive relationship between appraisal components and emotions."[60]

Lazarus was stung by Parkinson and Manstead's criticisms. He expressed puzzlement at their sniping and complained that, as they seemed to accept most of what he had argued about the role of appraisal in emotion, their discussion was "a bit odd." As they admitted at one point, they took issue with Lazarus only on the question of whether cognitive appraisal was the only route to the apprehension of the personal meaning of objective events or relationships, suggesting that emotions could be aroused without cognitive mediation, a position Lazarus saw no reason to accept. Lazarus conceded the methodological difficulties of separating appraisal and emotion empirically, using the self-report methods that were available. He even expressed a certain exasperation with Parkinson

59. Parkinson and Manstead, "Appraisal as a Cause of Emotion" (1992); Brian Parkinson and A. S. R. Manstead, "Making Sense of Emotion in Stories and Social Life," *Cognition and Emotion* 7 (1993): 295–323; Brian Parkinson, "Untangling the Appraisal-Emotion Connection," *Personality and Social Psychology Review* 1 (1) (1997): 62–79. In "Appraisals Are Direct, Immediate, Intuitive, and Unwitting . . . And Some Are Reflective," in line with Nisbett and Wilson's views about the unavailability of processes to awareness, Kappas complains of the methodological problems associated with the use of self-reports if dual- or multilevel models of appraisal are accepted, since unconscious, automatic appraisals are hidden from introspection. He suggests that it is therefore not surprising that Parkinson and Manstead had come to a negative conclusion about the essential role of cognition in emotion (968). Kappas cites Phoebe Ellsworth and Klaus Scherer's attempt to rebut this charge in their "Appraisal Processes in Emotion," in *Handbook of Affective Sciences* (2003), eds. Davidson, Scherer, and Goldsmith, 572–95.

60. Brian Parkinson, "Putting Appraisal in Context," in *Appraisal Processes in Emotion: Theory, Methods, Research*, 179.

and Manstead's demand for absolute proof of the causal role of appraisals.[61] In this context, he took heart from a recent paper by the German psychologist Rainer Reisenzein who, in response to a paper by Lazarus on the "vexing problems" posed by cognitive-mediational theories of emotion, argued that Lazarus's own standards for what could count as convincing evidence of the separate contributions of appraisal and emotion were too strict, and that the case for Lazarus's causal theory of appraisal based on self-report strategies could legitimately be made.[62]

I shall have more to say about Parkinson's contributions in the next chapter. For the present purposes, their 1990s work is of interest chiefly for their attempt to move the discussion toward a more dynamic and "transactional" account of the emotions. Their concern was that by emphasizing the mediating role of internal appraisal processes, Lazarus had individualized and privatized the meanings and interpretations generating the emotions, instead of treating meaning as embedded in the conventions and procedures of public life. In particular, they criticized Lazarus for treating meaning as if it was something that had to be "extracted" from the environment by constructing an interpretation of it, arguing instead on Gibsonian grounds that meanings are directly available to be picked up by the organism. It is as if the authors were almost but not quite able to offer an assault on the entire edifice of cognitive psychology, with its emphasis on the existence of internal mental mechanisms and representations to bridge the hypothesized gap between the individual and the world. Their objection was to Lazarus's picture of the individual or actor as "being always at least one critical step away from the social world, and never close to direct contact with social reality."[63]

In place of that picture, they proposed an emphasis on the public, per-

61. Richard S. Lazarus, *Stress and Emotion: A New Synthesis* (New York, 1999), 95–100. Lazarus is responding to Parkinson and Manstead's 1992 paper "Appraisal as a Cause of Emotion."

62. Rainer Reisenzein, "On Appraisal as Causes of Emotions," *Psychological Inquiry* 6 (1995): 233–37. Arguing that Lazarus's own standards for what could be regarded as convincing evidence for causality were "simply too strict," Reisenzein suggested that the case for Lazarus's causal theory of appraisal could be plausibly made by treating the theory "holistically" in terms of whether it could account better than any other cognitive or noncognitive theory for a broad range of relevant facts.

63. See Parkinson and Manstead, "Making Sense of Emotions in Stories and Social Life," where the authors criticize Lazarus for privatizing the appraisal process and pick up on one of Fridlund's early publications, on the sociality of solitary smiling, in order to suggest that the vignette methods used by Lazarus and his colleagues to study appraisal "miss out on these interpersonal factors by placing appraisal squarely in the realm of internal soliloquized representations" (314).

formative, and contextual nature of human affective interactions. It was a view of emotions that focused, more emphatically than in their opinion Lazarus had ever been able to do, on the real-time, ongoing, complex interpersonal gestural, bodily, and linguistic communications that directly structured people's emotional responses according to the situation and the (real or imagined) audiences involved. In this setting, it is not surprising that, beginning in 1993, Parkinson and Manstead began to pick up on and make use of the recent contributions of Alan Fridlund, who, starting in the late 1980s and on the basis of a rather different set of experimental and theoretical concerns, had begun to propose a "behavioral-ecological" or "communicative" approach to the emotions. Fridlund's work not only resonated with Parkinson and Manstead's evolving transactional-communicative position, it was one that offered a decisive correction to the mind-set that had dominated much scientific thinking until then, including Lazarus's. It is to Fridlund's work and its impact that I now turn.

[CHAPTER FIVE]

A WORLD WITHOUT PRETENSE?

Alan J. Fridlund's Behavioral Ecology View

In his *Last Writings on the Philosophy of Psychology,* Wittgenstein asks: "Could one imagine a world in which there could be no pretence?" Again he asks: "Can one imagine a people who cannot lie?—What else would these people lack?" and answers: "We should probably also imagine that they cannot make anything up and do not understand things that are made up."[1] Wittgenstein's point, or one of them, is that such people would be barely recognizable as human beings. And indeed, no one who has written about the emotions actually has envisioned a world in which humans entirely lack the capacity to pretend or simulate. But, in what might be viewed as a compensatory response to the seeming pervasiveness of lying, fraud, and deception in our daily lives, many emotion theorists *are* committed to the idea that the problem of cheating and dissimulation can be alleviated, if not completely eliminated, because there exist inbuilt mechanisms of reliable or truthful signaling. They adopt the view that honesty, or knowing whom to trust, is guaranteed in advance, as it were, because difficult or impossible-to-fake facial movements and other nonverbal signals have evolved through natural selection in order to provide accurate information about a person's internal, affective states. We might put it that such theorists believe they have found a solution to the philosophical problem of other minds by appealing to the idea of innately determined emotional "expressions" that unfailingly signal to us what other people are thinking and feeling because those expressions are hardwired to do so. These same theorists further believe, or proceed as if they believe, that without such an evolved, automatic system of communication to vouchsafe the genuineness and sincerity of emotional signals, trust, cooperation, and indeed genuine altruism are doomed to be undermined by the selfish human tendency to cheat. They therefore propose that the evolution of reliable emotional signals is at the very foundation of the social contract.

1. Ludwig Wittgenstein, *Last Writings on the Philosophy of Psychology: The Inner and the Outer*, vol. 2, eds. G. H. von Wright and Heikki Nyman, trans. C. G. Luckhardt and Maximilian A. E. Aue (Oxford, 1993), 37e, 56e.

Consider for example Robert Frank's influential book *Passions Within Reason: The Strategic Role of the Emotions*, published in 1988. In that work the author, who combines concerns with economics, game theory, and evolutionary psychology, set out to challenge the then commonly accepted view that people always act rationally, which is to say, selfishly, in pursuit of their own economic and other interests.[2] In opposition to such a "self-interest" model of human action, Frank proposed a "commitment" model of behavior. According to that model, our passions predispose us to act "irrationally." By this he meant that our emotions prompt us to act in ways that may be contrary to our own immediate interests but offer advantages to us because, if our commitments are trustworthy, we stand to benefit from the possibility of social cooperation that our credibility and trustworthiness elicit (*PR*, 47). Frank described the results of several one-shot Prisoners' Dilemma and other economic games in support of the idea that the majority of people put issues of fairness above self-interest, and that individuals are capable of communicating their honesty to others and discerning whom they, in turn, can trust. For Frank, then, our passions play an important role in our personal interactions because they are "commitment devices" that tell the truth about our inner emotional states, thereby encouraging others to count on us and work together with us.[3]

2. Robert H. Frank, *Passions Within Reason: The Strategic Role of the Emotions* (New York, 1988); hereafter abbreviated as *PR*. Frank equates rationality with self-interest. For the limitations of Frank's approach, see Jana Noel, "Review of *Passions Within Reason*," *Studies in Philosophy and Education* 10 (1990): 175–78, who suggests that for many philosophers rationality concerns the recognition of reasons. Of course, other explanations of cooperation have been proposed, such as the theory of kin selection, reciprocal altruism, and group selection.

3. "Commitment" in its game-theoretical sense "involves a player in a game taking an action that changes the configuration of the world such that, if the time comes when that agent would choose to renege on her promise or decline to follow through on her unsuccessful threat, she will be unable to do so. If the receiver of the threat or promise knows that such a commitment action has been taken, then she will take the promise or threat seriously." See Don Ross and Paul Dumouchel, "Emotions as Strategic Signals," *Rationality and Society* 16 (2004): 253. Frank contends that evolution has produced and sustains emotional responses in organisms to serve as psychologically endogenous vehicles of such commitment. In their paper, Ross and Dumouchel challenge Frank's commitment model of emotions on grounds not unlike those of Fridlund (whom they do not cite), namely, that Ekman-style affect programs cannot play the strategic role in commitment that Frank assigns them. The ensuing debate can be followed in Robert H. Frank, "In Defense of Sincerity: Response to Ross and Dumouchel," *Rationality and Society* 16 (3) (2004): 287–305; and Don Ross and Paul Dumouchel, "Sincerity Is Just Consistency: Reply to Frank," ibid., 307–18. For an analysis of this debate, see Leys, "A World Without Pretense? Honest and Dis-

Literary critic William Flesch has recently summarized Frank's argument this way:

> Honest signaling has a long-term advantage, and what makes honest signals honest is that they are difficult to fake and difficult to hide, even when doing so would yield a short-term advantage. Emotion commits us to doing things that might be against our short-term interest, and *palpable* emotion declares that commitment. To use Robert Frank's formulation . . . visible emotion solves commitment problems. You can trust someone who acts on the basis of passion rather than reason, and you can tell that they are acting on the basis of passion when the expression of their passions is counter to their own immediate interests. This is one of the things that make expression an honest—and so costly—signal. (his emphasis)[4]

As Flesch's remarks suggest, and as Frank also emphasized, if the passions are to function as commitment devices, trustworthy people must be discernibly different from those who are not, so that observers can accurately discriminate between honest signalers and imposters. Frank therefore argued that trust depends on the existence of visually salient "behavioral clues" (*PR*, 8) that are costly or difficult to feign or simulate because they involuntarily signal our internal affective states. Frank recognized that clues to behavioral dispositions are imperfect: lie-detection systems will sometimes fail, and double-dealers will exploit that fact. As he also suggested, in a world without pretense or lying, no one would be on the lookout for deception, with the result that cheaters would soon flourish and the whole signaling system would start to crumble. In any population consisting only of cooperators, he remarked, "no one would

---

honest Signaling in Social Life," *Philosophy of Education 2013*, ed. Cris Mayo (Urbana, IL, 2013), 25–42.

4. William Flesch, *Comeuppance: Costly Signaling, Altruistic Punishment, and Other Biological Components of Fiction* (Cambridge, MA, 2007), 92–93; hereafter abbreviated as *C*. Flesch is interested in the enforcement measures used by altruists to punish cheaters, arguing that the human fascination with fiction reflects these evolutionary concerns. He suggests that we want to see cheaters get their comeuppance and have evolved the ability to "track" the behavior of others and to learn their stories precisely in order to be able to punish non-cooperators and reward the virtuous. "[We] have specifically evolved an innate tendency toward what evolutionary theorists call 'strong reciprocity,'" he observes (21–22). Strong reciprocators mind everyone's business, not just the business of those who directly affect them, and they do this to ensure social cohesion. The basis for this strong reciprocity is the emotions.

be vigilant, and opportunities would abound for defectors. In a mixed population, cooperators can survive only by being sufficiently vigilant and skilled in their efforts to avoid good mimics."[5] In any mixed population, there thus exists an "uneasy balance between people who really possess the underlying emotions and others who merely seem to. Those who are adept at reading the relevant signals will be more successful than others. There is also a payoff for those who are able to send effective signals about their own behavioral predispositions" (*PR*, 10–11).

According to Frank, communication is made easier for us precisely because signaling our passions is not really a matter of choice but of the automatic discharge of internal states—states that manifest themselves in characteristic expressions of the face and the body that humans also recognize innately. In making this claim, Frank relied in part on the theories and findings of Paul Ekman, whose Basic Emotion Theory should by now be familiar from the discussion in chapter 2. As a reminder, the key features of Ekman's position can be summarized as follows:

1. There exists a small set of basic emotions, defined as pan-cultural categories or "natural kinds." Those basic emotions, which include fear, anger, sadness, disgust, joy, and surprise, are evolved, genetically hardwired, reflex-like responses of the organism.

2. Each basic emotion manifests itself in distinct physiological and behavioral patterns of response, especially in characteristic facial expressions that are automatically recognized.

3. According to Ekman's "neurocultural" model for explaining commonalities and variations in human facial movements, the face "expresses" the emotions, except when expressions are disguised by cultural or conventional norms ("display rules") for controlling and managing emotions in public, or are masked by deliberate deception. In other words, under pure or unfiltered conditions, facial displays are authentic "readouts" of the discrete internal states that constitute the basic emotions. Ekman called the muscles involved in the facial expression of the emotion "reliable" muscles because they are difficult to control and hence produce expressions that are hard or costly to fake.[6]

4. Each basic emotion is linked to specific neural substrates or subcortical "affect programs," an assumption that implies the embrace of some

5. Robert H, Frank, *What Price the Moral High Ground? How to Succeed Without Selling Your Soul* (Princeton, 2004), 11.

6. Paul Ekman, *Telling Lies: Clues to Deceit in the Marketplace, Politics, and Marriage* (New York and London, 2001), 132.

degree of modularity and information encapsulation in brain functions. The amygdala has been pinpointed as the neural seat of fear, while the insula has been implicated in disgust.[7]

5. Another of Ekman's assumptions is that, although the emotions can and do combine with cognitive systems in the brain, emotion and cognition are essentially separate. As he has put it: "[E]motional expressions are special . . . because they are involuntary, not intentional. Unlike the 'A-OK' or 'good luck' hand gestures, emotional expressions occur without choice . . . The communicative value of a signal differs if it is intended or unintended. Emotional expressions have such an impact; we trust them precisely because they are unintended."[8] Or, as Flesch has remarked: "The fact that emotional signaling is uncalculated ensures its honesty" (*C*, 97). According to this view, when we are alone, we are able to express our genuine feelings without the inhibition of culture because no one is watching us.

6. Not surprisingly, Ekman's ideas provide an explanation of lying according to which what is hidden in deception is detectable through the "leakage" of unmanaged emotional behavior, because emotions are unintentional, expressive states. He and his colleagues have performed numerous studies on the basis of this claim, ostensibly demonstrating the involuntary escape of emotion in situations of deceit.

Ekman's Basic Emotion Theory lay behind Frank's claim that our emotions are commitment devices because they are involuntary readouts of our internal affective states. Frank devoted a chapter of his book, entitled "Telltale Clues," to a summary of putative facts about involuntary facial expression largely drawn from Ekman's work.[9] Following Ekman, Frank reported that even when a facial expression involves muscles that are relatively easy to control, some of the most reliable clues to emotion come from so-called "micro-expressions," tiny, fleeting, involuntary movements of the muscles that are held to accurately convey an evanescent under-

7. For a critique of the arguments for the amygdala as the seat of the discrete fear emotion, see chapter 2. For a critique of experiments on the insula as the seat of the emotion of disgust, see Leys, "'Both of Us Disgusted in *My* Insula': Mirror Neurons and Empathy Theory," *nonsite.org* (February 2012); reprinted in a slighted shorter version in *Science and Emotions After 1945: A Transatlantic Perspective*, eds. Frank Biess and Daniel M. Gross (Chicago and London, 2014), 67–95.

8. Paul Ekman, "Afterword," in Charles Darwin, *The Expression of Emotions in Man and Animals*, 3rd ed., with an introduction, afterword, and commentaries by Paul Ekman (New York and Oxford, 1998), 373.

9. Frank's main source of information concerning how the body betrays lies was Paul Ekman's *Telling Lies*.

lying emotion. For Frank and Ekman, if deception is difficult to sustain, it is because those telltale micro movements tend to pierce our social masks and betray to others the truth of our authentic emotional states.[10]

THE "NEW ETHOLOGY" AND EMOTIONAL EXPRESSION

Frank's book on emotional commitment testifies to the widespread acceptance of the Ekman model of the emotions by the 1980s. And yet—and this is a point I especially wish to stress—well before 1988 a revolution in the theory of animal signaling had started to challenge the basic premises of Frank's and Ekman's arguments. Let me explain: according to the "classical" approach to animal communication associated with the influential work of the ethologists Konrad Lorenz and Niko Tinbergen that dominated the field until the 1970s, animal signals have evolved for the mutual benefit of signalers and receivers, which is to say that natural selection works for the good of the group or species.[11] As Searcy and Nowicki have observed in their valuable book on animal communication, the classical ethology view maintained that "'it is to the advantage of both parties that signals should be efficient, unambiguous, and informative.'"[12] Displays on this classical model were viewed as "a kind of fixed action pattern and the responses of the recipients were just as fixed."[13] Although both Lorenz and Tinbergen had given some weight to learning as a modifier of instinct, the "release" of innate patterns held sway and culture had little influence. A characteristic statement by Tinbergen claimed that

10. "[The face] lies all the time. It lies more than it tells the truth. But micro expressions, the very first signs of concealed emotion that occur in the 125th of a second, never lie. Most people miss them. But we are, to my surprise, learning that we can teach people to recognize them very quickly, even after an hour of training" ("Shamefaced: An Interview with Paul Ekman," by Christopher Turner, in *Cabinet* [31] [Fall 2008], http://www.cabinetmagazine.org/issues/31/turner.php).

11. Griffiths places the start of the "new ethology" and its rejection of the drive-discharge model of instinctive behavior even earlier, to the 1960s, suggesting a 1956 paper by Hinde as the origin of this new orientation. But Searcy and Nowicki follow the usual account that places the beginning of the new ethology in the 1970s, with the publication of Dawkins and Krebs's "seminal" paper of 1978. See Paul E. Griffiths, "Basic Emotions, Complex Emotions, Machiavellian Emotions," *Royal Institute of Philosophy Supplement* 52 (March 2003): 59; and William A. Searcy and Stephen Nowicki, *The Evolution of Animal Communication: Reliability and Deception in Signaling Systems* (Princeton, 2005), 6.

12. Searcy and Nowicki, *The Evolution of Animal Communication*, 7.

13. Alan J. Fridlund, *Human Facial Expression: An Evolutionary View* (San Diego, CA, 1994), 62–63; hereafter abbreviated as *HFE*.

> So far as our present knowledge goes, social cooperation seems to depend mainly on a system of releasers. The tendency of the actor to give these signals is innate, and the reactors' responses are likewise innate.
>
> We see therefore that a community functions as a result of the properties of its members. Each member has the tendency to perform the signal movements releasing the "correct" responses in the reactor; each member has specific capacities that render it sensitive to the species' signals. In this sense the community is determined by the individuals.[14]

Ekman's approach to the emotions conformed to this classical view because he treated facial expressions as the final common pathways of adaptive mechanisms that have evolved to convey accurate information to others about the organism's inner emotional state and therefore as useful signals for achieving social cooperation.[15]

Starting in the mid-1970s, however, evolutionary biologist Richard Dawkins launched the "selfish gene" theory by emphasizing that, in order to attain reproductive success, individuals do not act for the good of the group but in their own interests. Dawkins's selfish gene approach to natural selection immediately transformed the classical view of animal signaling. Rather than treating the relationship between signaler and receiver as one of harmonious cooperation, in which the signaler aims to transmit information efficiently and accurately to the receiver, Dawkins and the ethologists influenced by him began to view animal displays as the expression of a kind of coevolutionary "arms race" between animals. The emphasis now fell on the manipulative nature of animal signaling and on the corresponding selection pressures on the receivers to evolve vigilance and skepticism for displays in the form of finely tuned signal-detection abilities. On the classical model, animal communication was thought to be a cooperative process and in this sense good for the group, so deception was deemed unlikely. But when the classical model was re-

14. N. Tinbergen, *Social Behavior in Animals* (London, 1953), 85–86; cited in *HFE*, 63.

15. In Ekman's words, "it was central to the evolution of emotions that they inform conspecifics, without choice or consideration, about what is occurring: inside the person (plans, memories, physiological changes), what most likely occurred before to bring about that expression (antecedents), and what is most likely to occur next (immediate consequences, regulatory attempts, coping)." Paul Ekman, "Basic Emotions," in *Handbook of Cognition and Emotion*, eds. T. Dalgleish and P. Power (Chichester and New York, 1999), 46; hereafter abbreviated as "BE."

jected in favor of the new emphasis on the part played by the selfish gene in the evolutionary process, cooperative signaling was called into question and deception seemed more probable.[16]

Above all, the selfish gene approach to ethology stressed the idea that from the point of view of the individual or gene it would not be beneficial for the signaler to signal his or her intentions at all times. A "readout" view of signaling was thus rejected in favor of a manipulative or "exploitative" account of animal behavior, according to which signalers attempt to maximize their own interests by "mind-reading" their interactants—that is, by looking for the fine clues by which they can predict how the latter might act.[17] And of course the recipients of signals are not passive but have also evolved to exploit and mind-read as well, so that they too resort to concealment and active manipulation, including deliberately misleading their opponents. One of the important challenges for the new ethology then became how to understand and theorize the achievement of an ecological balance between honest and dishonest signaling. The problem posed was that if everyone starts to cheat, eventually no receiver will bother to respond to signals and the signaling system will collapse. But honest signaling cannot be the whole answer either, otherwise deception would disappear—and this had not happened. Plausible accounts were therefore needed to understand how an ecological balance between honest and dishonest signaling could be achieved. Much of the work on "Evolutionarily Stable Strategies" stimulated by the new ethology was

16. Richard Dawkins and John R. Krebs, "Animal Signals: Information or Manipulation?," in *Behavioral Ecology: An Evolutionary Approach*, eds. J. R. Krebs and N. B. Davies (Oxford, 1978), 282–309. Likewise, Peter G. Caryl reanalyzed in game-theoretical terms some classical ethological data on agonistic displays in order to challenge the traditional assumption that information about the probability of an attack should be conveyed by aggressive displays. The analysis tended to question the association that had been held to connect the expression of displays with internal motivational states. See P. G. Caryl, "Communication by Agonistic Displays: What Can Game Theory Contribute to Ethology," *Behavior* 68 (1979): 136–69; and P. G. Caryl, "Telling the Truth About Intentions," *Journal of Theoretical Biology* 97 (1982): 679–89.

17. Krebs and Dawkins describe the classical view of ethology as a "readout" view of motivation: "The classical ethological view emphasized the motivational state of the actor, and treated signals as formalized readouts of the actor's internal state . . . Natural selection is thought to favour actors who cooperate in having their intentions read" (Dawkins and Krebs, "Animal Signals: Information or Manipulation?," 306). The idea of facial signals or expressions as "readouts" of the signaler's internal emotional states has been especially emphasized by Ross Buck, *The Communication of Emotion* (New York and London, 1984), esp. 62–67.

(and continues to be) carried out by theorists expert in the use of game-theoretical models. As the huge literature in the field testifies, the issues at stake in this domain of research are still not resolved.

For my purposes, what was significant about the new ethology was its impact on the theorizing of the emotions. Almost for the first time, psychologists and ethologists found common ground in the need to rethink the phrase "the expression of the emotions" that had been inherited from Darwin. We might put it that the new ethology loosened the one-to-one correspondence that had been held to exist between internal states and external facial expressions by suggesting that the relationship was more labile than had been thought and by stressing instead the instrumental, cognitive, social-strategic dimension of facial displays.[18] The result was that quasi-reflexive theories of the human face, which were predicated on the classical ethological view of signaling, were "left stranded by the new understanding of the complexity of *non*human animal signals" (*HFE*, 30; Fridlund's emphasis).

There is an interesting story to be told about the work of various psychologists and ethologists who, stimulated in part by Dawkins's selfish gene approach to signaling, began during the 1970s to pose an alternative to the dominant "readout" view of emotional expression. In 1985, for example, the eminent British ethologist Robert Hinde suggested that Darwin's phrase "the expression of the emotions" had often pointed research in the wrong direction. "The implication of the title of Darwin's book is that emotional behaviour is the outward expression of an emotional state, and that there is a one-to-one correspondence between the two," he wrote. "Such a correspondence is assumed by much current work on emotional behavior . . . But if the internal state is complex one we must ask what it is that the emotional behavior expresses? . . . Several lines of comparative

18. Thus, the ethologist W. John Smith criticized the use of the term animal "expression" rather than "signal" or "display" on the grounds that its connotation was too narrow, because it was commonly used to imply that the referents of signaling are emotional states, a view he regarded as "untested." He cited Kraut and Johnston's 1969 study of audience effects on smiling behavior in support of his comments (see below for a further discussion of these audience effects). The net effect of Smith's approach was to emphasize the social-contextual character of signaler-recipient communicative interactions. W. John Smith, "Consistency and Change in Communication," in *The Development of Expressive Behavior: Biology-Environment Interactions*, ed. Gail Zivin (San Diego, 1985), 57. Smith's 1985 paper is briefly discussed by Fridlund, *HFE*, 74, where he emphasizes Smith's views on the development, in mammals and birds who live in close social groups, of "calibration" skills, or learned skepticism, by which their responses to a display depend on the identity of the displayer and the context of the display.

data . . . suggest a certain lability between aspects of the internal state and behavior . . . Thus the relation between emotional state and emotional behaviour is by no means simple." He offered evidence along these lines from the study of threat postures. From the point of view of natural selection, he observed with reference to Dawkins and Krebs's recent work that "it would not be adaptive for an individual to communicate to a rival the likelihood that he would attack or flee. Such signals make sense only if the threatening individual either is attempting to bluff, deceive, or manipulate the rival . . . or else is uncertain what to do next, because what he should do depends in part on the behavior of the other."[19]

Hinde made similar arguments in a volume published the same year in which several authors interpreted expressive behavior in humans and nonhumans in terms that raised questions about the readout view of emotions. But in fact neither Hinde nor any of the other contributors to that volume managed to break definitively with Ekman's paradigm. Instead, they tended to hold on to the notion of some kind of "internal state" as a necessary invoker of expressive behavior. They therefore proposed a compromise between "expression" and "negotiation" by positing a continuum between innately determined universal emotional "expressions" at one end (for example, infant expressions) and strategic signaling at the other end (for example, threat postures), and did so in terms that tended to repackage the dualism between innately driven facial signals versus learned cultural displays on which Ekman's emotion theory depended.[20]

There was one person, however, who at this juncture made a decisive break with Ekman. This was the psychologist Alan Fridlund, a former student of Ekman's who, in light of the new ethology, radically revised his

19. R. Hinde, "Was 'The Expression of the Emotions' a Misleading Phrase?," *Animal Behavior* 33 (1985): 988–89.

20. See R. Hinde, "Expression and Negotiation," in *The Development of Expressive Behavior: Biology-Environment Interactions*, ed. Gail Zivin (San Diego, 1985), 103–16, and related articles in this collection. "I remember what struck me as liberating," Fridlund recollects, "was that Hinde was a potential ally in trying to dislodge people from using Darwin as a reason to believe so devoutly in [Ekman's] emotions views of human expressions. After reading Dawkins-Krebs, though, I realized that Hinde had a real problem with the simple 'expression-negotiation' continuum, and that was deception. D-K [Dawkins-Krebs] recognized what seemed true to me, which was that 'deception' was as basic as 'honesty,' and that when you look at signals functionally rather than morally (as issues of 'trust' and 'fidelity' vs. 'lying'), the issue boiled down simply to one of signal reliability in both enaction and S/N [signal/noise] ratio of detection. Right, D-K's [Dawkins-Krebs's] later paper redressed an earlier inattention to receivers; changes in receivers' behavior have to be the driver of displayers' signals" (personal communication).

perspective. It was Fridlund who worked out the fundamental stakes of what he called the Behavioral Ecology View of Faces and who, in a deeply researched and brilliantly argued book, *Human Facial Expression: An Evolutionary View* (1994), laid out the new terms in which he thought expressive behavior should now be theorized and understood.

FRIDLUND'S BEHAVIORAL ECOLOGY VIEW OF FACES

Fridlund was born and raised in Mississippi. After pursuing an undergraduate degree at Northwestern University, in the late 1970s he started his graduate training in psychology at the University of Mississippi. In 1979 he transferred to Yale University in order to carry out electromyographic (EMG) studies of the face. In these studies, tiny electrodes were attached to the skin of the face in order to detect the micro movements that occurred when subjects alone in the laboratory watched movies or slides or when they were asked to imagine various social interactions. Introduced early in his graduate training to Tomkins and subsequently to Izard and Ekman, Fridlund saw no reason to challenge the by-then-well-established Basic Emotion Theory of facial expression. He enjoyed a close relationship with Ekman, who invited him to join him in the San Francisco area to continue his work, and during these years, as an expert on the demanding technicalities of electromyography, he published articles jointly with both Izard and Ekman.[21]

After obtaining his PhD in 1984, Fridlund was appointed Assistant Professor in Psychology at the University of Pennsylvania. The move was decisive for him. There he found himself in the company of several stimulating colleagues who held quite different views about the emotions and facial expression from those espoused by Ekman. Fridlund later recalled:

> [M]y first doubts were sown when I arrived at Penn and heard others' doubts about Ekman's theories for the first time. Up until then, all

21. A. J Fridlund and C. Izard, "Electromyographic Studies of Facial Expressions of Emotions and Patterns of Emotion," in *Social Psychophysiology: A Sourcebook*, eds. John T. Cacioppo, Richard E. Petty, and David Shapiro (New York, 1983), 243–86; A. J. Fridlund, G. E. Schwartz, and S. C. Fowler, "Pattern-Recognition of Self-Reported Emotional State from Multiple-Site Facial EMG Activity During Affective Imagery," *Psychophysiology* 21 (1984): 622–37; A. J. Fridlund and J. T. Cacioppo, "Guidelines for Human Electromyographic Research," *Psychophysiology* 23 (1986): 567–89; A. J. Fridlund, Paul Ekman, and Harriet Oster, "Facial Expressions of Emotion: Review of the Literature, 1970–1983," in *Nonverbal Behavior and Communication*, eds. A. Siegman and S. Feldstein, 2nd ed. (Hillsdale, NJ, 1987), 143–224.

> I had ever been exposed to was unquestioning reverence, and there was no doubt I was an acolyte. At Penn, though, the scholarly atmosphere was brutally critical and analytic and, at one time or another, I heard nearly every then-popular venture in psychology get knocked down by withering critique. When it came to Ekman, the criticism was that he tried to account for everything by a (neurocultural) theory that was full of weasel words and slip clauses such that . . . his view was non-disconfirmable. This slant on Ekman woke me up.[22]

Among the colleagues at Penn who "shook the scales" from Fridlund's eyes and forced him to take a "fresh look" at the face (*HFE*, xiii) were the psychologists John Sabini and Avery Gilbert (an olfaction smell expert who introduced Fridlund to behavioral ecology), with both of whom he would go on to perform experiments.[23] Fridlund's exposure to behavioral ecology was especially crucial for him. In particular, he "inhaled" the work of the ethologist W. John Smith (personal communication), whose "informational" approach to animal signaling in *The Behavior of Communicating: An Ethological Approach* (1977) offered Fridlund important insights into the function of facial and other displays, especially into their flexible, social, and contextual character. As he has recently observed, behavioral ecologists such as Smith "saw animal signaling not as vestigial reflexes, or readouts of internal state, but as adaptations that served the interests of signalers within their social environments."[24]

On this new model of animal signaling, the emphasis fell on the "formalization" of behavioral signals through the processes of "ritualization" and "conventionalization" for the purposes of display. Displays on this model were regarded as elaborations of two types of movements, "social instrumental habits" and protective reflexes. As Fridlund explained in his book, the social instrumental habits best preadapted for display were those behaviors that were likely to be performed in the course of an overt action toward others. The pioneering ethologist Oskar Heinroth, who had studied ducks and geese, had named these actions "intention movements," a concept that, as Fridlund noted, "has endured because in

22. Personal communication.

23. This is the same John Sabini whom we have already met in chapter 4 for his critique, with Maury Silver, of Nisbett and Wilson's claims about the reliability of self-reports.

24. A. J. Fridlund, "The Behavioral Ecology View of Facial Displays, 25 Years Later," *Emotion Researcher*, ed. Andrea Scarantino, http://emotionresearcher.com/the-behavioral-ecology-view-of-facial-displays-25-years-later/, accessed January 6, 2016; hereafter abbreviated as "BEV."

many cases it is self-evident, and not just in birds. A bird may peck grass just before it pecks at another bird, and it may flap its wings ineffectually just before it takes off. Similarly, we may yell at the dog just before we argue with a friend, and we may make repeated 'leaving' motions just before we end a conversation. The incipient acts we emit before we act in earnest announce our intentions" (*HFE*, 64). Intention movements had been widely studied in birds by Smith and others, as well as in nonhuman primates. Fridlund's goal became that of extending the analysis of intention movements to the actions of the human face. As for the common protective facial reflexes, such as gag or startle reflexes, the challenge for him was to figure out their role as preadapted components in the evolution of facial displays.

It is important to note that Fridlund's work involved a critique of the usual interpretation of Darwin's views on emotional expression. Darwin is typically thought by emotion theorists to have proposed that expressions are selected and adapted for the communication of emotion. But Fridlund has rightly pointed out that Darwin's conclusion was the opposite: he believed that most facial displays were not evolutionary adaptations, but vestiges or accidents. Darwin's aim in *Expression* was to overthrow the theological argument from design, as championed by William Paley, Charles Bell, and others, according to which human expressions are discontinuous with those of other animals and are divinely created. Instead, Darwin argued that expressions lacked adaptive functions. Fridlund thus suggests that modern theorists of emotion misrepresent Darwin's position when they treat emotional expressions as selected for the outward communication of internal emotional states.[25]

Convinced that modern behavioral ecology views of animal communication threw entirely new light on the nature of emotional-facial expression, Fridlund, in a series of experimental studies and position papers culminating in his book of 1994, offered a "manifesto" (*HFE*, xi) for a new Behavioral Ecology View of human facial behavior.[26] He explained that he

25. See A J. Fridlund, "Darwin's Anti-Darwinism in *The Expression of the Emotions in Man and Animals*," in *International Review of Emotion*, vol. 2, ed. K. Strongman (Chichester, 1992), 117–37; and idem, "How and Why Darwin Wrote *Expression*," *HFE*, 14–27. For a valuable, nuanced discussion of this point, see also Thomas Dixon, *From Passions to Emotions: The Creation of a Secular Psychological Category* (Cambridge, 2003), 159–79.

26. Fridlund published several articles in the early 1990s, especially a very long paper anticipating the argument of his 1994 book (Fridlund, "Evolution and Facial Action in Reflex, Social Motive, and Paralanguage," *Biological Psychiatry* 32 [4] [1991]: 3–100).

adopted the term "behavioral ecology" because it connoted "interdependence" and underscored the coevolution of formalization and vigilance in signaling, noting that he could have used the terms "interactional" or "communicative" instead (*HFE*, 124, n. 1). His aim was to provide an account of facial movements that squared with modern evolutionary theory by demonstrating the continuity between humans and nonhuman animals while also acknowledging the differences between the two (for instance, the difference that the evolution of language makes for humans).

Fridlund's fundamental insight was precisely that, in light of the new ethology, facial displays must be understood as intentional-communicative signals. As he argued:

> [A]ny reasonable account of signaling must recognize that signals do not evolve to provide information detrimental to the signaler. Displayers must not signal automatically, but only when it is beneficial to do so, that is, when such signaling serves . . . motives. Automatic readouts or spillovers of drive states (i.e., "facial expressions of emotion") would be extinguished early in phylogeny in the service of deception, economy, and privacy. Thus, an individual who momentarily shows a pursed lip on an otherwise impassive face is not showing "leakage" of anger but conflicting intentions . . . to show stolidity *and* to threaten. (*HFE*, 131–32; his emphasis)

From this it followed that emotional displays could not be regarded as readouts of internal states but as intentional movements serving various social motives. This meant not only that such displays are responsive to proximate elicitors but that they are also sensitive to those who are present, one's aims toward them, and the nature and context of the interaction.

Fridlund contrasted Ekman's Basic Emotion Theory of facial displays with his own Behavioral Ecology View in order to bring out the differences between them. Ekman's theory, as Fridlund described it, essentially proposed a model of the emotions involving two kinds of faces. First are the innate reflex-like faces that leak ongoing emotions: these are termed "facial expressions of emotion." Second are the learned, instrumental faces connoting emotion that is not actually being experienced: these instrumental faces reflect ordinary dissimulations, such as smiles of politeness. According to Ekman's two-factor model, the facial expressions of everyday life represent an interaction between emotional instigation and cultural adulteration or masking. The model thus assumes a dualism between felt or authentic displays on the one hand, and false or inauthentic displays on the other. For example, according to Ekman, the felt or so-called "Duchenne" smile is the smile we involuntarily produce when we are genuinely and honestly happy; the false or "non-Duchenne" smile is the smile we produce when we are merely pretending to feel joy. Ekman had claimed in this regard that the authentic smile cannot be produced on demand without betraying its inauthenticity.[27] Robert Frank in his 1988 book echoed that view when he observed that "we apparently know, even if we cannot articulate, how a forced smile differs from one that is heartfelt" (*PR*, 8). Ekman's model thus depended on a series of oppositions: between innate signals versus learned ones; between biology and culture; between involuntary signals and voluntary ones; between non-intentional emotional states versus intentional display; and between honest or reliable signaling versus dishonest or deceptive communication.

In contrast to Ekman's Basic Emotion Theory, Fridlund's Behavioral Ecology View did not treat facial displays as expressions of discrete, internal emotional states, or the outputs of modular affect programs. As Fridlund stated:

> For the contemporary ethologist or the behavioral ecologist, displays have their impact upon others' behavior because . . . vigilance for and comprehension of signals coevolved with the signals themselves. The balance of signaling and vigilance, countersignaling and countervigilance, produces a signaling "ecology" that is analogous to the balance of resources and consumers, and predator and prey, that characterizes all natural ecosystems.
>
> Displays are specific to intent and context, rather than derivatives or blends of fundamental emotional displays . . . Instead of there being six or seven displays of "fundamental emotions" (e.g., anger), there may be one dozen or one hundred "about to aggress" displays appropriate to the identities and relationships of the interactants, and the context in which the interaction occurs. The topography of an "about to aggress" display may depend on whether the interactant is dominant or nondominant, conspecific or extraspecific, and whether one is defending territory or young, contesting for access to a female, or retrieving stolen food or property. (*HFE*, 128–29)

27. P. Ekman and W. V. Friesen, "Felt, False, and Miserable Smiles," *Journal of Nonverbal Behavior* 6 (1982): 238–52; P. Ekman, R. J. Davidson, and W. V. Friesen, "The Duchenne Smile: Emotional Expression and Brain Physiology II," *Journal of Personality and Social Psychology* 58 (1990): 342–53; M. Frank, P. Ekman, and W. V. Friesen, "Behavioral Markers and Recognizability of the Smile of Enjoyment," *Journal of Personality and Social Psychology* 1 (1993): 83–93.

In comparing facial displays as interpreted by Ekman and his own Behavioral Ecology View, Fridlund did not depict prototype faces for each emotion category. This was because in his opinion there seemed to be *no* such prototype faces (*HFE*, 129).[28] Instead, Fridlund reinterpreted Ekman's emotion categories in the following way:

> [D]isplays exert their influence in the particular context of their issuance; a face interpreted as "contemptuous" in one context may be interpreted as "exasperated" or even "constipated" in another . . . Thus, in contexts in which one would try to appease another, any smile one issued would tend to be labeled a "false" smile in the Emotions View, which would connote a masking smile over some other emotion. For the behavioral ecologist, the same smile would likely be labeled an "about to appease" display, and it would deliver the same message as the word, "I give in" or "Whatever you say." (*HFE*, 129)

For Fridlund, the "facial expression of emotion" of Ekman's Basic Emotion Theory (or what Fridlund called Ekman's Emotion View) actually served the social motives of the displayer: "No distinction is made between 'felt' and 'false' displays issued by 'authentic' and 'social' selves; instead *all* displays are considered to arise from interaction, thus there is *only* an interactive self" (*HFE*, 130; his emphasis). He therefore contested Ekman's claim that when we are alone we display our natural expressions untainted by the contamination of culture; for Fridlund, solitary facial displays are social, too.

One of the significant changes in perspective that Fridlund's behavioral approach to facial displays brought about concerned precisely the nature and role of deception. In Ekman's Basic Emotion Theory, deception was regarded as a deliberate attempt to hide one's true feelings, an attempt that was betrayed by the "leakage" of unmanaged emotional behavior.[29]

28. As Fridlund noted, others at this time—notably Ortony and Turner—were also beginning to criticize the idea of emotions as discrete categories with prototypical faces, arguing that the idea of biologically or psychologically primitive "basic emotions" would not lead to progress in the field. See A. Ortony and T. J. Turner, "What's So Basic About Basic Emotions?," *Psychological Review* 97 (1990): 315–31.

29. See for example P. Ekman and W. V. Friesen, "The Repertoire of Non-Verbal Behavior: Categories, Origins, Usage, and Coding," *Semiotica* 1 (1969): 49–98; P. Ekman, "Mistakes When Deceiving," *Annals of the New York Academy of Sciences* 364 (1981): 269–78; P. Ekman, *Telling Lies* (1985); Paul Ekman, W. V. Friesen, and M. O'Sullivan, "Smiles When Lying," *Journal of Personality and Social Psychology* 54 (1988): 414–20; and P. Ekman and M. O'Sullivan, "Who Can Catch a Liar?," *American Psychologist* 46 (9) (1991): 913–20.

But Fridlund instead regarded deception as omnipresent in nature and "potentially highly advantageous for the displayer" (*HFE*, 138). In an original reinterpretation, he treated the results of lie-detection research, ostensibly demonstrating the existence of emotional leakage in subjects asked to deceive, as instead reflecting conflicts over intentions, specifically conflicts about lying (*HFE*, 137–38). "I know of no specific experiments demonstrating human skepticism," he wrote in this regard, "and they may be unnecessary. This is because nearly every human deception study . . . can be recast as a skepticism study, with the subjects who are the least deceived being, in effect, the best skeptics" (*HFE*, 75). He observed that the preponderance of investigations claiming to find leakage from unmanaged behavior in deceit had been poorly designed because true demonstrations of such leakage required not only that the "leaked" emotion be detectable, it also had to be "decodable," by which he meant that what was leaked must predict the specific emotion the subject was attempting to hide. Instead, by requiring subjects to deceive by stating counterfactuals or taking morally objectionable positions, deception had been "*confounded with conflict about deceiving*" (*HFE*, 137; Fridlund's emphasis).[30] Such conflicts could be moral conflicts over wanting to comply with experimental requirements by lying versus believing it is wrong to lie, or they could be pragmatic conflicts over wanting to comply by lying and believing there would be a price to pay. As Fridlund emphasized, however, citing an experiment by Bavelas, Black, Chovil, and Mullett (1996), when such conflicts were controlled for experimentally, subjects lied perfectly and showed no leakage.[31]

30. Fridlund cited M. Zuckerman, B. M. DePaolo, and R. Rosenthal, "Humans as Deceivers and Lie Detectors," in *Nonverbal Communication in the Clinical Context*, eds. B. Blanck, R. Buck, and R. Rosenthal (University Park, PA, 1986), 13–35; and M. Zuckerman and R. E. Driver, "Telling Lies: Verbal and Nonverbal Correlates of Deception," in *Multichannel Interpretations of Nonverbal Behavior*, eds. A. W. Siegman and S. Feldstein (Hillsdale, NJ, 1985), 129–47.

31. Thus, in an experiment by Bavelas et al., subjects had to lie in order to keep secret a friend's surprise birthday party. As Fridlund reported: "They lied perfectly but showed no 'leakage.' Moreover, the authors discovered that when subjects were faced with the choice to lie or tell a hurtful truth, the subjects rarely lied. Rather, they *equivocated,* supplying 'meanings implied but not claimed' . . . For example, when friends ask us to appraise their wretched new piece of art, we tend neither to damn nor to deceptively praise it. Instead, we utter equivocal comments like, 'Very interesting. Who's the artist?'" (*HFE*, 138; his emphasis). The details of the experiment can be found in Janet Bavelas, Alex Black, Nicole Chovil, and Jennifer Mullett, "Truths, Lies, and Equivocations: The Effects of Conflicting Goals on Discourse," *Journal of Language and Social Psychology* 9 (1990): 135–61. On the assumption that "deceptive" or "pretend" displays might elicit the same visceral-autonomic changes as "honest" ones, Fridlund

At the same time—and this is another point I wish to stress—Fridlund characterized Dawkins and Krebs's somewhat cynical model of the evolution of signal systems as "regrettably oversimplified" (*HFE*, 138) because of its exploitative view of animal communication. On the one hand, Fridlund welcomed the latter's approach to signaling because its emphasis on manipulation and deception helped turn attention to the flexibility and sociality of nonhuman displays. On the other hand, like the ethologist W. J. Smith and others who influenced his approach, Fridlund stressed that the relationships between manipulator and mind reader or receiver are often cooperative. He therefore posited the existence of a dynamic equilibrium between cooperative signals on the one hand, and exploiters who devalue the signals by imitating them (i.e., by cheating) on the other. It is worth noting here that for Fridlund and the new ethology theorists, the notion of "cooperation" was not understood psychologically, but in behavioral-ecological terms as a situation in which both parties to an interaction, not excluding in some cases predator and prey, stand to gain from the same outcome. Viewed in this way, cooperation applied not only to the relationship between conspecifics but also to certain predator-prey encounters. What Fridlund denied, however, was that cooperation comes about because some facial signals function indexically and causally to produce automatic readouts of internal states, as if nature had provided an automatic guarantee of reliability. For Fridlund, facial movements are strategic signals through and through; they are, as he stated, simply messages that influence the behavior of others because vigilance for and comprehension of signals have coevolved with the signals themselves (*HFE*, 105).[32] As he has recently put it,

suggested that all studies assaying physiological changes (or indeed other behavioral responses held to be cues to deception) require simulated-emotion controls. The control procedure would require trained actors briefed about the experiment, and told to lie without "real emotion" the reactions of experimental subjects. In the Behavioral Ecology View, the same autonomic patterns should be obtained in the subjects who pretended (*HFE*, 170–71). Pretend controls of this kind had been fairly common in hypnosis studies to test whether any effects were due to the "trance" state. Fridlund has commented that the experiment would require measures ensuring that the non-method pretend actors did not fall into role, so the control procedure would not be without problems (personal communication).

32. An interesting recent game-theory study of human deception, informed by Fridlund's views, shows that the trade-off between the short-term benefits of successful deception and the long-term costs to deceivers' reputations often favors signalers who produce imperfectly dishonest human facial expressions rather than perfectly honest or perfectly dishonest signals. See Paul W. Andrews, "The Influence of Postreliance Detection on the Deceptive Efficacy of Dishonest Signals of Intent: Understanding Facial

> The "truth" of a display inheres neither in the display nor its displayer, but in the moving average by which the recipient continually calibrates and recalibrates the reliability of the signals issued in that context by that displayer. Greater predictability of displayers' signals and lower skepticism by recipients toward them naturally co-evolve with repeated cooperation, else breaches occur that force recalibration, confrontation, or termination of interaction. Likewise, the "leakage" professed by BET theories [Basic Emotion Theories] is not the breakout of "genuine emotion" through an outer mask, but simply momentary conflict in intentions in social negotiation. ("BEV")

In other words, facial movements primarily serve social motives, with the result that the costs and benefits of signaling vary with the momentary social context and the animal's intentions. For Fridlund, there is no way to know in advance if another human being is trustworthy just because her facial expression is thought to provide surefire signs of the truth of her inner state, regardless of context or receiver, as Ekman maintained. Rather, human signals, like the signals of many other animals, are dependent on motive and situation. They are social tools that aid in the negotiation of social communications. For him, displays are "declarations that signify our trajectory in a given social interaction."[33] In short, Fridlund criticized Ekman for exaggerating the automaticity and involuntary nature of facial movements and neglecting their strategic and performative character.[34]

AUDIENCE EFFECTS AND SOLITARY EXPRESSIONS: THE STATE OF THE EVIDENCE

In his book, Fridlund not only proposed a theoretical intervention in the debate over the nature of the emotions and signaling. He backed up his ar-

---

Clues to Deceit as the Outcome of Signaling Tradeoffs," *Evolution and Human Behavior* 23 (2002): 103–21. As Andrews notes, both the Basic Emotion Theory and Fridlund's Behavioral Ecology View admit constraints on deception. But whereas the Basic Emotion Theory proposes that the nervous system is wired in such a way that signalers must leak clues of their true emotional states when they experience strong emotions, Fridlund's theory proposes a much more flexible wiring in the nervous system that pays attention to the costs and benefits of deception across different situations. The author concludes that "people's nervous systems are not as constrained in their ability to deceive" as the Ekman model assumes (115).

33. A. J. Fridlund, "The New Ethology of Human Facial Expressions," in *The Psychology of Facial Expression*, eds. James A. Russell and José-Miguel Fernández-Dols (Cambridge, 1997), 107–8.

34. For some further comments on signaling and cooperation, see appendix 1.

guments by exhaustively reviewing the extensive literature in the field, by reporting new experimental results, and by making critical reinterpretations of some of Ekman's canonical experiments. Especially striking was the evidence Fridlund adduced on "audience effects," that is, on the ways in which human and indeed nonhuman displays vary with the presence of interactants and with the relationship between those interactants and the displayer. These findings supported Fridlund's model of display behavior as strategic moves in the context of social transactions. As Fridlund argued, if—as the Ekman model of emotion maintained—

> displays simply read out fundamental emotions (deception notwithstanding), then across species the displays should be largely a function of emotional elicitors . . . Alternatively, if displays serve social motives throughout phylogeny, then across species their occurrence should be a function not only of the proximal elicitors, but of those who are present, one's aim toward them, and the context of the interaction. (*HFE*, 145)

Moreover, because displays can enhance fitness, the behavioral ecology view predicted that, controlling for elicitors, individuals would display more when nearby interactants are genetic relatives than when they are unrelated.

Research findings provided confirmation of the Behavioral Ecology View of signaling. Fridlund cited studies of the behavior of an array of nonhuman animals, including cats, squirrels, spider monkeys, vervet monkeys, and domestic chickens, all of which demonstrated such audience effects. For instance, in an important investigation, Moore and Stuttard (1979) demonstrated that cats, in the classic Thorndike "puzzle box," made their characteristic flank-rubbing "greeting display," interpreted as simply an instrumental escape behavior by Thorndike and later by Edwin Guthrie, only when the experimenter was visible outside the cat box door (*HFE*, 147–48).[35] Similarly, in an influential series of field studies, ethologists Cheney and Seyfarth (1990) showed not only that vervet monkeys produced predator-specific calls that did not conform to the emotional view of vocalizations, but that their calls referred to the identity of the recipients and the place of the displayer in the social structure (*HFE*, 148–49).[36] Systematic studies of audience effects in domestic chickens yielded similar results. In several studies, Marler and Evans demonstrated that

35. B. R. Moore and S. Stuttard, "Dr. Guthrie and *Felis domesticus* or: Tripping over the Cat," *Science* 205 (1979): 1031–33.

36. See D. L. Cheney and R. M. Seyfarth, *How Monkeys See the World* (Chicago, 1990).

the food and predator vocal calls of these chickens were sensitive to the presence of conspecifics. Cockerels preferentially emitted food calls in the presence of females, fewer food calls in the presence of familiar rather than strange females, fewer still when alone, and fewest of all when in the presence of another male. As Fridlund noted, other data suggested that male chickens emitted deceptive food calls when food was absent and females were distant; these calls tended to elicit female approach, suggesting that such calls were strategies that had evolved for that purpose. The same kinds of audience effects were identified in the case of the cockerel's alarm calls. Males sounded more alarm calls in the presence of their mate or that of another female, a selectivity that worked just as well with a televised image.

In addition, Marler and Evans had carried out experiments that addressed the question of whether enhanced calling was the result of increased "emotional arousal," due to a female's presence. They arranged for a male to receive food only after pecking an operant key, and then only during signaled periods of food availability. In Fridlund's account of the findings:

> When males were placed in their cages adjacent to a female, they tended to produce a stereotyped courting dance. However, they did not issue food calls until the signal light came on and indicated food availability. Still, food calls might be interpreted as independent conditional responses from previous key-pecking trials, but one other manipulation counters this explanation. Males made substantially more food calls when females were present, but their operant rates of key pecking for food were unchanged. The obvious interpretation is referential; the males' food calls "referred" to the food independent of their level of "emotional arousal." (*HFE*, 150)

As Fridlund also reported, Marler and Evans had gone on to compare and contrast the "affect hypothesis" of animal signaling versus the "referential" hypothesis. According to Marler and Evans, the two approaches should differ in certain important ways. For example, the two models should differ in their "input specificity," since "emotional" outbursts should be elicited by a wide range of stimuli, whereas signals conforming to the referential model (or Behavioral Ecology View) should have much more specific elicitors or referents. The suggestion was that referents "would only influence the animal to signal if, in context, it would be functional to do so." In addition, the two kinds of signals should differ in their "conditions of production," that is, in the pattern of behavior comprising the signal. Signals that were emotional outbursts should consist of stereotyped and invariant patterns of behavior, whereas the pattern of behavior comprising referential signals should be more variable. In Fridlund's summary:

> Thus a signal of fear might include piloerection [erection of the hair of the skin], agitation, and vocalization—and it should always include these if fear has been evoked. In contrast, the pattern of behavior comprising referential signals should be more variable, with the animal capable of producing some components while withholding others. Thus an animal that is threatened by a predator may indeed piloerect and become agitated, but it may not vocalize if it is disadvantageous to do so—if, for example, a pack of similar predators is nearby. (*HFE*, 151)

Fridlund was intrigued by Marler and Evans's model because, as he stated, it was "so pragmatic." It permitted researchers to test the involvement of "emotion" in displays without needing to define emotion. This was because the model was "connotative rather than essentialist." As he stated: "That is, whatever 'emotion' is, most researchers from an Emotions View would probably endorse the proposition that affect-based signaling would be elicited by a range of stimuli and manifest as a nondissociable configuration of responses. Likewise, most would probably agree that the combination of specificity of referents and context-dependent dissociability of responses would not follow from an affect-based account" (*HFE*, 152). But Fridlund observed that the model's strength was also its Achilles heel, because it relied on agreements about what the term "emotion" connoted. As he remarked, advocates of the basic emotion position had failed to define "emotion" and instead had rendered the term so elastic that it could encompass nearly any behavior. He therefore anticipated "no failures of imagination in attempts to retain affect-based explanations even of signals that unequivocally belong on the 'referential' side" [of Marler and Evans's model] (*HFE*, 152). As we shall see in the next chapter, this is exactly what subsequently happened in the case of Griffiths, because he found in the research described by Marler and Evans the evidence he was looking for in order to defend an affect-based theory of animal signaling.

As for human audience effects, here too recent studies provided support for Fridlund's position. For example, in a landmark study, Kraut and Johnston (1979) demonstrated that bowlers rarely smiled when they faced the bowling pins or even when they obtained a strike, but smiled frequently when they turned to face their friends and others. These findings suggested that smiling was not the result of an inner emotional state of happiness but was a function of social interactions. Kraut and Johnston presented their work as an ethologically inspired alternative to the readout view of emotions, so there is an important sense in which their general orientation to the problem of emotional behavior converged with that of Fridlund (*HFE*, 152–55).[37]

In a related development, even before Fridlund began proposing his Behavioral Ecology View of faces, researchers in the field of nonverbal communication—a field tracing its lineage back to the anthropological work on communication undertaken in the 1950s and 1960s by Roy Birdwhistell who, as I mentioned in chapter 2, denied the existence of universals in nonverbal gestures—had recently started to publish findings on motor mimicry that also contradicted the readout view of emotional expression. Motor mimicry, such as wincing when we see someone else being hurt, or cringing when we hear of another's fear, had previously been interpreted as the expression of a vicarious inner emotional state owing to the imitation of the observed person's feelings. But as Bavelas and her colleagues demonstrated experimentally, the pattern and timing of wincing at a confederate's pain was strongly dependent on the experimental subject's eye contact with the ostensible sufferer. The authors' conclusion was that motor mimicry was not a reflex readout of the subject's vicariously experienced inner emotional state but a form of *representation* by which the observer intended to communicate to another:

> Expressive behaviors . . . are not an inadvertent by-product of a private experience but are primarily and precisely interactive; they are constant evidence that in our social behavior we are intricately and visibly connected to others. For these reasons we propose that the overt behavior of motor mimicry is primarily communicative and that, moreover, it conveys a message fundamental to our relationships with others: I am like you, I feel as you do. Thus the centuries-old puzzle of motor mimicry—and perhaps many other behaviors—may be solved by looking beyond the individual to the immediate interpersonal context in which the behavior occurs.[38]

37. R. E. Kraut and R. E. Johnston, "Social and Emotional Messages of Smiling: An Ethological Approach," *Journal of Personality and Social Psychology* 37 (1979): 1539–53. Kraut and Johnston tended to think a compromise was possible between the "emotional" and the "social" hypotheses about the causation of smiling, even suggesting the existence of two kinds of smile, not unlike Ekman's distinction between genuine "Duchenne" smiles and false "non-Duchenne" smiles.

38. J. Bavelas, A. Black, C. R. Lemery, and J. Mullett, "'I Show How You Feel': Motor Mimicry as a Communicative Act," *Journal of Personality and Social Psychology* 50 (2) (1986): 328, cited also by Fridlund, *HFE*, 157. Bavelas and her colleagues did not repudiate the idea of intrapersonal emotion processes, but suggested that proving a direct correlation between those internal processes and communicative ones would be dif-

In another investigation, Bavelas and her colleagues demonstrated that when subjects listened face-to-face to a storyteller who punctuated her storytelling with brief movements to her left, the listeners did not imitate the storyteller by performing "rotational" imitative movements (that is, by moving to *their left,* as if they had occupied the same "role" as her and had adopted her inner state), but performed "reflective" or mirroring movements, by moving to *their right* (in order, as Bavelas and her coauthors suggested, to maintain an empathic relationship with her). Once again, Bavelas et al.'s conclusion was that

> motor mimicry is not the manifestation of a vicarious internal state but a *representation* of that state to another person . . . We do not agree with theories that, in effect, assume that the overt behavior is a "spill over" from an inner vicarious experience. We propose that it has a function of its own, a communicative one.[39]

Likewise:

> We propose that the observer is not having the other's reaction but is portraying it. He or she is sending the message. "It is *as if* I feel as you do." In other words, the motor mimicry is an encoded representation quite different from, say, actual pain.[40]

Bavelas et al.'s findings and conclusions about mimicry contrast strikingly with those of many of today's theorists who commonly emphasize the uncontrolled, "dumb," and noncognitive nature of motor imitation or "mimesis."[41] But on the basis of their results, Bavelas and her colleagues in-

---

ficult. "[W]e conclude that overt motor mimicry is best explained as a communicative act," the authors wrote, "controlled by interpersonal processes and independent of any intrapersonal processes that may accompany it. It is important to emphasize that we do not reject the existence of the latter. Instead, we propose a *parallel process* theory: Both communicative and internal psychological processes can be elicited by the same stimuli but thereafter proceed independently. Any interdependencies would have to be demonstrated, and both the present data and Kraut and Johnston's (1979) earlier data on smiles suggests that this may be a difficult isomorphism to establish" (8).

39. J. Bavelas, A. Black, N. Chovil, C. R. Lemery, and J. Mullett, "Form and Function in Motor Mimicry: Topographic Evidence That the Primary Function Is Communicative," *Human Communication Research* 14 (3) (1988): 278–79.

40. J. Bavelas, A. Black, C. R. Lemery, and J. Mullett, "Motor Mimicry as Primitive Empathy," in *Empathy and Its Development*, eds. Nancy Eisenberg and Janet Strayer (New York, 1987), 329.

41. In a large literature on automatic imitation, see for example Tanya L. Chartrand and John A. Bargh, "The Chameleon Effect: The Perception-Behavior Link and Social Interaction," *Journal of Personality and Social Psychology* 76 (1999): 893–910, in which

stead underlined the symbolic or cognitive-intentional character of such mimetic actions and did so in terms that were congruent with Fridlund's emphasis on the intentional-communicative nature of signaling.

In their work, Bavelas et al. raised a particularly important topic that Fridlund tackled head-on. If facial movements and displays are a function of the relationship between interactants and the social context, why do people make faces when they are alone?[42] It will be remembered from my discussion in chapter 2 that, ever since the work of Lavater in the eighteenth century, the belief that the emotional truth shows itself on a person's face when he or she is alone had been the default position of many emotion theorists, including Ekman and his followers. From this it followed that "spontaneous" faces should occur in the alone condition because societal or "display" rules would be inoperative. Solitary facial expressions could thus be regarded as the authentic expression of the subject's inner emotional state.

But in fact, the opposite finding was obtained by Fridlund. In an experiment carried out with Avery Gilbert and John Sabini, he showed that solitary subjects displayed *few* spontaneous faces to odors they had rated as smelling good, bad, or neutral (*HFE*, 158). In the experiment, the facial responses of female subjects were videotaped while they smelled six odors in each of three experimental conditions (spontaneous, posing to real odors, and posing to imagined odors). Videotaping was covert in the spontaneous condition (in other words, the subjects were unaware of being filmed) but overt in the posed condition. Raters were then asked to identify the type of odor—good (cloves, roses), bad (urine, rancid sweat), and neutral (mineral oil only)—from the poses. The findings demonstrated that subjects exhibited few facial responses to the odors when smelling them in pri-

---

the authors cite Bavelas et al.'s work but reinterpret mimicry as a form of unconscious "behavioral coordination" that occurs automatically in the absence of cognition or any intentional-communicative goal. For a discussion and critique of Chartrand and Bargh's work, see Leys, "The Chameleon Effect and Mirror Neurons," *nonsite.org* (forthcoming).

42. To the question of what explanations can be offered for motor mimicry when an individual is alone, Bavelas et al. anticipated Fridlund's notion of an implicit audience by observing that "participants watching a video in an experiment are not alone but in the presence of both an experimenter and a camera," and that to an individual alone in front of a TV, the persons depicted are sufficiently real psychologically to elicit a display. They added: "The occurrence of motor mimicry . . . when alone is the nonverbal equivalent of talking to oneself. Just as we represent our thoughts in words that sometimes spill over into muttering to ourselves, we might represent some thoughts in nonverbal actions that are not always suppressed." Bavelas et al., "'I *Show* How You Feel,'" 328.

vate, despite dramatic differences in their hedonic ratings of the odors. In short, patterned faces did not automatically accompany the hedonics of odor, but were influenced by the social demands of the setting.[43]

Nevertheless, as Fridlund also observed, solitary faces do occur, and this fact had been routinely mentioned as definitive evidence that faces are spontaneous readouts of internal emotions (*HFE*, 158). Thus, Fridlund cited the psychologist Ross Buck as stating: "'When a sender is alone . . . he or she should feel little pressure to present a proper image to others, and any emotion expression under such circumstance should be more likely to reflect an actual motivational/emotional state'" (*HFE*, 158).[44] Ekman's distinction between the so-called "Duchenne" and "non-Duchenne" smiles was premised on that claim: according to him, the spontaneously occurring or involuntary Duchenne smile is the genuine "felt" or "enjoyment" smile that occurs when people are alone; it combines a contraction of the muscles at the corner of the eyes (wincing) with a contraction of the muscles at the sides of the mouth (smiling). Smiles performed without the eye contraction were regarded by Ekman as "fake" or deceitful. In a recent paper, he had presented evidence in favor of this claim based on the study of smiles performed by subjects who watched films when they were alone in the laboratory, the alone condition being treated by Ekman as enabling purely spontaneous behavior uncontaminated by display rules or conventions. On this basis, Ekman had declared: "'Facial expressions do occur when people are alone . . . and contradict the theoretical proposals of those who view expressions solely as social signals'" (cited in *HFE*, 158).[45]

But in a searching discussion, Fridlund suggested that just because we are alone in a room or a laboratory space does not mean that we are alone *psychologically*. Rather, he considered facial movements in the "alone" or solitary condition a function of an implicit sociality because of the presence of imagined interlocutors and interactants of various kinds. In other words, we often act as if others are present even when they are not. In an experiment designed to document the importance of an implicit audience as a mediator of solitary faces, Fridlund set up a situation in which subjects watched an amusing videotape in one of four viewing conditions:

43. Avery N. Gilbert, Alan J. Fridlund, and John Sabini, "Hedonic and Social Determinants of Facial Displays to Odors," *Chemical Senses* 12 (2) (June 1987): 355–63; and *HFE*, 158.

44. Fridlund was citing Ross Buck, *The Communication of Emotion*, 20.

45. Paul Ekman, Richard J. Davidson, and Wallace V. Friesen, "The Duchenne Smile: Emotional Expression and Brain Physiology II," 351.

1) alone; 2) alone, but with the belief that a friend nearby was engaged in an irrelevant task; 3) alone, but with the belief that a friend nearby was viewing the same videotape in another room; and 4) when a friend was physically present. Viewers' smiles were measured with electromyography over the *zygomatic major* muscles at the side of the mouth that are responsible for smiling. As Fridlund reported, the results showed that

> smiling among solitary viewers who believed a friend was viewing nearby was equal to that shown in the actual presence of the friend, but greater than that shown by subjects who simply viewed alone. Self-ratings of happiness did not differ among the viewing conditions, and within conditions rated, happiness correlated negligibly with smiling. The findings confirmed audience effects for human smiles, demonstrating that the effects do not depend upon the physical presence of the interactant, and suggested that the smiles did not result from altered emotional experience. (*HFE*, 161)[46]

In another experiment, Fridlund and his colleagues asked subjects to imagine situations that they enjoyed either alone (low-social condition) or with other people (high-social condition). Their smiling during the imagery was again measured by using facial EMG overlying the *zygomatic major* muscles. Following each trial, the subjects used emotion scales to report how happy they had felt during the imagery. Controlling for these happiness ratings statistically, the results showed more EMG activity in the zygomatic major sites in high-social imagery than in low-social imagery, even when the happiness ratings were equalized. This experiment not only corroborated the hypothesis that solitary facial displays are mediated by imaginary interactants but also demonstrated the separability of emotional ratings and sociality.[47]

In a related study, as Fridlund reported, psychologist Nicole Chovil, a coauthor with Bavelas on the studies of the communicative function of motor mimicry, presented the results of an experiment in which she controlled for differing degrees of sociality. In Chovil's experiment, subjects heard stories about a harrowing close situation call in one of four conditions: 1) alone, from an audiotape recording; 2) from another subject

46. A. J. Fridlund, "Sociality of Solitary Smiling: Potentiation by an Implicit Audience," *Journal of Personality and Social Psychology* 60 (1991): 229–40.

47. A. J. Fridlund, J. P. Sabini, L. E. Hedlund, J. A. Schaut, J. I. Shenker, and M. J. Knauer, "Social Determinants of Facial Expressions During Affective Imagery: Displaying to the People in Your Head," *Journal of Nonverbal Behavior* 14 (1990): 113–37; and *HFE*, 164.

separated by a partition; 3) alone, over the telephone; and 4) talking to another subject face-to-face. Subjects' facial displays—largely wincing and grimacing—were scored using visible facial coding. Chovil then obtained sociality ratings for the four conditions by asking a separate group of raters to estimate the "psychological presence" afforded by each condition, where the conditions were ordered according to their estimated sociality. Her results showed that the mimicry or facial displays increased with rated sociality, thereby providing evidence in favor of a communicative view of facial displays.[48]

In an interesting exchange over the interpretation of Chovil's findings, Buck offered what would emerge as one of the most frequent counterarguments used to defend the Basic Emotion Theory. His claims reveal how easy it was for adherents of the readout view to fall back on received ideas according to which facial displays simply serve communication by functioning as innately determined signals the meaning of which the receiver innately knows. As he saw it:

> [T]he primary role of the display is . . . to "inform other members of the species about the sender's emotional state" . . . The display is innate—a phylogenetic adaptation—and it must have evolved in concert with pre-attunements so that the display constitutes a "social affordance" in the terms of Gibson's . . . perceptual theory. The receiver "knows the meaning" of the display directly as a kind of inherited knowledge.
>
> The result is a spontaneous communication process that is, in effect, a conversation between limbic systems. It answers the problem of "other minds" posed by philosophers: we know certain feelings and desires in others directly because they are constructed to display those states and we are constructed, given attention, to know the meaning of those displays just as directly as we know pain when we stub our toe. In spontaneous communication the sender and receiver constitute literally a biological unit.[49]

There could hardly be a franker acknowledgment that for Buck the readout theory of emotional signaling solved the skeptical problem of how

48. Nicole Chovil, "Social Determinants of Facial Displays," *Journal of Nonverbal Behavior* 15 (1991): 141–54. Chovil thanks Fridlund for his help in analyzing the data and for providing comments on earlier drafts of her paper (141).

49. Ross Buck, "Social Factors in Facial Display and Communication: A Reply to Chovil and Others," *Journal of Nonverbal Behavior* 15 (3) (1991): 158–59; hereafter abbreviated as "SF."

one person knows the mind and feelings of another person by proposing that such knowledge is the result of the direct transmission of inherently truthful signals between brains: the body does not lie. As Chovil and Fridlund in their reply to Buck succinctly and critically put the point: "Buck's idea of limbic systems in intimate communication precludes the possibility of deception or misinterpretation."[50]

In her paper, Chovil had suggested that the readout paradigm and a communicative view of facial displays offered different predictions about the impact of interpersonal contexts and audiences on facial behavior. On the readout model, the presence of other people might make us moderate or inhibit an otherwise "spontaneous" expression of emotion of the kind we make when we are alone, suggesting that in the presence of others the intensity and frequency of facial displays would be *reduced*. By contrast, the communicative view predicted that on the whole there would be fewer facial movements in unobserved situations than in those in which an amenable addressee or audience is present, suggesting that in the presence of others the intensity and frequency of facial displays would be facilitated or *enhanced* ("WE," 143). This latter prediction is what her findings confirmed.

Buck took Chovil to mean that sociality *always* potentiates facial displays, and offered evidence that contradicted that view. But as Chovil and Fridlund observed in their reply to Buck, the communicative view of displays did not predict that facial displays must always increase in number or intensity with increasing sociality. In Chovil's study, increased sociality led to a greater frequency of displays because, the authors suggested, the more social situations were closer approximations to face-to-face encounters ("WE," 163). However, as Fridlund especially would go on to emphasize, in other social contexts—for example, in the social setting of a game where being able to maintain a poker face can be important—the intensity of facial displays might decrease (*HFE*, 164). In short, it was important not to be too literal about the relationship between facial behavior and sociality: the nature of facial displays always depended on the context of the interaction as well as on the relationship of the interactants, with the result that in some intensely social conditions facial movements might be reduced but in other ones they might be enhanced.[51]

50. N. Chovil and A. J. Fridlund, "Why Emotionality Cannot Equal Sociality: Reply to Buck," *Journal of Nonverbal Behavior* 15 (3) (1991): 166; hereafter abbreviated as "WE."

51. In other words, social role is an important determinant of facial behavior. Fridlund has observed in this connection that it is hard to strip an experiment of the impact of such social roles. The result is that social influences are difficult to control

Chovil and Fridlund characterized Buck's related argument that sociality simply equaled emotionality as the strongest criticism of the communicative view of faces. "[M]anipulations of sociality will naturally be judged by subjects to be manipulations of emotion (although the opposite would not, of course, be the case)," Buck argued. "In general, I suggest that any situation that is judged to be 'social' will also be judged to be 'emotional'" ("SF," 160). Chovil and Fridlund replied by mentioning counterexamples showing that emotion and sociality are not covariant "but independent or reciprocal," as when people are actually happier when others leave the room, or when children cry when their caretakers leave and stop when they return ("WE," 164). They also referred to two of Fridlund's experiments, briefly discussed above, demonstrating the separability of sociality and reported emotional experience. In the first experiment, sociality affected facial displays when emotion was statistically controlled; in the second, the manipulation of sociality did not affect reported emotion ("WE," 164).

But crucially, Chovil and Fridlund also called attention to the problematic status of the "display rule" thesis itself. They noted in this regard that the display rule idea required the separation between emotion and sociality, with faces instigated by the former and mediated by the latter. As they observed: "The display rule concept stipulates that individuals can alter their spontaneous faces by faking or exaggerating an absent emotion, minimizing an actual one, or masking one's natural expression with a substitute. Consequently, display rules can only be operative if two individuals show different facial displays while having the same emotion. To

---

or handle experimentally. In any situation, the number of variables potentially influencing behavior is enormous; even unconscious influences can't be ruled out. The risk of being too literal about the "alone" (or "spontaneous") condition is especially high, with researchers too often assuming that just because an experimental subject is alone in a laboratory setting he or she loses all cultural influences. The same criticisms apply to experiments, all too frequent, that operate on the assumption that facial electromyography, the electrical recording of facial muscle activity through the use of tiny electrodes attached to the skin, is uncontaminated by social roles and contexts. Such studies often wrongly assume that electromyography reflects pure emotion because the experimental subject is "alone" while viewing or imagining, and because electromyographic activity is below or at the edge of visibility. But from a behavioral ecology point of view, electromyography techniques don't offer the chance to penetrate below the social, but instead reveal the covert sociality of what cannot be seen (see *HFE*, 171–72). Likewise, Despret has observed that leaving a subject alone in a room and thinking he or she is not aware of being observed borders on the naïve—and thinking that subjects are naïve (Vinciane Despret, *Our Emotional Makeup: Ethnopsychology and Selfhood* [New York, 2004], 85).

our knowledge, this criterion has never been demonstrated in any display-rule study (see reanalysis of Ekman-Friesen's 'Japanese–American' study by Fridlund, 1991)" ("WE," 165). In making this statement, with its reference to Fridlund's very recent criticism of Ekman's Japanese–American study, Chovil and Fridlund would have been well aware that they were raising an explosive issue. The whole point of that famous Japanese–American study was its claim to have shown the differential effects of display rules on the same emotion experienced by Japanese and American students. If that canonical study was flawed, Ekman's display rule thesis was in serious trouble.

EKMAN'S JAPANESE–AMERICAN STUDY: A MISREPORTED EXPERIMENT

It is a remarkable fact that the Japanese–American experiment that has been considered so decisive for the validity of Ekman's neurocultural theory of the emotions has been repeatedly misreported by him. The purpose of that study was to test the validity of Ekman's "display rule" concept by comparing the emotional experiences and facial expressions of subjects from two different cultures under conditions in which social conventions could be manipulated and controlled. Ekman and Friesen first reported the study in two separate reports published in 1972.[52] Japanese students were chosen for the comparison with Americans because of the popular idea that the Japanese were inscrutable to the Westerner, suggesting the operation of specific cultural display rules controlling for polite behavior when a Japanese person was in the presence of others, especially an authority figure.

Here is a standard description of the experiment given by Ekman, this one from a statement in 1975 as cited by Fridlund (*HFE*, 286):

> Research conducted in our laboratory played a central role in settling the dispute over whether facial expressions are universal or specific to each culture. In one experiment, stress-inducing films were shown to college students in the United States and to college students in Japan. Part of the time, each person watched the film alone and part of the time the person watched while talking about the

52. P. Ekman, "Universals and Cultural Differences in Facial Expressions of Emotion," in *Nebraska Symposium on Motivation*, ed. J. Cole (Lincoln, NE, 1972), 207–83; hereafter abbreviated as "UCD"; W. V. Friesen, *Cultural Differences on Facial Expressions in a Social Situation: An Experimental Test of the Concept of Display Rules*, unpublished doctoral dissertation, University of California–San Francisco, University of California Microfilm Archives; hereafter abbreviated as *CD*.

> experience with a research assistant from the person's own culture. Measurements of the actual facial movements, [secretly] captured on videotapes, showed that when they were alone, the Japanese and Americans had virtually identical facial expressions. When in the presence of another person, however, where cultural rules about the management of facial appearance (display rules) would be applied, there was little correspondence between Japanese and American facial expressions. The Japanese masked their facial expressions of unpleasant feelings more than did the Americans. This study was particularly important in demonstrating what about facial expression is universal and what differs for each culture. The universal feature is the distinctive appearance of the face for each of the primary emotions. But people in various cultures differ in what they have been taught about managing or controlling their facial expressions of emotion.[53]

The experiment therefore purported to demonstrate the "leakage" of the Japanese students' basic negative emotions in response to the stressful films prior to the covering over of those emotions by the Japanese cultural display rule controlling for polite, smiling faces. In Ekman's words: "In this single experiment we had shown how facial expressions are both universal and culturally different. In private, when no display rules to mask expression were operative, we saw the biologically based, evolved, universal facial expressions of emotion. In a social situation, we had shown

53. P. Ekman and W. V. Friesen, *Unmasking the Face* (Englewood Cliffs, NJ, 1975), 23–24. In this and other publications, Friesen published two video frames side by side of the facial expression of a Japanese and American student taken when the students were alone and ostensibly exhibiting spontaneous negative feelings. Subsequently, Ekman claimed that slow-motion videotape analysis showed the actual sequencing in which the Japanese students covered over their negative feelings with a false smile when in the presence of the graduate assistant. "Measurement showed that the facial movements were no longer the same. The Japanese looked more polite, showed more smiling than did the Americans. Examining these video-tapes in slow motion it was possible to observe sometimes the actual sequencing in which one movement (a smile for example) would be superimposed over another muscle action (such as a nose wrinkle, or lower lip depressor)." See Paul Ekman, "Biological and Cultural Contributions to Body and Facial Movement," in *The Anthropology of the Body*, ed. J. Blacking (London, 1977), 70; idem, "Biological and Cultural Contributions to Body in the Expression of Emotions," in *Explaining Emotions*, ed. Amélie Oksenberg Rorty (Berkeley, CA, 1980), 94; idem, "The Argument and Evidence about Universals in Facial Expressions of Emotion," in *Handbook of Social Psychophysiology*, eds. Hugh Wagner and Antony Manstead (Chichester and New York, 1989), 151. But so far as I am aware, the sequence of video frames showing this transition was never published.

how different rules for the management of expression led to culturally different facial expressions."[54]

On the basis of this experiment, to this day Ekman and Friesen have been widely understood to have proved the validity of their display rule thesis, a linchpin of their neurocultural version of the Basic Emotion Theory. The experiment is considered an iconic or founding study in the emotion field, routinely cited throughout the literature on the emotions. But in a brilliant dissection and dismantling of this "paradigmatic 'demonstration' of display rules" (*HFE*, 286), Fridlund showed that the experiment and its findings had never been reported completely or accurately by Ekman and his colleagues. He also demonstrated that, when the experiment was reconstructed correctly, the opposition between natural, genuine emotional expressions versus culturally coded displays could not be sustained.

In the statement quoted above and in many other publications, Ekman over the years had described the experiment as involving two experimental conditions or phases. In the first phase, the Japanese and American students watched neutral and stress film clips when they were alone in the viewing room; in the second phase, the students watched stress film clips when in the company of an authority figure. According to Ekman, it was during the second phase that the facial expressions of the Japanese and American students diverged and the Japanese covered over their negative feelings by polite smiles. But as Fridlund discovered on the basis of a close scrutiny of the unpublished doctoral thesis by Ekman's laboratory collaborator Wallace V. Friesen, based on work carried out in the Department of Psychology at the University of California–San Francisco under Ekman's supervision, Friesen's detailed description of the display rule study deviated considerably from the published reports. The Japanese–American experiment actually involved three phases or conditions, not two, and this made a decisive difference. The three phases were (*HFE*, 287–88):

Phase 1. Twenty-five Japanese and twenty-five American male undergraduates, each in his own country, individually viewed a total of four movies lasting twenty minutes while they were alone in the room, and after they had been fitted with devices to measure galvanic skin conductance and heart rate during the film viewing. The first film clip was designed to be mildly pleasant, and showed two men taking a canoe trip. The other three clips were intended to be stressful; they included clips of *Subincision*, the ritual circumcision film already used by Lazarus in his

54. Paul Ekman, "Expression and the Nature of Emotion," in *Approaches to Emotion*, eds. P. Ekman and L. Scherer (Hillsdale, NJ, 1984), 321, cited in *HFE*, 289.

cross-cultural studies of stress, as discussed in chapter 3, a suction-aided infant delivery, and nasal sinus surgery. Friesen did not include this solitary viewing phase in his dissertation, but instead referred to it as one previously carried out by Ekman, Friesen, and Malmstrom in an unpublished study first reported by Ekman in 1972 (*CD*, 11). As Fridlund noted, Ekman's 1972 chapter contained the only published listing of any of the solitary viewing data, and this listing was fragmentary.

Phase 2. When all the films had been shown, a research assistant—actually a graduate student from the experimental subject's own culture, but dressed in a white coat to give the appearance of authority—entered the viewing room and engaged the subject in a one-minute face-to-face interview about the latter's experience while viewing the films. The data for this second phase consisted of the subject's videotaped facial behavior lasting for approximately 10–20 seconds as the research assistant began the interview with the question "How are you feeling right now?" (*CD*, 75; cited in *HFE*, 288). This was the "first episode" studied by Friesen in his dissertation.

Phase 3. In the third phase, the research assistant sat facing the viewing subject with his back to the viewing screen while the most unpleasant portions of the final stressful clip (a clip showing nasal sinus surgery) were replayed. As the replay began, the research assistant resumed his interview with the question "Tell me how you feel right now as you look at the film" (*CD*, 75; cited in *HFE*, 288). The data for this third phase included about 20–30 seconds of the subject's facial behavior following that question. This was the "second episode" studied by Friesen.

It will be seen that Phase 1 involved solitary viewing and both Phase 2 and Phase 3 involved the presence of the research assistant or authority figure. As Fridlund observed: "This fact is crucial to the interpretation of the results for the three conditions" (*HFE*, 288). Fridlund's inspection and summary of the available data led to three main conclusions, as follows (*HFE*, 288–89):

1. Subjects in the first, solitary viewing phase of the experiment produced a preponderance of putative "negative-affect" faces, equivalent for the two cultures. Eleven of the fifty subjects made no faces at all despite viewing such gruesome films.

2. In the second phase, about two-thirds of the faces made during the interview with the research assistant consisted of smiles, while the remainder were interpreted as "negative-affect" faces. Statistical analysis showed *no* differences between the two cultures (*CD*, 51–53).

3. In the third and final phase, the Japanese did not change either in smiling or "negative" affect faces compared to the preceding interview.

In contrast, the Americans smiled less and made more "negative-affect" faces. In other words, in Phase 3 it was less a matter of the Japanese changing their facial expressions from negative-affect expressions to polite smiles than of their maintaining the polite faces they had already displayed during Phase 2. It was the Americans whose facial movements changed during the switch from Phase 2 to Phase 3 as they smiled less and made more negative faces. This aspect of the findings was obscured by Ekman's omission of Phase 2 in his descriptions of the experiment.

Ekman's standard interpretation of the findings was that both the Americans and the Japanese felt revulsion at watching the films. The Americans, however, were more authentic in showing their negative feelings, whereas the Japanese masked their revulsion with a false smile. But Fridlund suggested that Ekman's interpretation was unsupportable on several counts. Thus, Fridlund remarked on the absence of any direct comparisons between social and solitary viewing, rendering any statement about the "masking" of emotion by the Japanese subjects untenable. It could well have been that the American students were merely histrionic. He also wondered why about twenty percent of the subjects in the solitary viewing condition (six Americans and five Japanese) displayed no discernible facial activity at all, leading him to ask: "Is it conceivable that the stressful films left all these subjects entirely unaffected?" Moreover, he asked, if the Japanese "'masked their revulsion with a false smile,'" then why did the five Japanese students who made no observable facial movements during solitary viewing make smiles during the interview?[55] Furthermore, differences in facial displays were found only in Phase 3 of the experiment, the interview-while-viewing condition, yet "there was no prior basis for believing that any display rules would operate in one condition but not in another" (*HFE*, 289).

"Most critically" (*HFE*, 291), Fridlund argued—as he had argued against Buck—that attributing the Japanese-American differences to display rules would have required, as the definition of display rules stipulated, verification that both cultures had equivalent emotions in the condition

55. Ekman continued to claim that the Japanese students masked their negative expressions after Fridlund's critique. For instance, in 1997 he observed: "Display rules are learned, should vary across cultures, and may differ among distinct social groups, within a culture. Our study of display rules (Ekman, 1972) found that Japanese more than Americans attempted to conceal negative emotional expressions in the presence of an authority figure, using a masking smile." P. Ekman, "Expression of Communication About Emotion," in *Uniting Psychology and Biology: Integrative Perspectives on Human Development*, eds. Nancy L. Segal, Glenn E. Weisfeld, and Carol C. Weisfeld (Washington, DC, 1997), 327–28.

in which their faces differed (Phase 3). Fridlund noted that Ekman himself had made that stipulation when he discussed facial behavior at funerals. He quoted Ekman as having written:

> "All too often a common emotional state is inferred simply because the same event was compared. For example, at funerals Culture Y might show down-turned, partially open or trembling lips, inner corners of the brows drawn together and up, and tightened lower lids (the sad face), while Culture X might show up-turned, partially opened lips, deep nasolabial folds, wrinkling in the corners of the eyes, and bagging of the lower eyelid (the broad smiling face). Before declaring that the facial expression of sadness varies across these two cultures, it would be necessary to verify that the stimulus *funeral* normatively elicits the same emotion in the two cultures rather than being an occasion for sadness in one culture and happiness in another." (*HFE*, 291; "UCD," 215; Ekman's emphasis)

By this reasoning, Fridlund contended, a prerequisite to any comparison of facial behavior in the Japanese–American study would have been to ensure the equivalence of emotion in the two groups of students. But it was precisely such independent evidence about emotional states that was lacking in the study. This was the "pitfall" of the neurocultural model, Fridlund wrote. Put simply, "here the problem comes home to roost" (*HFE*, 291).

Fridlund drew attention to the fact that Ekman himself had admitted the lack of independent evidence about emotional state as a shortcoming of the experiment, and had suggested self-report as a possible source of validation. "The question might be raised as to whether we have actually shown that the emotional expressions were the same in the two cultures, or only that the facial behavior was the same," Ekman had written. "How do we know that the behavior we called fear, for example, is not actually behavior unrelated to emotion, or behavior relevant to anger? . . . We have no direct evidence to show that the items we call fear correlate with either a subject's immediate self-report, or with the elicitation of that one emotion. (We are in the process of obtaining such validation data)" ("UCD," 259–60). Fridlund noted that self-report data were in fact collected and available, because at the conclusion of the experiment, the subjects were asked to fill out questionnaires concerning their feelings about each stress film and were also requested to complete an entire California Psychological Inventory (*CD*, 25–26). But these data were not reported or considered by either Ekman or Friesen.

It will come as no surprise to the reader that, in light of his Behavioral

Ecology View, Fridlund treated the differences observed between the facial displays of the Japanese and American students in Phase 3 of Ekman's experiment as the effects of cultural differences in the subjects' stance toward the viewing conditions and hence to differences in the management of facial behavior. He pointed out that "the experimenter himself is always an implicit audience, and his or her laboratory is always the stage for the directorial effort known as an 'experiment.' Thus the 'alone' phase of the study was implicitly social, and the two interview phases were simply more explicitly social. Thus contrasting the facial behavior in the 'alone' versus interview phases as authentic versus managed is over-statement at minimum" (*HFE*, 291). In short, Fridlund treated Ekman's Japanese–American experiment as a study of audience effects.

On this basis he offered a simpler explanation than masking via display rules to account for the results by suggesting that the Japanese students smiled out of politeness to the research assistant interviewer. When the stress film was being replayed, the positioning of the interviewer left the subjects the choice to look at him or at the film. Although the purpose of the film replay was to obtain reactions to the film when the interviewer was present, it would be rude for the Japanese students to ignore the interviewer who was addressing them, which is why they were unaffected by the replay. "It would be far less rude for the American student to view the film while being addressed. Thus the cultural difference may not have been managing facial behavior, but in attending to the film. In the Behavioral Ecology View, the faces issued by the Japanese and Americans were equally authentic displays of social motives. The Americans were authentically moved to comment facially on the film, whereas the Japanese were authentically moved to show *politesse* to the experimenter" (*HFE*, 292). As Fridlund's alternative explanation helped make clear, in any cross-cultural study social roles and the context of the interaction were potential confounds in assessing display rules: not only was it necessary to control the emotions of the subjects, but also their relation to those to whom they are displaying, in terms, for instance, of power, status, obligation, and so on.

EKMAN'S ATTEMPTED RESPONSES TO FRIDLUND: TRYING TO HAVE IT MANY WAYS

Fridlund in his 1994 book and related papers gave the first, and to this day the only, detailed analysis and critique of the Ekman–Friesen Japanese–American experiment in all its reconstructed details, to devastating effect. Understandably, Ekman felt called upon to answer his criticisms. Yet nowhere have the contradictions inherent in his neurocultural theory been more in evidence than in his various efforts to rebut Fridlund. In general, over the years Ekman's responses to criticisms have involved a number of different, sometimes competing strategies, which can be enumerated as follows:

1. Ekman's most commonly used strategy has been to restate his basic theoretical and empirical claims while simply ignoring criticisms of his work. For instance, so far as I have been able to ascertain, he has admitted only once that the Japanese-American study involved more than two phases, but instead has repeated his original description of the experiment as if it had never been challenged. This strategy has sometimes been accompanied by an acknowledgment that, for some of his empirical claims—for example, those concerning the link between discrete emotions and specific facial expressions or specific patterns of physiological activity—the evidence is incomplete, although this disclaimer is frequently accompanied by the assertion that nevertheless the preponderance of the data supports his position.

2. Ekman mentions criticisms but dismisses them by suggesting that they are trivial or, in Fridlund's case, merely "technical," and by implying that they have been already answered elsewhere—even though those answers, when examined closely, turn out to be no answers at all.[56]

3. Ekman appeals to evidence that is said to favor his position but that has not yet been published.

4. Ekman makes concessions and revisions to his neurocultural theory in ways designed to make it appear that the points his critics have raised have in some sense always been his own, when in fact they completely undermine his position. At the same time, he later quietly returns to his previous position, as if the revisions and concessions he has just made are of no consequence and therefore do not count.

Ekman deployed this last strategy early on in a paper that involved a

56. For Ekman's claim that Fridlund had raised "technical questions" that Ekman has answered elsewhere, see Paul Ekman, "Afterword: Universality of Emotional Expression? A Personal History of the Dispute," in Charles Darwin, *The Expression of the Emotions in Man and Animals,* 3rd ed., introduction and commentaries by Paul Ekman (Oxford, 1988), 389, where Ekman refers the reader to his paper "Expression or Communication about Emotion" (1997), discussed further in n. 58 below. For the claim that the alternative views proposed by Fridlund and Russell are "ideologically, not empirically based," see J. Cacioppo, J. Hager, and P. Ekman, "The Psychology and Neuroanatomy of Facial Expression," in *Final Report to the NSF of the Planning Workshop on Facial Expression Understanding*, eds. P. Ekman, T. S. Huang, T. J. Sejnowski, and J. C. Hager (1993), cited by James Russell, "Facial Expressions of Emotion: What Lies Beyond Minimal Universality?," *Psychological Bulletin* 118 (3) (1995): 389.

startling revision of his entire neurocultural theory. In a statement that clearly alluded to Fridlund's implicit sociality thesis but without mentioning Fridlund by name, in 1997 Ekman recognized that "I expect that some display rules are so well-established that some people may follow them even when they are alone. And some people when alone may imagine the reactions of others, and then follow the appropriate display rule, as if the others were present. And finally, there may be display rules that specify the management of expression not just with others but when alone."[57] (As I have already mentioned in chapter 2, Ekman had made a similar statement even earlier when, in 1982, he had discussed the existence of "miserable smiles," understood by him as governed by display rules, in the solitary condition where display rules were supposed to be inoperative.) In other words, instead of it being the case that when we are alone we are free of the cultural display rules that govern our behavior in public, as he had previously argued, Ekman now stated that display rules might carry over into private or solitary behavior—precisely Fridlund's point. But in making this concession, Ekman dissolved the distinction between universal expressions of emotion and the cultural conventions that moderate them, a distinction on which his entire neurocultural theory of emotion had depended. As Fridlund has recently observed of this moment of slippage, "notably, Ekman did not specify how one might ascertain when such 'solitary display rules' were in effect and when they were not" ("BEV").[58]

And of course, as Fridlund has also observed, if Ekman's 1997 turnabout appeared to solve one problem for his neurocultural theory—by implicitly recognizing the validity of Fridlund's views about the sociality of even so-called "solitary" conditions—it opened up a larger difficulty concerning Ekman's Japanese-American study. As we have seen, in that experiment Ekman and Friesen had contrasted the facial responses of Japanese and American students to watching stress films when they were alone in the

57. Paul Ekman, "Expression or Communication about Emotion," 328.

58. Nevertheless, in an apparent contradiction of what he had just conceded about the operation of display rules in the alone condition, in the same 1997 paper Ekman went on to accuse Fridlund, among others, of failing to grasp the asocial, involuntary nature of expressions of emotion, stating that "we can and do have emotions when we are neither in the presence of others nor imagining that we are," as when we respond to thunder, a sunset, or a tornado, or to other animals, to the loss of physical support, autoerotic activity, and other situations, the implication being that the facial displays occurring in such situations are authentically felt expressions stripped of social conventions (Ekman, "Expression or Communication About Emotion," 334). But he does not say how the differences between authentic expressions and display rule or conventionalized expressions in the solitary condition can be distinguished.

viewing room and when they were in the presence of an authority figure. The authors had claimed that, whereas in the solitary viewing situation both the Japanese and American students had experienced the same negative emotions, in the presence of an authority figure those expressions had diverged. The Japanese students had displayed more polite faces than the Americans, owing to the intervention of the Japanese cultural display rules controlling for polite smiles in the presence of another, display rules that did not apply to the American students. Ekman had offered that experiment as decisive proof of the idea that solitary facial behavior was free of cultural display rules. As he had written: "In private, when no display rules to mask expression were operative, we saw the biologically based, evolved, universal facial expressions of emotion. In a social situation, we had shown how different rules about the management of expression led to culturally different facial expressions."[59]

But with Ekman's 1997 expansion of his neurocultural theory to include *solitary* display rules, Fridlund rightly asked: "Is it now certain that the solitary faces observed in the Japanese-American study were in fact display-rule free and thus 'biologically based, evolved, universal facial expressions of emotion'? If so, how would that be verified?" Indeed, as Fridlund also pointed out, there are wider repercussions to Ekman's concession about the existence of solitary display rules, in that by making "private" behavior potentially conventional or cultural, he in effect placed himself in a position that is virtually indistinguishable from that of his rivals, such as Margaret Mead or Ray Birdwhistell, who emphasized the pervasiveness of cultural learning in all aspects of life ("BEV"). Nevertheless, in a gesture characteristic of what I have called Ekman's fourth strategy for dealing with criticisms, a strategy whereby concessions are quietly followed by retractions, in 2005 Ekman reversed his position by denying Fridlund's implicit sociality thesis all over again, even going so far as to complain that the thesis was "not subject to disconfirmation, and therefore of no use"—a complaint that of course in all logic applied to his own 1997 statement about the operation of cultural display rules in the solitary condition—a statement he had now apparently abandoned.[60]

The same incoherencies, inconsistencies, and slippages are on exhibit

59. P. Ekman, "Expression and the Nature of Emotion," in *Approaches to Emotion*, eds. P. Ekman and L. Scherer (Hillsdale, NJ, 1984), 321; cited in Fridlund, *HFE*, 158.

60. Paul Ekman, "Conclusion: What We Have Learned by Measuring Facial Behavior: Further Comments and Clarifications," in *What the Face Reveals: Basic and Applied Studies of Spontaneous Expression Using the Facial Action Coding System (FACS)*, eds. Paul Ekman and Erika L. Rosenberg (Oxford, 2005), 613.

in an especially revealing fashion in another of Ekman's attempts to reply to Fridlund, one from 1999.[61] We have seen that Fridlund had objected that attributing Japanese–American differences in the presence of the authority figure to display rules would, as the display rule concept stipulated, have required verification that both sets of students experienced equivalent emotions in the third phase of the experiment when the facial movements of the two groups of students diverged. He had wondered why the data from the self-reports obtained at the end of the experiment were not published. In a response to Fridlund, Ekman now rectified his previous descriptions of the experiment by acknowledging for the first time that the study had indeed included, as Fridlund has observed, an interview phase in which both the Japanese and the American students had shown similar facial behavior.[62] Ekman also attempted to explain why the evi-

61. Paul Ekman, "Facial Expressions," in *Handbook of Cognition and Emotion*, eds. T. Dalgleish and M. Power (New York, 1999), 310–20; hereafter abbreviated as "FE."

62. "Rather than regarding the similarity when being interviewed as further evidence of universality," Ekman observed, "Fridlund viewed it as a challenge to our findings of differences in the third condition, when watching the film with the authority figure present. Why did they not show differences in the second condition when being interviewed? Fridlund asked. The answer is straightforward. The differences occur when negative emotions were being aroused—by a film—and managed, masked by smiling. The interview did not elicit sufficiently strong negative emotion, and was not intended to. It is only when they were viewing the very unpleasant films with the authority figure present, that the differences emerged" ("FE," 313).

Further details about this phase emerge in Friesen's dissertation. There, he acknowledged that the smiles in Phase 2 of the experiment were a complicating factor in the findings, and attributed them to the influence of yet another display rule, one that controlled for greetings and initial exchanges between two individuals. Citing Erving Goffman's description of the function of such greeting and exchange rituals in Western culture, rituals that included smiles of recognition and other salutation displays, Friesen extended Goffman's findings to Japanese culture in order to suggest that, because of the presence of the research assistant-interviewer, all the components of this greeting ritual were part of the context for both the Japanese and the American students in Phase 2 and accordingly accounted for their smiling behavior. But Friesen further complicated matters by supplementing this interpretation with an additional explanation for the similarities in the facial displays during Phase 2. According to that supplementary explanation, at least some of the Japanese and American subjects were actually experiencing positive emotion during the episode. The source of the positive affect, Friesen suggested, "could be the relief experienced with the termination of the stressful films . . . This explanation could easily account for at least some of the observed positive displays, and since some of the subjects verbalized their relief, this source of positive affect was almost certainly a part of the findings." He acknowledged, though, that the two interpretations could not be separated satisfactorily by the data presented in his discussion (*CD*, 70). Thus, according to Friesen, the smiling faces in

dence concerning the students' emotional feelings during the experiment was not included in the final reports. He justified the omission on the grounds that the data from the questionnaires about how the students felt would have been contaminated: the same display rules that caused the Japanese students to mask their negative expressions in the presence of an authority figure would have also led them not to report their negative feelings when filling out questionnaires. "[Fridlund] asked why we did not report the data we collected on what the students said after the experiment about how they felt," Ekman wrote. "But these reports should also be influenced by cultural differences. The same display rules which cause the Japanese to mask negative expressions in the presence of an authority figure, would lead them not to report as much negative emotion in questionnaires given to them by that very same authority figure. For that reason we never analyzed those reports" ("FE," 313).[63] Unexplained by Ekman is why he presumed that his subjects' self-reports actually *were* affected by display rules which, as the logic of the entire experiment resting as it does on the neurocultural theory stipulates, should have affected "display" but not emotion, and if so, how he would have verified that they were or were not affected by such rules.

Instead, Ekman stated, he and Friesen had used a "very different strategy": they knew that the films used in the experiment had had the same emotional effect on both sets of students from the prior research of Lazarus and his colleagues, who in their own cross-cultural studies had found the same physiological response to these films among both the Japanese and the Americans. "The films we showed to these subjects we already knew had the same emotional impact, from prior research by Richard Lazarus and his colleagues, which found the same physiological response to these films in Japanese and American subjects. We selected

---

Phase 2 represented *both* the intervention of putative greeting display rules, whose function was by definition to modulate or cover over the students' emotional state, in this case presumably the students' negative experiences from watching the gory films, *and* the presence of authentic positive feelings owing to the relief the students felt at the termination of the stress films. As Friesen indicated, in the absence of independent data about the actual emotional experience of the experimental subjects, there was no principled way to choose between the two interpretations on the basis of the facial displays alone.

63. But then why were the self-reports collected in the first place? Moreover, Ekman must have forgotten that concern about the value of the verbal data had not prevented Friesen from making a brief reference to them when, in his thesis, he appealed to the relief expressed by some of the students in Phase 2 as evidence that some of the students must have experienced positive affect in that episode.

these films precisely because of that fact, because we could be certain that they would arouse the same emotions" ("BE," 313).

But Ekman's reply to Fridlund exposed the incoherence of his position. In two experiments published in 1983 and 1990, respectively (already mentioned in chapter 2), Ekman and his colleagues had asked actors and scientists familiar with Ekman's laboratory protocols to pose faces by following precise coaching instructions as to which muscles to contract. The same actors had also been asked to "re-experience" each of the six emotions by reliving past emotional experiences. During both of these tasks, facial behavior had been videotaped and autonomic responses had been recorded; the claim was that voluntarily contracting the facial muscles into universal emotion signals produced emotion-specific autonomic activity.[64] Of course, if it was true that each specific emotion expression was accompanied by signature patterns of visceral-autonomic response, then Ekman's Basic Emotion Theory would receive valuable reinforcement, which is why experiments of this kind have had an important place in the defense of his position.

In those experiments, Ekman and his group had claimed that *the face leads physiology*, since the different facial poses were said to generate distinct physiologies. According to that logic, then, the Japanese students who displayed different faces from those of the Americans in Phase 3 of the Japanese–American study must also have experienced a *different* physiological response, because facial behavior was said to determine physiology. But, according to the logic of Ekman's display rule concept, the Japanese students actually felt the same negative emotions as the Americans even as they masked their feelings with polite smiles, and must therefore have had the *same* physiological response as them. He could not have it both ways.[65]

Finally, it is worth pointing out that Lazarus himself never claimed that "emotional impact" could be determined by physiology, since he was known for emphasizing the role of appraisal in eliciting affective states and for believing that physiology was just a backdrop or neutral context for the emotion experience. So, Ekman was on weak footing if he wanted to claim by reference to Lazarus's work that the Japanese and the Americans

64. P. Ekman, R. W. Levenson, and W. V. Friesen, "Autonomic Nervous System Activity Distinguishes Among Emotions," *Science* 221 (1983): 1208–10; R. W. Levenson, P. Ekman, and W. V. Friesen, "Voluntary Facial Action Generates Emotion-Specific Autonomic Nervous System Activity," *Psychophysiology* 27 (1990): 363–84.

65. For Fridlund's criticisms of Ekman et al.'s 1983 experiment, focusing on the confounds between sociality and emotion, see *HFE*, 171.

"felt" the same because they showed the same physiology. As a matter of fact, in their studies of the physiological responses of Japanese and American students to stress films, Lazarus and his colleagues had reported one striking difference between the two groups. Unlike their American counterparts, whose levels of autonomic arousal as measured by skin conductance fluctuated in response to the sequence of threatening or less threatening film scenes, the Japanese students not only showed consistently elevated physiological arousal when viewing both the benign and stress films but also exhibited very little variation corresponding to the different scenes. Thus, the Japanese subjects appeared to be in a continuous state of physiological arousal throughout the film viewings, showing an overall pattern of arousal that was not reflected in their self-reports. Lazarus and team hypothesized that the reason for the difference was that, compared to the Americans, the Japanese were not used to experiments of this kind with their requirement of being observed and evaluated, a situation they experienced as highly stressful. In short, the observational aspect of the entire experimental situation was threatening to the Japanese, rather than the specific contents of the stress film.[66]

Ekman's various replies to Fridlund's critique exhibited both the cunning and the weakness of his neurocultural model. Its cunning was its great flexibility and hence its ability to explain virtually any finding. But its weakness was precisely that same flexibility, because it led to arbitrary interpretations of the evidence, to ad hoc labeling of faces as genuine or inauthentic depending on the needs of the experimental findings, and to contradictory arguments in order to "save the phenomena" (or rather, to save the neurocultural theory).[67] It took Fridlund, who knew the theory inside out and had become immune to its seductions, to recognize what was wrong with it. As he observed:

> The two-factor neurocultural model was constructed explicitly to reconcile putative universalist data . . . and cultural-relativist obser-

66. Richard Lazarus, Masatoshi Tomita, Edward Opton Jr., and Masahisa Kodama, "A Cross-Cultural Study of Stress-Reaction Patterns in Japan," *Journal of Personality and Social Psychology* 4 (1966): 622–33; and James Averill, Edward M. Opton Jr., and Richard Lazarus, "Cross-Cultural Studies of Physiological Responses During Stress and Emotion," *International Journal of Psychology* 4 (2) (1969): 83–102.

67. As Fernández-Dols and Crivelli have likewise commented, the concept of display rules was underspecified and could always be invoked post hoc whenever the expected facial expression predictions failed to occur. José-Miguel Fernández-Dols and Carlos Crivelli, "Emotion and Expression: Naturalistic Studies," *Emotion Review* 5 (1) (2013): 25.

> vations, mostly by anthropologists and linguists. It was successful in leading researchers to attend to both genetic and epigenetic features in facial display.
>
> But even a cursory analysis reveals that the model was markedly *over*inclusive. Any facial display, in any circumstance, could be explained by one of the model's provisos. For example, if a facial display was observed to fit one of the proposed prototypes, then it *was* an expression of a primary emotion. If the face was not prototypic but had features of more than one prototype, then that face *was* a blend of the prototypes. If the face was in no way prototypic, then it represented the action of a cultural display rule. If the face was prototypic but did not accord with expectations given the situational elicitors, then the single face might have emerged from an emotion about an emotion, with the face that would normally be linked to the primary emotion blocked by a display rule. These assignations would be no problem if there were independent criteria for determining when the facial action program was activated, when a blend of emotions was being elicited, and when a cultural display rule was operative. Unfortunately, the neurocultural model provided no such criteria. (*HFE*, 274)

Or, as he also remarked: "One might cynically remark that the neurocultural model . . . [had] so many provisos and escape clauses that it was untestable from the start" (*HFE*, 276).

• • •

Much, much more could be said about the confrontation between Fridlund's Behavioral Ecology View and Ekman's Basic Emotion Theory. As further evidence of what was wrong with Ekman's claims about the universality of the "basic emotions" and the role of cultural display rules, Fridlund included in his book a long chapter by the psychologist James Russell offering a superb critique of Ekman's cross-cultural judgment recognition studies. These studies, equally famous in the emotion field, purported to show that facial expressions, as manifested in photographs of posed or spontaneous faces, were universally recognized by literate and illiterate people alike as expressing the basic emotions. But in a masterly assessment, Russell demonstrated that the results were artifactual because they depended on the use of forced-choice labeling formats and other problematic methods that begged the questions to be proved in ways that radically undermined Ekman's claims for the universal nature of the basic emotions. Russell's criticisms thus emphasized the poor link or relationship between internal emotional states and facial displays on which so much of Ekman's position depended. The net result of Fridlund's and Russell's criticisms was to dramatically challenge the theoretical and empirical validity of Ekman's by-now-well-entrenched views.[68]

The stage seemed set for a brand-new orientation in the field. But that was not to be—it has not happened, at least not yet. In spite of these developments, including more recent challenges to Ekman's work that build on Fridlund's and Russell's critiques, since 1994 the Basic Emotion Theory has remained the dominant paradigm in psychology and the affective neurosciences and has continued to inform and inspire the majority of studies in the emotion field. But if one is convinced, as I am, that Fridlund, Russell, and other critics are right and that Ekman is mistaken about the emotions, then the continued success of his paradigm cries out for explanation. That topic is the subject of the next chapter.

68. More recent studies and analyses challenging Ekman's universality thesis include N. L. Nelson and James A. Russell, "Universality Revisited," *Emotion Review* 5 (1) (2013): 8–15; José-Miguel Fernández-Dols and Carlos Crivelli, "Emotion and Expression: Naturalistic Studies"; N. L. Nelson and James A. Russell, "Universality Revisited," ibid., 24–29; C. Crivelli, P. Carrera, and J. M. Fernández-Dols, "Are Smiles a Sign of Happiness? Spontaneous Expressions of Judo Winners," *Evolution and Human Behavior* 36 (2015): 52–58; Carlos Crivelli, Sergio Jarillo, James A. Russell, and J. M. Fernández-Dols, "Reading Emotions from Faces in Two Indigenous Societies," *Journal of Experimental Psychology-General* 145 (7) (2016): 830–45.

[CHAPTER SIX]

THE DEBATE CONTINUES

Paradigm Change or Status Quo?

In this chapter, I plan to assess various responses to Fridlund's Behavioral Ecology View of facial expression with a view to throwing light on the history of research on emotion since 1994, the year his book appeared. This is not to deny that others prior to 1994 had already raised important criticisms of the basic emotions position. For some time social constructionists, such as James Averill and Rom Harré, had been raising questions about the basic emotions concept in terms that resonated with aspects of Fridlund's arguments.[1] The philosopher Robert Solomon had also voiced strong opposition to the prevailing Ekman model by emphasizing the cognitive status of affective life.[2] Moreover, in a well-known paper, Ortony and Turner had questioned the concept of "basic emotions" on the grounds that the belief in such entities might be "an article of faith rather than an empirically or theoretically defensible basis for the conduct of emotion research."[3] As we shall see in the epilogue, in retrospect Fridlund has suggested that the Ortony and Turner paper precipitated a major change in Ekman's position, when he introduced the term "emotion families" in order to widen BET's criteria for the expression of emotion. But the significance of this shift in Ekman's views was not fully appreciated at the time.

I think it is fair to say, though, that none of those authors had Fridlund's potential to upend Ekman's Basic Emotion Theory. They were not

1. Rom Harré, ed., *The Social Construction of Emotions* (Oxford, 1989), with contributions by James Averill and others. See also James R. Averill, "The Future of Social Constructionism: Introduction to a Special Section of *Emotion Review*," *Emotion Review* 4 (3) (2012): 215–20.

2. Solomon's more recent publications include "Emotions, Thoughts and Feelings: What Is a 'Cognitive Theory' of the Emotions and Does It Neglect Affectivity?," in *Philosophy and the Emotions*, ed. Anthony Hatzimoysis (Cambridge, 2003), 1–18; idem, "Emotions, Thoughts, and Feelings: Emotions as Engagements with the World," in *Thinking and Feeling: Contemporary Philosophers on Emotions*, ed. Robert C. Solomon (Oxford, 2004), 76–88.

3. A. Ortony and T. J. Turner, "What's Basic About Basic Emotions?," *Psychological Review* 97 (1990): 315.

internal to the daily routines of the laboratory practices that informed Ekman's approach to the emotions, whereas Fridlund was a well-regarded scientist who, as an expert in the electromyography of the face, had belonged to Ekman's research community for several years before breaking away. His credentials alone made it more likely that his criticisms of Ekman's position would carry weight. Moreover, because Fridlund's claims were at least as grounded in Darwinian theory as were Ekman's, their competing approaches met on the same evolutionary turf, so to speak, making the dispute between them a particularly sensitive and loaded matter, although, as emerged in chapter 5, they occupied different Darwinian positions on a number of crucial issues. Fridlund knew that he had to present a biologically sound alternative to the Basic Emotion Theory, in part because that theory had coasted so long on its claims to represent Darwin. His biologism may have scared off anthropologists and cultural theorists even though, as I shall argue, his arguments actually give them a safe and comfortable harbor. But Fridlund's claim to represent Darwin's thought was perhaps one reason why the pushback from the basic emotion theorists against his intervention was so immediate. Indeed, there was such a strong reaction in Ekman's camp against Fridlund's "apostasy" that within a few years, he more or less withdrew from active involvement in the emotion field, for a long time leaving it mostly to others to sort things out.

The debate precipitated by Fridlund's work has been ongoing ever since. After the appearance of his book, he was joined in his questioning of the validity of the Basic Emotion Theory by a number of scientists. Indeed, as we saw in chapter 5, already in his contribution to Fridlund's book James Russell offered a decisive critique of numerous cross-cultural judgment studies of emotion, an analysis that seriously damaged Ekman's claims as to the universality of facial expressions. Moreover, Russell and his colleagues also demonstrated the powerful effects of context on the perception of facial expression in terms that not only supported Fridlund's line of argument, but according to the latter "broke the paradigm lock BET [Basic Emotion Theory] had on facial-expression research."[4] Especially important in forwarding the idea that something was amiss with Ekman's views was Russell and José-Miguel Fernández-Dols's collection *The Psychology of Facial Expression* (1997), which not only gave a prominent place to Fridlund's Behavioral Ecology View but included a series of wide-

4. A. J. Fridlund, "The Behavioral Ecology View of Facial Displays, 25 Years Later," *Emotion Researcher*, ed. Andrea Scarantino, http://emotionresearcher.com/, accessed January 6, 2016; hereafter abbreviated as "BEV."

ranging articles that opened up Ekman's position to critical discussion.[5] Indeed, in that book and elsewhere, Fernández-Dols and others have powerfully reinforced Fridlund's claims about the sociality of emotional expressions by showing that, contrary to Ekman's account, so-called Duchenne smiles are at least as affected by sociality as are non-Duchenne ones, and that such smiles can be produced deliberately.[6] More generally, since 1994 there has been a steady flow of increasingly influential research and literature-review articles in a variety of books and journals by Russell, his former student Lisa Feldman Barrett, psychologists Brian Parkinson and Rainer Reisenzein, and others who have raised significant objections to the Ekman paradigm of the emotions by suggesting that the empirical evidence concerning the correlations between emotions and facial expressions claimed by Ekman and his school do not stand up to scrutiny.[7]

5. James A. Russell and José-Miguel Fernández-Dols, *The Psychology of Facial Expression* (Cambridge, 1997).

6. J. M. Fernández-Dols and M. A. Ruiz Belda, "Are Smiles a Sign of Happiness? Gold Medal Winners at the Olympic Games," *Journal of Personality and Social Psychology* 69 (1995): 1113–19; M. A. Ruiz-Belda, J. M. Fernández-Dols, P. Cerrera, and K. Bachard, "Spontaneous Facial Expressions of Happy Bowlers and Soccer Fans," *Cognition and Emotion* 17 (2003): 315–26; M. Mehu, K. Grammer, and R. I. M. Dunbar, "Smiles When Sharing," *Evolution and Human Behavior* 28 (2007): 415–22; C. Crivelli, P. Cerrera, and J. M. Fernández-Dols, "Are Smiles a Sign of Happiness? Spontaneous Expressions of Judo Winners," *Evolution and Human Behavior* 36 (2015): 52–58.

7. In addition to publications cited elsewhere in this chapter, the following sampling from a large literature offers an overview of the new ways of viewing the emotions: Brian Parkinson, "Do Facial Movements Express Emotions or Communicate Motives?," *Personality and Social Psychology Review* 9 (4) (2005): 278–311; Lisa Feldman Barrett, "Are Emotions Natural Kinds?," *Perspectives on Psychological Science* 1 (2006): 28–58; Feldman Barrett, Kevin N. Ochsner, and James J. Gross, "On the Automaticity of Emotion," in *Social Psychology and the Unconscious: The Automaticity of Higher Mental Processes*, ed. John A. Bargh (New York and Hove, East Sussex, 2007), 173–207; Lisa Feldman Barrett, Batja Mesquita, Kevin N. Ochsner, and James J. Gross, "The Experience of Emotion," *Annual Review of Psychology* 58 (2007): 373–403; *The Mind in Context*, eds. B. Mesquita, Lisa Feldman Barrett, and E. Smith (New York, 2010); *Categorical Versus Dimensional Models of Affect: A Seminar on the Theories of Panksepp and Russell*, eds. P. Zachar and R. Ellis (Amsterdam, 2012); José-Miguel Fernández-Dols, "Advances in the Study of Facial Expression: An Introduction to the Special Section," *Emotion Review* 5 (1) (2013): 3–7; A. Nelson and James A. Russell, "Universality Revisited," in ibid., 8–15; Rainer Reisenzein, Markus Studmann, and Gernot Hortsmann, "Coherence Between Emotion and Facial Expression: Evidence from Laboratory Experiments," in ibid., 16–23; José-Miguel Fernández-Dols and Carlos Crivelli, "Emotion and Expression: Naturalistic Studies," in ibid., 24–29; Brian Parkinson, "Contextualizing Facial Activity," in ibid., 97–103; *The Psychological Construction of Emotion*, eds. Lisa Feldman Barrett and James A. Russell (New York and London, 2015); Brian Parkinson

Predictably, though, Ekman and his colleagues have continued to vigorously defend the correctness of his position—in Ekman's case, as we have seen, sometimes in disconcertingly slippery ways. It is not surprising that many scientists trained in Ekman's experimental methods and theoretical presuppositions have likewise stuck with his research program, even if important critiques have had to be ignored and cracks in the edifice papered over. Ekman's methodologies and assumptions fit so well with the reigning paradigms in experimental psychology and with modern imaging technologies, such as fMRI, that they have proven hard to relinquish. Nevertheless, it is striking how few advocates of Ekman's Basic Emotion Theory appear to have taken the trouble to read the original descriptions of his Japanese–American study or Fridlund's critical reconstruction of it. It is as if they are so strongly in the grip of Ekman's picture of the emotions that they have been unwilling to evaluate its merits dispassionately. Also puzzling is why certain scholars in the humanities and social sciences who are swept up in today's general "turn" to affect should likewise be attracted to Ekman's ideas. One answer is that, for reasons to be discussed in chapter 7, there is at the present moment a strong trend toward theorizing affect in the same noncognitive and non-intentional terms as in Ekman's model, which is to say that there exists a natural affinity between the ideas of the new affect theorists and Ekman's emotion views.

Especially interesting in this regard are certain well-informed commentators who, in spite of the fact that they have read and in a certain sense appreciate Fridlund's critique, nevertheless continue to seek ways to accommodate Ekman's views. In particular, the philosopher of science Paul Griffiths, whose arguments I briefly discussed in the introduction, and his former student Andrea Scarantino have offered sophisticated defenses of the basic emotions program by revising it in ways intended to make it less vulnerable to criticisms of the kind launched by Fridlund and others.

But even some psychologists sympathetic to Fridlund's work have found it difficult to accept the implications of his approach. So, for example, in recent years the "Psychological Constructionist" movement launched by Russell and Feldman Barrett has played an important role in setting the terms of the debate over emotions, specifically over the existence of discrete emotion categories. But, as I will suggest, there is an important sense in which their efforts to come up with new approaches to emotion have been less than fully successful. This is because, in spite of their commitment to cognitivism and intentionalism in emotion sci-

---

and A. S. R. Manstead, "Emotion Research in Social Psychology: Thinking About Emotions and Other People," *Emotion Review* 7 (2015): 371–80.

ence, in attempting to propose alternatives to the Basic Emotion Theory, Russell, Feldman Barrett, and their colleagues have ended up reproducing the very dichotomy between biology and culture, or affect and cognition, that they claimed to be attacking in rejecting Ekman's program. This is a fascinating development, one that reveals the difficulty of handling issues of cognition and intentionality in psychology, a topic that all along has been a central concern of this book.

I will have more to say about this aspect of the recent history of research on emotion as I proceed. But it needs to be stated at the outset that it has been all too easy for scientists, even those generally supportive of Fridlund's work, to repeat the very conceptual mistakes that, as he had tried to show, have been inherent in Ekman's picture of the emotions from the first. There is therefore a sense in which the new approaches to the emotions have ended up tending to displace from view the radical implications of Fridlund's proposals. The question that then arises is whether such new approaches constitute a progressive research program, as advocates believe, or whether in the end they amount to business as usual, with the result that the emotion field remains to this day in a state of virtual stasis.

In order to set the stage for my discussion of these issues, I want to describe certain features of Fridlund's approach that were implicitly in play in the previous chapter but that I have not yet made explicit, features that raise a number of important questions about how to conceptualize and study the emotions. The features of Fridlund's approach that I will single out for discussion I consider strengths, but they also help explain why so many researchers have remained resistant to his ideas and findings. Fridlund has recently described his position as a "tough sell" ("BEV"). This chapter is in part an attempt to say why that has been the case.

SOME GENERAL FEATURES OF FRIDLUND'S BEHAVIORAL ECOLOGY VIEW

Among the general features of Fridlund's position that I wish to highlight are:

1. *Fridlund's "Behaviorism" and the Question of Intentionality.* Fridlund labels his position a "Behavioral Ecology View," so the question arises as to what is implied by this name. Fridlund's behavioral orientation does not connote a commitment to the behaviorist movement associated with the work of John Broadus Watson or B. F. Skinner, a movement that attempted to explain the behavior of persons and nonhuman animals without appeal to psychic or mental states. Rather, Fridlund's stance entails adopting a "functional" or "externalist" approach to psychology, one that emphasizes the idea that the organism's mindedness shows itself to others di-

rectly, in publicly observable actions and performances.[8] He thus views intentionality as a form of behavior characterized by the "aboutness" of the actions involved (*HFE*, 61, n. 1), and he does so in terms that gesture toward the philosopher Daniel Dennett's "intentional stance" (*HFE*, 134) on the one hand and the philosopher Gilbert Ryle's "neo-behaviorist" or Wittgensteinian views on the other (*HFE*, 186), while prudently leaving to others the task of trying to solve the formidable philosophical problem of the nature of intentionality as such.

"The issue of intentionality is a thorny one," Fridlund has observed. "These issues have been explored by Dennett (1987), and [the ethologists] Cheney and Seyfarth (1990) paid them special attention in the context of audience effects. In their discussion they question just how much vervets and other animals 'know' about what they are doing, just what they 'mean' when they display, and whether receivers and audiences 'know' the intended effects of displays. I have circumvented these 'levels of intentionality' issues in the interests of space, and use intentionality in a purely functionalist sense, that is, one discerns the intent from the effect of displays, and from the efforts of the displayers to obtain those effects" (*HFE*, 146). As he has stated more recently, many of the classic expressions of the Basic Emotion Theory

> can be recast in functional terms. In suitable contexts, so-called "happy faces" solicit affiliation or play, whereas those for "sadness" recruit succor, "anger" faces threaten or deter, "fear" faces predict submission or withdrawal, "disgust" faces indicate rejection or intent to spew, and so on. When, in counterpart to classical cross-cultural matching studies, participants are asked to match iconic . . . expressions to functional redescriptions, such redescriptions (e.g., "back off or I'll attack") achieve matching rates equal to emotion terms . . . Unlike emotion terms, however, such functional descriptions imply no particular internal state, nor any moral assignations

8. A. J. Fridlund, *Human Facial Expression: An Evolutionary View* (San Diego, CA, 1994); hereafter abbreviated as *HFE*. Fridlund's functionalism is not a version of the philosophical functionalism that informs the cognitive sciences and adopts a view of the mind as made up of internal or mental representations. For a useful contrast between that kind of functionalism and what the philosopher Descombes calls "holistic" functionalism, in which the psychic life of a system is "not the reflection of its internal construction but of the problems posed by the environment in which it lives"—a functionalism compatible with Fridlund's approach—see Vincent Descombes, *The Mind's Provisions: A Critique of Cognitivism*, trans. Stephen Adam Schwartz (Princeton and Oxford, 2001), 141.

> about which signals are "honest" or "genuine." They are predicated merely on the view that facial displays are only probabilistic signals of social intentions that would, in everyday life, be accompanied by words, vocal prosidies, and gestures congruent with intent. ("BEV")

By adopting such a behavioral-ecological approach to the emotions and to the question of intentionality, Fridlund has been able to avoid certain important theoretical-conceptual pitfalls:

a) By pursuing a behavioral ecology view of facial expression, Fridlund rejects the idea of the emotions as private internal states. We might put it that he has attempted to "undo the psychologizing of psychology" in ways that can be seen to be compatible with Wittgensteinian arguments of the kind advocated by the philosopher Anthony Kenny against the idea of a private language. (The phrase in quotation marks is Stanley Cavell's, meant to characterize Wittgenstein's project in the *Philosophical Investigations*.) The critique of private mental states does not disqualify us from appealing to our sense of having an inner life: metaphors of inside and outside are irresistible; they cannot be legislated away and must therefore represent something significant. But it turns out that it is very difficult to operationalize those metaphors for scientific purposes without raising hard problems about the status of emotional and other mental states, and it is those problems Fridlund's behavioral ecology approach avoids.[9] Indeed, there is an interesting sense in which his position converges with recent Wittgensteinian–Cavellian attempts to theorize the intentionality of the various "forms of animal life" in ecological-ethological terms.[10]

9. Bernie Rhie has recently suggested that Fridlund fails to treat the meaning of facial expressions as immediate and direct but regards them as readouts of internal mental states, and hence as a matter of inference and interpretation. But this seems to me a misunderstanding of Fridlund's project. See Bernie Rhie, "Wittgenstein on the Face of a Work of Art," *nonsite.org*, no. 3 (2011).

10. See Jason Bridges, "The Ecology of Reasons," unpublished paper (my thanks to the author for allowing me to read this text). Fridlund (*HFE*, 61) also cites without comment two papers on intentionality in animal signaling: Marc Bekoff and Colin Allen, "Intentional Icons: Towards an Evolutionary Cognitive Ethology," *Ethology* 91 (1992): 1–16; and M. D. Hauser and D. A. Nelson, "'Intentional' Signaling in Animal Communication," *Trends in Ecology and Evolution* 6 (1991): 186–89. Without fully endorsing Ruth Millikan's teleosemantic attempt to naturalize intentionality, Bekoff and Allen explore the value of Millikan's views concerning the evolution of "proper functions" for framing empirical research on intentional behavior in ethology, especially the study of play behavior. But see Jason Bridges's valuable critique of teleosemantics in "Does Informational Semantics Commit Euthyphro's Fallacy?," *Nous* 60 (2006): 522–47; and idem, "Teleofunctionalism and Psychological Explanation," *Pacific Philosophical Quarterly* 87 (2006): 403–21.

b) By assuming that mentation or psychic life shows or reveals itself directly, in publicly observable facial actions and other behaviors, Fridlund takes the existence of intentionality as a basic fact that is not reducible to non-intentional concepts. He therefore avoids positing the existence of internal mental mechanisms of the kind assumed by nearly all cognitive psychologists, especially those attracted to computational models of the mind. As a result, he has circumvented difficult questions about the validity of those computational assumptions and about the empirical-causal status of intentionality, questions that, as has been seen in previous chapters, haunted the work of Richard S. Lazarus and others. Fridlund's is an externalist, functional view of mentation, a view that he has described as "Ryle, applied to faces."[11]

c) By concentrating on the behavior of intact nonhuman and human animals in their natural-social settings and by denying the existence of "basic emotions" or discrete "affect programs" in the mind-brain, Fridlund has dodged the fallacies and reductionisms so often involved when researchers attempt to locate emotions in the "head." He has also fended off the related tendency of emotion researchers to attribute functions and properties to parts of the brain that properly belong to the whole person or organism. This of course puts his position at odds with all those affective neuroscientists today who, making use of brain imaging technologies such as PET-scanning and fMRI, believe it is possible to locate specific emotions in certain parts of the central nervous system by using Ekman's photographs of posed expressions or other stimuli to establish correlations between the perception of external events and particular anatomical-cerebral sites. The fact that, as recently documented by Feldman Barrett and several other researchers, strong evidence in favor of such putative correlations between specific emotions and specific brain localizations has failed to materialize, suggests that Fridlund has been correct in declining to take that route.[12]

11. Fridlund, personal communication, January 19, 2016.

12. See Lisa Feldman Barrett et al., "Of Mice and Men: Natural Kinds of Emotions in the Mammalian Brain? A Response to Panksepp and Izard," *Perspectives on Psychological Science* 2 (2007): 297–312; Kristen A. Lindquist, Tor D. Wager, Eliza Bliss-Moreau, and Lisa Feldman Barrett, "The Brain Basis of Emotion: A Meta-Analytic Review," *Behavioral and Brain Sciences* 35 (3) (2012): 121–43. See also S. A. Guillory and K. A. Bujarski, "Exploring Emotions Using Invasive Methods: Review of 60 Years of Human Intracranial Electrophysiology," *Social Cognitive Affective Neuroscience*, Epub: March 2014. In her writings, Barrett has offered a valuable critique of the widespread mistake in emotion research of relying on a concept of the hierarchical organization of brain functions, such as Paul MacLean's "triune" brain concept, according to which in hu-

Moreover, Fridlund has explicitly rejected the idea that in humans the basic emotions are located in a phylogenetically "primitive" or "proto-reptilian" part of the brain that, according to the now somewhat discredited "triune" or "limbic" theory of Paul MacLean, controls the older and more fundamental affective processes—processes that can be partly overridden or modulated by the newer, neo-mammalian brain controlling for intellectual-cognitive or cultural "display rule" processes. This is a hierarchical view of brain function that is still held in a revised form by basic emotion theorists and their allies, such as LeDoux, Craig DeLancey, and others. It is a picture that permits the continuation of what Fridlund has labeled the "two-factor" emotion model positing two kinds of faces: the innate reflex-like faces that express or read out ongoing emotional states controlled by subcortical sites in the brain, and the learned, instrumental faces that connote the everyday modulation and adulteration of facial expressions owing to the requirements of convention and deception and that are controlled by neocortical centers.

But, as Fridlund remarked in 1994, the human brainstem is "not reptilian, it is distinctly human, as is the human 'limbic system'; all parts of the human brain, like all extant animals, are equally evolved" (*HFE*, 110). As he also stated: "All parts of our brains are equally evolved, and act in coordinated fashion to promote our survival amid the social matrix into which we are born" (*HFE*, 294). Feldman Barrett has recently made the same point: "There have been changes to the human brain (particularly the expansion of the neocortex and its reciprocal connections to affect-related subcortical areas) . . . that have been shaped by evolutionary pressures related to social functioning and group size. These changes make strict homologies between humans and other mammals unlikely."[13] The implication is that researchers such as Griffiths and Scarantino are wrong to assume that, in spite of observable differences between humans and nonhuman animals, such as the human acquisition of language, those differences do not matter when it comes to the basic emotions because of the "deep" homology between their respective brains. This is a topic to which I will return.

d) A related implication of Fridlund's approach is that it is an error to believe that the concepts or cognitions involved in emotional behavior are

---

mans the neocortex sits atop a more ancient and primitive subcortex. This is an assumption that leads to the idea that the phylogenetically older subcortical structures, such as the amygdala, subserve the "lower" or more primitive emotional functions independently of higher cortical processing.

13. Lisa Feldman Barrett, "Are Emotions Natural Kinds?," 46.

added or tacked on to primitive biological-affective processes that themselves lack such concepts and cognitions. His rejection of a dualism that opposes the raw, biological "givens" of non-intentional affect programs to cultural conventions taking the form of "display rules" or learning means that he avoids having to explain how that add-on occurs—a difficult problem, to say the least. Fridlund's position is compatible with the "embodied world-taking cognitivism" advocated by Hutchinson and some others today. This is a cognitivism that, as I observed in the introduction, accepts as valid the philosopher John McDowell's neo- or post-Kantian critique of the idea of nonconceptual perception or experience and maintains that emotions are conceptual through and through. In short, Fridlund's Behavioral Ecology View of the emotions is at odds with any theory, of the kind so widely accepted today in cognitive science, that attributes to the "basic emotions" some sort of nonconceptual "mental" content to which conceptuality or meaning is subsequently tacked on in order to explain the so-called "higher" emotions.

Of course, as we shall see in a moment in the case of Griffiths and Scarantino, those who defend a theory of the basic emotions as containing no conceptual content believe they have the philosophical resources to meet objections to their position. But it is important to realize that there are powerful arguments against their views, and that the question of how to theorize cognition and conceptuality, which is to say intentionality, is one of the fundamental issues at stake in responses to Fridlund's behavioral ecology approach to the question of emotional behavior.

2. *What Is "Emotion"?* Another feature of Fridlund's Behavioral Ecology View is that it puts pressure on the value of the very concept of "emotion" as deployed by Ekman and his school and indeed by the majority of emotion researchers. This is because in Fridlund's opinion "emotion" has become such a nebulous construct as to defy anyone's ability to pin it down. There is a superficial sense in which Fridlund shares this estimation with Griffiths and Scarantino, who have recently argued that the everyday or "Folk Psychology" term "emotion" contains such a heterogeneous mixture of meanings that it has become useless for the purposes of scientific research. But Fridlund takes his concerns in an entirely different direction from those authors.

As we shall see, Griffiths and Scarantino argue that what needs to be done is to separate a properly scientific use of the term "emotion" from its vernacular or everyday deployment in order to promote a study of the basic emotions along Ekman-inspired or -related lines. What worries Fridlund, however, is not the everyday use of emotion terms but precisely the problems involved when "emotion" is used according to the demands of Ekman's Basic Emotion Theory or related scientific projects. In this context, Fridlund is skeptical about the term "emotion" because it has become so ineffable as to make dealing with it a form of shadowboxing. Indeed, one could argue that for him, uncertainty about what the word "emotion" connotes makes the idea that a behavior or a face provides "non-emotional information" equally vague.

Fridlund's Behavioral Ecology View is designed to avoid reliance on such tenuous constructs. As he has commented, proponents of the Basic Emotion Theory "provide no agreed-upon way to determine whether one has an emotion or is being emotional" (*HFE*, 134, n. 7). And: "I suggest that facial displays can be understood without recourse to emotion or emotion terms . . . [W]hat cannot be done is to show that emotions have no role in facial displays, because excluding emotion would require a definition that allows it. At present, arguing against 'emotion' in any form is shadow boxing" (*HFE*, 186). Fridlund's point is that without criteria determining what it includes and excludes, the term "emotion" lacks a clear meaning. As he stated in a response to Buck's objection that he had not defined emotion: "*Human Facial Expression* is agnostic about the existence of emotion (how can I deny what cannot be defined?). My claim is utilitarian; our expressions can be understood without recourse to causation by ineffable entity. It leaves to hermeneutics the shell game of what is and what is not 'emotion.'"[14]

One can see why Fridlund says these sorts of things. The uncertainty as to whether "emotion" represents a separate and distinguishable condition, combined with a failure to specify criteria for the presence or absence of "emotion" independent of facial movements, means that the distinction between emotion and display rules or culture, central to the basic emotion model, is so unstable that it accommodates arbitrary interpretations of the data as needed. We have already seen how Ekman has exploited that instability to push back against Fridlund's criticisms, to the point of going so far as to concede the operation of display rules in the "alone" condition—a move that, without his acknowledging the point, undoes the very opposition between affect programs and conventional display rules on which his Basic Emotion Theory depends. Fridlund has forcefully expressed his concerns about the elasticity of the term "emotion" in a 1997 summary of his position that is worth quoting at length:

14. Alan J. Fridlund, "Reply to Buck's Review of *Human Facial Expression: An Evolutionary View*," *Communication Theory* 5 (4) (1995): 398. Fridlund was replying to Ross Buck's review of *Human Facial Expression*, in ibid., 393–96.

Throughout this chapter, I use the word *emotion* only as a prop, because proponents of the Emotions View all use the term but mean different things by it. This is evident in the contributions to this volume by Izard and by Frijda and Tcherkassof. Despite their excellent defenses of two variations on the Emotions View, both illustrate the difficulties inherent in trying to retain "emotion" as an explanation for human facial expression. The pivotal question to ask of each is, "When should facial expressions occur and when should they not?"

For Izard . . . "facial expression . . . is conceived as an evolved, genetically influenced but highly modifiable and dissociable component of emotion," but he then states that "observable expression is not viewed as a necessary component of emotion." Thus, emotion can include facial expression, but it can occur without an expression, and furthermore, expression can occur without emotion. What are the conditions of linkage and dissociation that would produce each? Izard is unclear. He appends the term *motivational* to *emotional* to describe states that might elicit facial expressions, while also suggesting that emotion can occur without occasioning motives, after having stated earlier that "emotions motivate and organize perception, cognition, and action." When, then, should a facial expression occur: When emotion motivates, when it does not, or doesn't it matter? This I cannot glean from Izard's chapter.

Fridlund continues:

The chapter by Frijda and Tcherkassof contains a corresponding set of semantic tangles. After a quite literate introduction to the problems faced by a traditional Emotions View of facial expression, the authors conclude that facial expressions are not adequately described in terms of "emotion." Instead, facial expressions usually point to a motivational state . . . which we call the individual's *states of action readiness*. Facial expressions express states of action readiness. After these declarations, do the authors forsake emotion? Far from it, because emotions are states of action readiness. In fact, they constitute the "core" of emotions. Here this theory parallels Izard's on emotion and motivation. Emotion can engender action readiness, but action readiness is only one of the moderately correlated components of emotion, and can occur without emotion. On the question of when a facial expression occurs, Frijda and Tcherkassof logically concluded, where Izard demurred, that facial expressions of emotion occur if and only if the emotion is accompanied by a state of action readiness. In taking this step, the authors remove emotion from the direct causal chain, since action readiness is only one moderately correlated component of emotion, and since many factors other than emotion can occasion action readiness.

What is also notable from the two chapters—and is frankly pandemic among expositors of the Emotions View—is that the differences in how they use the term *emotion* indicate that they are talking about two different constructs. Neither offers a proper technical definition of emotion, one which would stipulate criteria—independent of facial expression—for when an emotion is occurring (an inclusion criterion) and when it is not (an exclusion criterion). How can we entertain a theory whose critical term cannot be defined concretely and consensually? (his emphasis)[15]

In this context, it is not surprising to find Fridlund complaining that "the logical question is what *isn't* emotion" (*HFE*, 185). Nor is it strange to find him arguing that because emotion terms might be "reified" social motives to begin with, attempts to dissect them in order to segregate out the associated displays are a dead end, on both conceptual and utilitarian grounds. In this connection, he has remarked:

> Conceptually, hermeneutical dissection of emotions will inevitably lead to the level of explanation by social motive. This is because emotion terms, when dissected sufficiently, usually become statements of motives and context. For example, "happiness," an emotion term without obvious motives and context, includes "amusement," "serenity," and "relief," to pick a few. Yet the definitions of each of the putative subtypes denote motives, roles, or context: "amusement" implies affiliation and play behavior with another, "serenity" implies repose and preference for solace, and "relief" implies termination of an aversive interaction.
>
> On utilitarian grounds, the intentional analysis of displays will lead to a display taxonomy that predicts (and would be validated by) classes of observable behavior, rather than depending on progressively finer distinctions among emotions, distinctions that tend to become more dubious phenomenologically. (*HFE*, 279–80)

Nevertheless, Fridlund's agnosticism regarding the term "emotion" has been difficult to accept for many scientists both within and outside

15. Alan J. Fridlund, "The New Ethology of Human Facial Expressions," in James A. Russell and José-Miguel Fernández-Dols, *The Psychology of Facial Expression* (1997), 124–25.

the Ekman camp, who feel that they cannot do without it. For one thing, dispensing with the term seems to go against the grain of phenomenological experience, according to which we feel our emotions welling up inside us and prompting us to act unless we are able to control them by inhibiting or modulating them in some way. Parkinson has recently expressed that intuition about the phenomenology of affect when, in a review comparing and contrasting Ekman's and Fridlund's positions, he comments:

> In Fridlund's (1994) view, social convention does not lead to the suppression of authentic emotion expressions. Instead, conflicting facial control processes are at an equivalent level, both reflecting social motives. On the one hand, we want to communicate an intention to someone else; on the other hand and for other reasons, we do not want to communicate it (or there are two conflicting intentions, both of which we want to communicate). However, this provides no direct explanation of the phenomenal experience that the first impulse (to express something) is at a more basic level . . . Our sense is often that the expressive impulse comes from deep within, whereas expressive control is imposed on this impulse from elsewhere.

Parkinson acknowledges that this subjective perception may be false in the sense that its origins may lie elsewhere than in the automatic welling up of emotions from deep inside us, against our will and control, but he nevertheless seems to endorse the apparent implications of that perception:

> Of course, this subjective perception may represent a form of false consciousness arising from our implicit socialization into romantic ideologies. Alternatively, the felt priority of one impulse may simply reflect its earlier origins in infancy when desires were unregulated. A third possibility is that our deeply felt urges may reflect closer identification with the relevant identity positions or roles rather than any natural origins . . . However, any such implication that one set of motives comes from a more personally felt place takes us closer to an acknowledgement that these motives are more emotional. Any distinction between motives based on their priority or proximity to the self implies a two-factor analysis (not unlike Ekman's) in which expressive urges are controlled by culturally determined regulation strategies.[16]

16. Parkinson, "Do Facial Movements Express Emotions or Communicate Motives?," 292.

As Parkinson notes, this is the sort of thinking that animates Ekman's model, which is why in response to Fridlund's criticisms of the Basic Emotion Theory, Ekman's coauthor and colleague Erika Rosenberg has defended Ekman by retorting that "explicit in Fridlund's [argument] is the notion that the face has nothing to do with emotion—that it is not a[n] emotional response system per se and it has no meaningful relationship to the experience of emotion."[17] In fact, as we have seen, Fridlund remains noncommittal on the emotion-facial movement or "expression" link precisely because in the research on offer, "emotion" remains undefined.

But even researchers sympathetic to Fridlund's approach seem to find his skepticism as to the usefulness to science of the emotion concept hard to accept. Indeed, some researchers appear to succumb to the lure of the concept at the very moment when they are trying to defend Fridlund's thinking on this issue. In a recent paper reviewing research findings on the relation between "emotion" and facial expression, Russell, Bachorowski, and Fernández-Dols start out by observing that the claim that emotional expressions express emotion "is a tautology but may not be a fact." In line with Fridlund's arguments, they then proceed to document the ways in which modern evolutionary thought has cast doubt on the traditional idea that the function of facial expressions and other nonverbal signals of the basic emotions is to cooperatively or altruistically transmit veridical or "honest" information about the sender's internal state that the receiver then automatically decodes. Instead, as the authors report, researchers now separate the study of the sender's signaling intentions from those of the receiver, on the grounds that the sender's and receiver's interests may conflict. They observe: "As the interests of sender and receiver only sometimes coincide, it is not always in the sender's interest to provide veridical information. EEs [Emotional Expressions] are thus as capable of being deceptive as honest."[18] As a result, and as the authors document, the quasi-reflexive or readout theory of facial expression, according to which there is a tight correlation between the hypothesized internal discrete emotions and characteristic facial expressions, has given way to a new appreciation of the looseness of fit between facial movements or signals and any hypothesized internal emotional states.

17. Erika L. Rosenberg, "Emotions as Unified Responses," in *What the Face Reveals: Basic and Applied Studies of Spontaneous Expression Using the Facial Action Coding System (FACS),* 2nd ed., eds. Paul Ekman and Erika L. Rosenberg (New York, 2005), 88.

18. James A. Russell, Jo-Anne Bachorowski, and José-Miguel Fernández-Dols, "Facial and Vocal Expressions of Emotion," *Annual Review of Psychology* 54 (2003): 331; hereafter abbreviated as "FV."

Nevertheless, toward the end of the same paper, Russell, Bachorowski, and Fernández-Dols refer to Owren and Bachorowski's 2001 speculations about the evolution of the smile in terms, suggesting that for the latter authors the hold of the "emotion" concept has not been broken. Thus, in their article Owren and Bachorowski had suggested that smiling involved two different but related systems: a phylogenetically older and simpler reflex-like system that produces "spontaneous" smiles as reliable signals of positive feelings toward a specific receiver; and a more recently evolved version of the first system in which volitional smiles are produced in a more controlled process. "In contrast to spontaneous smiles," Russell, Bachorowski, and Fernández-Dols paraphrase the argument, "volitional smiles are emancipated from affect in that they can occur during the experience of any affective state. Sometimes thought of being 'deceptive' or 'dishonest,' the power of volitional smiles lies in their inherent unreliability as a cue to the sender's state." To which they add:

> Fridlund's (1994) evolutionary account places a similar emphasis on the smile being directed at a receiver but substitutes "friendly intentions" for '"positive feelings." Although not denying that emotions and feelings exist or are correlated with EEs, Fridlund argued that the most coherent causal story can be told in more behavior-relevant or functional terms.[19] Fridlund applied the same analysis to other EEs as well, centered on other social intentions, including aggression, appeasement, and help-seeking. Perhaps because this account strays from the traditional assumptions associated with EEs and maintained in previous accounts, it has been frequently misunderstood. For example, "intentional" should not be taken to mean conscious state, but simply involving a behavioral disposition aimed at a specific receiver. Fridlund's account does not require a simple correlation between the amount of signaling and the degree of sociality of the situation. Nor does Fridlund's account deny that EEs can occur when the sender is alone. Indeed, he offered evidence that EEs produced when alone are directed at an imaginary, implicit, or animistic audience. Like Owren and Bachorowski's (2001), Fridlund's account suggests the power of modern evolutionary theory to overturn long-held assumptions and open the door to fresh perspectives on EEs. ("FV," 342)

19. In fact, Fridlund does not offer a causal theory of the production of facial expressions, but a descriptive or functional account. See also my epilogue for some further comments on this point.

What is striking about this statement is that while it defends Fridlund's behavioral ecology or communicative theory of animal behavior, including facial behavior, as an approach that does not have to appeal to an ineffable concept of "emotion," it nevertheless seems to imply that there is little to choose between Fridlund's position and that of Owren and Bachorowski, because they are more or less equivalent. Yet this is not the case. For, in their article, Owren and Bachorowski employ various hypotheses about the role of "handicaps" or costs to signalers and receivers in the evolution of the smile in order to reinstate an Ekman-like dualism between emotionally determined, reflex-like, reliable happy smiles and culturally determined, voluntarily produced "deceptive" or dissimulative smiles, a dualism that has been the target of Fridlund's criticism all along. By appearing to remain neutral regarding the choice between Fridlund's account of facial movements and Owren and Bachorowski's different approach, Russell, Bachorowski, and Fernández-Dols seem to believe, or proceed as if they believe, that Fridlund's "friendly intentions" can be substituted for Owren and Bachorowski's non-intentional "positive feelings" without a fundamental change of meaning or loss of intellectual clarity.

It is worth noting in this regard that if one asks what motivates theorists such as Owren and Bachorowski who, in spite of damaging criticisms of Ekman's position, nevertheless continue to reproduce its basic premises, it seems plausible that the answer lies in their concern about deception. In other words, it seems reasonable to suggest that it is because they wish to secure facial signaling from the threat of dishonesty and lying that they continue to support a readout concept of internal emotional states and hence implicitly to support Ekman's questionable views. However, Fridlund would presumably respond to Owren and Bachorowski's arguments by reminding them that on behavioral-ecological grounds, their belief that an "emotion-dependent" signaling system guarantees the reliability and authenticity of certain displays, such as the genuine smile, simply postpones the ecological problem, since the next evolutionary step would be the introduction of "deception" by the signaler's mimicking of authenticity cues (see appendix 1 for a further comment on these issues).[20]

20. For an attempt by a philosopher to reconcile the dispute between Fridlund and Ekman by positing a continuum between emotions defined as involuntary, hard-to-fake readouts of internal states and emotions as strategic signals, see Mitchell S. Green, *Self-Expression* (Oxford, 2007). For a somewhat similar effort to reconcile Ekman and Fridlund's views by a specialist in primate behavior, see also Marc D. Hauser, *The Evolution of Communication* (Cambridge, MA, and London, 1998), 490–96.

This last point brings me to a further issue worth emphasizing, which concerns Fridlund's explanation of why the readout theory of internal emotional states has proven so attractive over the years. His suggestion is that such a readout view appeals to many researchers because it is a legacy of Rousseauean romanticism, according to which faces that are held to be the product of automatically generated internal emotional states are thought to signify authenticity, whereas rule-governed social faces denote "the inevitable loss of innocence forced by society" (*HFE*, 293). No doubt there is something to this suggestion, which also accounts for the pull of a series of related oppositions—between "private" and "public" faces, nature and culture, the endogenous and the exogenous, and the authentic self and the dissimulating self—that are inherent in Ekman's neurocultural approach and that Fridlund rejects.

In addition, like the social constructionists with whose views his own are sometimes compared, Fridlund has made the suggestion that by regarding our own emotions as ineluctable eruptions of our inner passions, we exonerate ourselves from responsibility for our intentions and actions.[21] "Instead of calling someone who has hurt us a 'worthless piece of excrement,'" he has proposed, "we can say . . . 'I can't help but have these angry feelings toward you.' This lets us both off the hook. Our emotions just happened to us . . . And if *we* do something we wished we hadn't, we can just claim that we were 'being emotional'" (*HFE*, 185–86). Fridlund has made use of this argument to suggest why it is that even if according to his analysis faces do not express "emotions" as the classical emotion theorist construes them, we nevertheless constantly attribute emotion categories to faces. He proposes that one reason we do so is that emotion language is more palatable to us than the language of social intent, because it offers us excuses, rather than requiring us to give reasons, for our actions (*HFE*, 281).

I note in this connection that emotion theorist Feldman Barrett offers a different explanation for our tendency to employ discrete emotion categories when interacting with others. Agreeing with Fridlund's rejection of the basic emotions approach, she argues that if we nevertheless continue

21. Thus, social constructionists have similarly emphasized the role of "emotions" as offering excuses for our actions instead of demanding that we give reasons for them—that is, as ways of moving socially sanctioned behavior into the realm of passive, involuntary, and thus excusable passionate behavior. See for example James R. Averill, "Emotions Unbecoming and Becoming," in *The Nature of Emotion: Fundamental Questions* (New York and Oxford, 1994), 265–69.

to deploy discrete emotion categories, it is not because they are natural kinds. Rather, it is because it is through the ineluctable process of categorization that our emotional responses are given meaning. According to Feldman Barrett, it is only as a result of conceptualization and interpretation that a hypothesized primitive "Core Affect," defined by her as internal bodily and incoming sensory input, becomes "bound to the object" through the application of learned, situated category knowledge, with the result that we become "angry" about something or someone, or "afraid" of something, and so on.[22] Thus, in spite of her opposition to Ekman's basic emotion view, Feldman Barrett ends up proposing a theory according to which meaning and intention are added on to a meaningless "Core Affect" by a process of conceptual construction and classification, which is why we end up with the familiar, discrete categories of emotion.[23] In short, Feldman Barrett shares Fridlund's emphasis on the collective intentionality of the interactions between humans and many nonhuman animals. But unlike Fridlund, she understands intention as itself the product of a process of meaning-making by which significance is imposed on internally or externally generated sensory input that is itself without meaning. In this regard, in spite of her commitment to cognitivism, Feldman Barrett's effort to provide an "alternative" to the basic emotion approach comes close to reinstating Ekman's two-factor model according to which

22. See for instance Lisa Feldman Barrett, "Solving the Emotion Paradox: Categorization and the Experience of Emotion," *Personality and Social Psychology Review* 10 (1) (2006): 20–46; Lisa Feldman Barrett, Kristen A. Lindquist, and Maria Gendron, "Language as Context for Emotional Perception," *Trends in Cognitive Science* 11 (8) (2007): 327–32; Kristen A. Lindquist and Lisa Feldman Barrett, "Constructing Emotion: The Experience of Fear as a Conceptual Act," *Psychological Science* 19 (9) (2008): 898–903; Christine D. Wilson-Mendenhall, Lisa Feldman Barrett, W. Kyle Simmons, and Lawrence W. Barsalou, "Grounding Emotion in Situated Conceptualization," *Neuropsychologia* 49 (2011): 1105–27; and Lisa Feldman Barrett, "The Conceptual Act Theory: A Précis," *Emotion Review* 6 (4) (2014): 292–97. Daniel Gross and Stephanie Preston offer somewhat similar arguments based on notions of "situated cognition," in "Emotion Science and the Heart of a Two-Cultures Problem," in *Science and Emotions After 1945: A Transatlantic Perspective*, eds. Frank Biess and Daniel M. Gross (Chicago and London, 2014), 96–117. For some further comments on these views, see my epilogue.

23. Similarly, Russell defines non-intentional core affect—the consciously accessible neurophysiological or somatosensory signals of hedonic and arousal values emanating from the individual's internal body—as well as the signals from the external world as the bottom-up "raw data" out of which emotional "metaexperiences" (or the intentional mental states of fear, anger, and other emotional interpretations) are built up. See Russell, "My Psychological Constructionist Perspective, with a Focus on Conscious Affective Experience," in *The Psychological Construction of Emotion*, eds. Lisa Feldman Barrett and James A. Russell (New York and London, 2015), 195.

cognition or conceptuality is tacked on to a biologically primitive nonconceptual affective process—a model that Fridlund refuses.[24]

3. *Is the Behavioral Ecology View Disconfirmable?* Another general feature of Fridlund's work that some have found hard to accept is his frank admission that at the limit, his position cannot be disconfirmed. As we have seen, according to his view of facial behavior, in humans implicit or imagined interactions with others can never be excluded. This suggests that the implicit sociality view is not disconfirmable, since it is difficult to imagine any situation short of complete narcotization when we are not social in some way. "In fact," he says of the implicit sociality view, "it probably is unfalsifiable *in extremis*" (*HFE*, 167).

But Fridlund does not regard this admission as a fatal flaw or as an impediment to the viability of the implicit sociality view as a scientific concept. As he notes, the absence of absolute proof is a feature shared by many well-established theories whose limiting cases are unattainable—for example, superconductivity and special relativity theories share the same trait. What this means, he suggests, is that the implicit sociality concept can be studied quite easily within experimentally manipulable ranges. "The strict falsifiability objection is also instructive," he comments in this regard, "in that it reveals the extent to which we view most animals (especially humans) atomistically, instead of as components of, and agents within, an encompassing web of social relations—even when alone (see [George Herbert] Mead, 1934). Indeed, 'solitude'—not implicit sociality—may be the odder concept" (*HFE*, 167). He points out that the implicit sociality view is hardly novel, since several of the pioneers of psychology, such as Wundt, Piderit, Gratiolet, Ribot, and even Darwin (in *The Descent of Man*), had all proposed "imaginary object" accounts of solitary "facial expressions" made toward absent others (*HFE*, 167–68).[25]

24. Jan Plamper briefly takes me to task for including the work of Lisa Feldman Barrett and James A. Russell under the rubric of "good science," arguing that in spite of their overall intentionalism, the notion of "core affect" is ultimately on the side of basic emotions "à la Tomkins" in its dissociation between affect and intention. As my discussion of Barrett and Russell in this book (including the epilogue) goes to show, I do not disagree. But this is not to say that there is nothing of value or importance in their work, and I would certainly resist Plamper's suggestion that my appeal to "good science" even partly masks a move to "naturalize contingent political positions." Jan Plamper, *The History of Emotions: An Introduction* (Oxford, 2015), 263–64.

25. Indeed, as Fridlund himself recognizes, his position resonates with the American philosopher and social theorist George Herbert Mead's earlier suggestion that even private actions are performed with someone else in mind, even if that person is simply the "'generalized other,'" a point noted by Brian Parkinson in his paper "Do Facial Movements Express or Communicate Motives?," 285.

However, many researchers complain that Fridlund's behavioral ecology or social motive view of facial expression is not falsifiable. Ekman himself brought up the issue when in 1997 he commented that Fridlund's suggestion that the reason people sometimes make facial expressions when alone is because they imagine people are present "makes Fridlund's position not subject to disconfirmation and therefore of little use."[26] There is a considerable irony in Ekman's reproach since, as emerged in the previous chapter, in the same year, he conceded that cultural display rules might carry over into private behavior—an admission which meant that logically the non-disconfirmability complaint must apply to his own position.

In fact, the problem of the lack of absolute proof haunts the entire field of emotion research today, whatever thesis is being pursued. For instance, and as we shall see in the epilogue, many researchers today attempt to keep the concept of "emotion" alive and scientific by proposing that emotions are made up of a number of quite loosely associated neurochemical, behavioral, and cognitive components, including facial expressions. By positing a "componential" or "package" approach to emotional behavior, such researchers are able to hold on to the nebulous concept of "emotion" while finding a way to explain (or explain away) the variability and multiplicity of responses obtained in their experiments. On this view, the presence or absence of "emotion" cannot be determined by the presence or absence of any single component or subset of components, with the consequence that, as Fridlund has recently put it, the position is reduced to "no more than hand waving about the knottiness of the phenomena and *ad hoc* choices of stipulated emotion 'measures,' with the result that surveys research and formal meta-analyses continually find disappointing links between 'emotion' and 'expression'" ("BEV").

The same problem of proof comes up in Brian Parkinson's social-constructionist approach to the emotions. We briefly encountered Parkinson in chapter 4 when considering his and Manstead's 1990s critical responses to Lazarus's work. I noted there that in an effort to overcome what they diagnosed as an untenable separation between the individual subject and the world implicit in Lazarus's appraisal theory, Parkinson and Manstead emphasized the public, social, and performative or transactional character of human affective interactions. This soon led Parkinson to pay attention to Fridlund's behavioral ecology views, especially the latter's emphasis on the intentional, communicative nature of facial signaling.

26. Paul Ekman, "Conclusion: What Have We Learned by Measuring Facial Behavior?," in *What the Faces Reveals: Basic and Applied Studies of Spontaneous Expression Using the Facial Action Coding System*, 613.

In 2005, Parkinson in a cautious review undertook a comparison between Ekman's and Fridlund's approaches, concluding that the preponderance of the argument and the evidence spoke in favor of Fridlund's position.

Yet, even as the logic of his own arguments seemed to deny the possibility of doing so, in the same 2005 article Parkinson retained a distinction between "social motives" and what he called "emotion" or "emotion itself." Thus, on the one hand, his general point appeared to be that it is impossible to know what "emotions" are outside the interpersonal contexts that constitute them as such. This seemed to be what he was saying when he wrote that "My central aim is to point the way toward a more comprehensive account of facial movement in social interaction that properly situates emotional communication in its everyday functional context." But he also stated that "a fuller articulation of dynamic emotion processes in their interpersonal context may render many (but not all) of the distinctions between emotions and social motives redundant."[27] It is the "but not all" in this last sentence that is puzzling. It is as if Parkinson already adhered to a definition of emotion in opposition to the social motive thesis but chose not to tell the reader what that definition was, thereby begging the very question at stake in the Ekman–Fridlund confrontation.[28]

At the same time, in his paper Parkinson complained more than once that Fridlund's social motive-communicative account of facial expression was partly undermined by its under-specification of the central concepts of social motive (or behavioral intentions) and audience attunement. In particular, Parkinson reproved Fridlund for not making consistent or systematic predictions about audience effects on facial movements, predictions applying across all conceivable situations. But this was a very demanding requirement, one that inevitably applied to Parkinson's own position, which—decrying Fridlund's notion of "social motive" as not specific enough and rejecting the notion of "audience effects" as too narrow—stressed the need to evaluate all the dynamic factors that moderated and influenced the impact of interpersonal contexts and behavior on facial movement. How, after all, could one begin to control for all the possible factors influencing human behavior—especially if unconscious interper-

27. Brian Parkinson, "Do Facial Movements Express Emotions or Communicate Motives?," 279.

28. The same distinction comes up in an even starker form in a more recent article in which Parkinson distinguishes between "emotions themselves" and "emotional meanings," implying that in spite of his argument that culture and biology are inseparable, he thinks emotions are meaningless states to which signification is then added. Brian Parkinson, "Contextualizing Facial Activity," *Emotion Review* 5 (1) (2013): 100.

sonal factors were included in the mix, as it appeared from Parkinson's text that they were?[29] In short, Parkinson's own transactional-social model of emotional behavior was so expansive that it was not obvious what experimental limits he himself would be able to set: his worry that Fridlund's theory was capable of explaining just about everything, and therefore in some sense nothing, inevitably rebounded against his own position.

This is not to downplay the fact that Fridlund's social motive position and denunciation of the Basic Emotion Theory raise intriguing questions about the direction that future research in the field would have to take if his arguments were to be fully accepted as valid, a topic to which I will return at the end of this book.

DEFENDING THE BASIC EMOTION THEORY

Against this backdrop of Fridlund's critique of Basic Emotion Theory and the features of his proposed Behavioral Ecology View of facial signaling, it is time to evaluate some recent work in the emotion field. More than any other figures, the philosopher and historian of science Paul E. Griffiths and his former student and sometime coauthor Andrea Scarantino have offered the most spirited defense of the basic emotion approach in psychology, and it is to their ideas and claims that I now turn.

As I noted in the introduction, in 1997 Griffiths burst on the scene with a book provocatively titled *What Emotions Really Are: The Problem of Psychological Categories*. In this hugely influential text, which has been widely cited not only within the field of the emotion sciences but also by scholars in the humanities and social sciences who have become interested in affect, Griffiths argued against the idea that all the emotions can be explained by a single mechanism. He maintained instead that it was necessary to accept a theoretical pluralism by differentiating between Ekman's basic emotions and the higher, so-called "cognitive" emotions that required a different psychological explanation. The challenge for Griffiths was then to suggest how the cognitive emotions and the basic emotions are connected—not an easy problem to solve.[30]

29. As Parkinson implies, remarking that when in an experiment solitary viewers watch a film, their sad expressions might be due to their identification with the movie characters and "immersion" in the imagined interpersonal interactions depicted in the film clips. Parkinson, "Do Facial Movements Express Emotions or Communicate Motives?," 299.

30. Paul E. Griffiths, *What Emotions Really Are: The Problem of Emotional Categories* (Chicago and London, 1997); hereafter abbreviated as *WERA*.

Griffiths's book gave an enormous boost to Ekman's affect program theory at the very moment Fridlund and Russell were advancing their critiques. His approach was motivated by several considerations. Claiming that cognitivists or "propositional attitude" theorists in philosophy, such as Anthony Kenny and Robert Solomon, had reached a dead end by focusing on conceptual analysis to the neglect of empirical research, he proposed a scientific approach to the emotions based on psychoevolutionary considerations. His book included the following fundamental claims:

1. Vernacular, everyday, or "folk psychology" emotion terms of the sort studied by the cognitivist philosophers do not constitute natural kinds corresponding to real distinctions in nature and for that reason should be eliminated from the sciences of emotion. Griffiths's position on this topic rested on various assumptions about natural kind semantics associated especially with the views of the philosopher Richard Boyd (see *WERA*, 173–75, for Griffiths's brief discussion of Boyd's views).

2. Vernacular emotion terms should therefore be replaced by two natural kind terms corresponding to the two identified categories of emotion, Ekman's "affect programs" category on the one hand, and the higher, so-called "complex emotions" or "cognitive emotions" category on the other. Griffiths asserted that Ekman's affect program theory was correct but had only limited application. It covered the "basic emotions," such as anger, fear, disgust, contempt, surprise, and joy, but not the more complex emotions. Therefore, a quite different account was needed for emotions such as envy, guilt, jealousy, and love. The implication of Griffiths's position was that one theory alone could not cover all the empirical facts because underlying the superficial similarities between various emotional behaviors were fundamental differences whose significance could only be determined by considering evolutionary factors. From Griffiths's perspective, as Hutchinson in an acute critique has commented, "any likeness between the 'basic emotions' and the 'complex emotions' is purely analogous (a correspondence only in purpose or function) and not homologous (correspondence in terms of essential structural form). Therefore, these emotions, basic and complex, are *really* different and only *superficially* similar or of the same kind."[31] Privileging homology over analogy, Griffiths claimed that only similarities based on homology or shared evolutionary history pointed to real correspondences in nature.

31. Hutchinson, "Emotion-Philosophy-Science," in *Emotions and Understanding: Wittgensteinian Perspectives*, eds. Ylva Gustafsson, Camilla Kronqvist, and Michael McEachrane (Basingstoke and New York, 2009), 64–65.

3. It followed from the argument on homology that common evolutionary origins justified treating the basic emotions as the same in both humans and nonhuman animals, thereby eliminating the worry that claims about the nature of certain emotions in humans do not apply to the "lower" animals. In a recent paper, Griffiths argues:

> Suppose that two animals have homologous psychological traits, such as the basic emotion of fear in humans and fear in chimpanzees. We can predict that, even if the function of fear has been subtly altered by the different meaning of "danger" for humans and for chimps, the computational methods used to process danger-related information will be very similar and the neural structures that implement them will be very similar indeed. After all, Joseph LeDoux's widely accepted account of fear processing in the human brain is largely, and legitimately, based on the study of far more distantly homologous processes in the rat . . . Now suppose that two animals have psychological traits that are analogous—fear in the rat and fear in the octopus, for example. It is a truism in comparative biology that similarities due to analogy (shared adaptive function) are "shallow." The same problem can be solved in different ways, and so the deeper you dig, the more likely it is that mechanisms will diverge . . . In contrast, similarities due to homology (shared ancestry) are notoriously "deep": even when function has been transformed, the deeper you dig, the more similarity there is in the underlying mechanisms. Threat displays in chimps look very different from anger in humans, but when their superficial appearance is analyzed to reveal the specific muscles whose movements produce the expression and the order in which those muscles move, it becomes clear that they are homologues of one another.[32]

In making statements like this, however, it could be argued that Griffiths begs numerous significant questions. These questions include whether his methodological preference for homology over analogy is justified; whether characterizing homology as "deep" and analogy as "shallow" is anything more than a rhetorical ploy designed to make the analogy appear superficial and less real; whether it makes no substantive difference that the meaning of danger in humans is "subtly different" from its meaning in nonhuman animals, such as a chimp or a rat; whether arguments based on scientific consensus, as in Griffiths's claim that Le-

32. Paul E. Griffiths, "Is Emotion a Natural Kind?," in *Thinking About Feeling: Contemporary Philosophers on Emotions*, ed. Robert C. Solomon (Oxford, 2004), 237–38.

Doux's account of fear processing in humans has been "widely accepted," are relevant to the evaluation of the correctness or validity of those same arguments; whether computational psychology is the right approach to the study of mind; and so on. Indeed, although Griffiths confidently presents his views as vindicated by advances in the philosophy of natural kind semantics and by up-to-date empirical findings, his arguments have not gone uncontested. Philosophers such as Hutchinson and psychologists have disputed, often tellingly, all of Griffiths's fundamental premises, including his understanding of cognitivism, his account of recent developments in natural kind semantics, and his privileging of homology over analogy in the treatment of the emotions.[33]

For my purposes, what is especially interesting is the way in which Ekman's affect program theory matches Griffiths's theoretical needs by offering a solution to the ostensible problems posed for cognitivism in psychology. According to Griffiths, those problems include the existence of so-called objectless emotions, that is, states of depression, anxiety, elation, and other clinical cases that appear to lack intentional objects and hence to lack cognition or propositional attitudes; the existence of emotions that are said to be triggered without the kinds of judgment cognitivists usually consider central to the affects, as in the examples of emotional preferences primed by subliminal stimuli documented especially by Zajonc; the suggestion that the basic emotions are "sources of emotion not integrated into the system of desires and beliefs" (*WERA*, 243), which is to say that the basic emotions are not intentional states; and the idea that emotions are accompanied by characteristic autonomic changes, facial expressions, muscle movements, and expressive vocal changes, signature changes and behaviors that Griffiths accuses the cognitivists of being unable to explain.

Ekman's Basic Emotion Theory fits Griffiths's theoretical requirements perfectly, since it suggests the existence of discrete affects that are automatically and unintentionally triggered by unlearned or learned stimuli with at best minimal cognitive input and are discharged through characteristic facial expressions and signature patterns of physiological

33. For criticisms along these and related lines, see especially Hutchinson, "Emotion-Philosophy-Science," and idem, *Shame and Philosophy: An Investigation in the Philosophy of Emotions and Ethics* (Basingstoke, 2008), 7–41. Hutchinson's arguments are in line with the position of scientists such as Feldman Barrett, Fridlund, and others who, on what they consider properly evolutionary and other evidentiary grounds, in the case of emotions deny the idea of absolute commonalities between humans and nonhuman animals.

changes. As Griffiths puts it: "The work of Ekman and his collaborators is of particular interest because they seem to treat emotion terms as the names of categories of psychological events in the manner I advocated" (*WERA*, 77). In sum, although Griffiths acknowledges that Ekman's claims have been controversial, he defends them on the grounds that once it is recognized that the latter's empirical work applies to only a limited range of instances of what is meant by anger or sadness and not to every vernacular use of these terms, his theoretical formulations and empirical findings, including his cross-cultural judgment studies and his iconic Japanese–American study, can be accepted as sound.

In his book, Griffiths made no reference to Fridlund's criticisms of Ekman's Basic Emotion Theory. But within a few years he began to acknowledge Fridlund's work. In 2001, in an article on "Emotion and Expression," Griffiths for the first time described Fridlund's behavioral ecology (or "paralanguage") view of facial expression as the "main contemporary alternative to the affect program theory."[34] Nevertheless, in a sign of the continued allure of Ekman's picture of the emotions, in the same article Griffiths continued to cite without critical comment the disputed Ekman–Friesen Japanese–American study as decisive evidence in favor of Ekman's neurocultural theory. He also defended the validity of the affect program theory by arguing that some facial signals have evolved in nature to provide reliable or trustworthy signals of underlying emotions. Admitting Fridlund's point that evolution would not favor direct readouts of emotional state, Griffiths nevertheless pushed back by asserting:

> Fridlund argues that an evolutionary perspective actually favors the paralanguage view, since involuntary signals of true emotional state would be subverted by the evolution of dissimulation and deceit. While this is an important perspective on the evolution of emotional expressions, it does not constitute the decisive argument that Fridlund seems to suppose. Veridical signals do evolve in nature, often by making use of so-called *hard to fake* signals. The cost of being unable to suppress a signal of emotional state may be balanced by the advantage of being believed. A purely theoretical argument based on evolutionary dynamics of signaling systems seems more likely to support the view that "examples of emotional behavior lie along a continuum from expression to negotiation" (Hinde, 1985, p. 989)

34. Paul E. Griffiths, "Emotion and Expression," *International Encyclopedia of the Social and Behavioral Sciences*, eds. Neil J. Smelser and Paul B. Baltes (Amsterdam and New York, 2001), 4436; hereafter abbreviated as "EE."

than a purely expressive or purely manipulative picture of emotional expression. ("EE," 4436–37; his emphasis)

In other words, Griffiths here assumed, as the affect program theorists did, that there is a "state" inside the individual that in the natural course of things will show itself on the outside, especially the face, and that when it does it is a true or reliable readout or exhibition of an emotion—precisely the point at issue in Fridlund's critique.

Griffiths's dismissive response to Fridlund's arguments does not seem to have been founded on a genuine understanding of the latter's criticisms of Ekman's experiments or on a thorough examination of the burgeoning literature on the question of the costs of veridical signaling. Instead, his reaction appears to have been driven by his prior commitment to the affect program theory, a commitment that was so fundamental that it prevented him from being troubled by the flaws in Ekman's experimental evidence or in Ekman's reasoning. Nevertheless, Griffiths was impressed by Fridlund's emphasis on the strategic nature of animal signaling, with the result that he began to try to combine the affect program view with theories designed to explain strategic signaling or what he calls "situated" or "transactional" emotional responses.[35] As this was exactly what the animal ethologist Robert Hinde had earlier attempted (see chapter 5), it is not surprising that Griffiths accepted Hinde's views on emotion and expression rather than Fridlund's more "radical" position. Griffiths therefore treated Hinde, rather than Krebs and Dawkins, W. John Smith, or Fridlund himself, as the true source and origin of the new ethological ways of thinking about the situated emotions. On Hinde's continuum model, emotional displays considered to be emotional were viewed as "unconditional predictors" revealing a motivational state that explained future behavior, while expressive behaviors seen as forms of negotiation were viewed as "conditional predictors" forecasting how the signaler would behave according to the responses of receivers.[36]

Thus, in keeping with his commitment to theoretical pluralism, Griffiths has attempted to unite the two apparently competing approaches of

35. Paul E. Griffiths and Andrea Scarantino, "Emotions in the Wild: The Situated Perspective on Emotion," in *Cambridge Handbook of Situated Cognition*, eds. P. Robbins and M. Aydede (Cambridge, 2009), 437–53.

36. Paul E. Griffiths, "Basic Emotions, Complex Emotions, Machiavellian Emotions," *Royal Institute of Philosophy Supplement* 52 (March 2003): 52. Griffiths has even suggested that Hinde's work inspired Fridlund, whereas in fact Fridlund was critical of Hinde's attempted solutions and paid far more attention to the work of W. John Smith, Krebs, and Dawkins (see chapter 5, n. 20).

Fridlund and Ekman by proposing that emotional behaviors lie on a continuum between purely expressive, non-contextual, involuntary readout responses at one end and strategic "negotiations" or "situated emotions" at the other. This appears to amount to little more than a rephrasing of Ekman's "two factor" theory.[37] One of Griffiths's proposals in this regard has been that the affect programs and the "situated" emotions operate on different time scales, such that the basic emotions are restricted to short-term, stereotyped, quasi-reflex short bursts, while the situated emotions involve longer emotional episodes (*WERA*, 241). Let me say that I have never found convincing the argument for the noncognitive status of emotional responses based simply on the speed with which they are said to occur. I therefore find implausible Griffiths and Scarantino's view that just because an emotional response is rapid it must be noncognitive and must therefore conform to Ekman's now-discredited views.[38]

In another argument that harks back to Lazarus's discussion of levels of appraisal, Griffiths has made use of claims that an organism's appraisals of stimulus inputs take place at several different levels, including crucially at subpersonal levels. He therefore proposes that many stimulus assessments can occur on the basis of very crude or "raw" sensory inputs without conceptual content of any kind, thereby dissociating basic emotions from cognition and stipulating the cognitive impenetrability of many emotional responses.[39] Figuring prominently in Griffiths's dis-

37. For Griffiths's laudatory review of Paul Ekman's book, *Emotions Revealed*, see "Smile and the Whole World Smiles with You," *New Scientist* 178 (2003): 56.

38. My view of this matter of speed of response finds confirmation in a recent paper that challenges the "standard hypothesis" according to which there is a dedicated, modular system that operates rapidly, automatically, and subcortically to process basic emotions. On the basis of the evidence, the authors make the case that visual processing of emotion stimuli occurs no faster than visual processing in the cortex. Luiz Pessoa and Ralph Adolphs, "Emotion Processing and the Amygdala: From a 'Low Road' to 'Many Roads' of Evaluating Biological Significance," *Nature Reviews* 11 (December 2010): 773–82.

39. Griffiths cites John D. Teasdale, "Multi-Level Theories of Cognition-Emotion Relations," in *Handbook of Cognition and Emotion*, eds. Tim Dalgleish and Mick J. Power (London, 1999), 665–81. Teasdale acknowledges as a weakness of his and other multilevel theories of cognition-emotion relations that it may be difficult to distinguish empirically between a multiplicity of explanatory accounts for any given phenomenon that the different models, or indeed the same model, can provide. "Put bluntly," he writes, "if we can 'explain' all aspects of cognition-emotion relations, simply by an unconstrained re-description of those relations in the language of a complex framework which does not yield verifiable predictions, then we may have really explained very little" (675).

cussions is LeDoux's analysis of fear, according to which there are two separate neuronal pathways, a "quick and dirty" subpersonal "low route" through which crude aspects of stimuli can activate emotion production, and a slower, less direct "high route" through which more complex perceptual features of stimuli can elicit affects. Griffiths also makes use of the related suggestions about the role of "automatic appraisal" and affective primacy put forward by Ekman, Zajonc, Öhman, and others.

What is striking about Griffiths's recent publications on these topics is how much theoretical-philosophical effort he has expended in order to deflate claims about the intentional nature of "emotional" responses. It is as if he is so wary of attributing intentionality, agency, or cognition to animals that he has done his utmost to theorize even strategic signaling behavior of the kind emphasized by Fridlund in non-intentional terms. It is not enough for Griffiths that the so-called "basic emotions" are non-intentional and noncognitive: even the more strategic or so-called "Machiavellian" emotions must be viewed as non-intentional too. In an original but disputable interpretation, he goes so far as to suggest that "strategic" responses can best be understood as repertoires that have been built into the organism by evolution. He has therefore proposed the following "Machiavellian Emotion Hypothesis":

> *The Machiavellian Emotion Hypothesis*. Emotional appraisal is sensitive to cues that predict the value to the emotional agent of responding to the situation with a particular emotion, as well as cues that indicate the significance of the stimulus situation to the agent independently of the agent's response.
>
> Put in the language of appraisal theories, the hypothesis is that the appraisal hyperspace has "strategic" dimensions. Current appraisal theories identify multiple dimensions that assess the organism-relative significance of what has happened. The Machiavellian emotions hypothesis predicts that there will also be dimensions that assess the payoff to the organism of having an emotion. Putting the hypothesis in more philosophical terms, the emotional appraisal ascribes to the environment the property of affording a certain strategy of social interaction.[40]

Focusing on studies showing audience effects or "Machiavellian factors" in animal behavior, Griffiths now suggests that Ekman's concept of a display rule needs to be amended "in a way that makes even Machiavellian expression a more integral part of the actual emotion process

40. Griffiths, "Basic Emotions, Complex Emotions, Machiavellian Emotions," 54.

than it at first seems." In other words, he proposes that "display rules" in nonhuman animals are not learned during individual development, as Ekman's display rule concept had proposed, but are themselves "built in" or inherited as evolved emotional responses.[41] On the basis of this revision, and with reference to Marler's 1997 research, Griffiths argues against Fridlund:

> Fridlund frames these results [on audience effects] as a refutation of basic emotion theory, but it is not clear that the results support this interpretation. It is true that Ekman has argued that the "display rules" that modulate emotional behaviors according to social context are acquired, culturally specific, and do not interfere with the actual working of the automatic appraisal mechanism and the affect programs . . . *But there is nothing to prevent an affect program theorist from building audience effects into the evolved "emotion module" itself. Emotional behavior exhibits audience effects in many organisms in which it seems much more likely that they are part of the evolved emotion system itself rather than that they are acquired behaviors*—organisms such as domestic chickens. (my emphasis)[42]

It is worth noting that in making the claim in this passage for the unlearned character of audience effects or strategic signaling in chickens, Griffiths refers to a 1997 paper by Marler on this topic (actually an article by Marler and Evans) that in fact contradicts him on this point. In their article, Marler and Evans had undertaken an evaluation of Ekman's Basic Emotion Theory in the light of Fridlund's Behavioral Ecology critique. Citing their experiments on the sensitivity of chicken predator and food calls to the presence or absence of conspecifics—experiments I discussed in the previous chapter and that Fridlund had earlier cited as favoring his Behavioral Ecology View of facial signaling—Marler and Evans had cautiously accepted Ekman's neurocultural explanation of such audience effects rather than Fridlund's more skeptical conclusions. In short, like Hauser, whose work on animal communication Griffiths also cited, Marler and Evans had proposed that the findings on audience effects in non-

41. Apparently without Griffiths being aware of it, as early as 1992 Ekman himself had begun to incorporate Tooby and Cosmides's view of emotions as evolutionary adaptations to life tasks, with the result that he, too, had started to treat strategic expressions as themselves inherited, built-in forms of display that enhanced fitness. The result was that the role of cultural display rules appeared to be diminished or made redundant, a point to which I will return in the epilogue.

42. Griffiths, "Emotions," in *Blackwell Guide to Philosophy of Mind*, eds. Stephen P. Stich and Ted A. Warfield (Malden, MA, 2003), 298–99.

human animals could be combined with Ekman's Basic Emotion Theory by suggesting that audience effects were the result of social cues (or display rules) governing the modulation of endogenously driven behaviors.

Nevertheless, in an effort to discover whether sensitivity to audiences in animal signaling required cognitive skills or could be explained as hardwired operations, Marler and Evans had obtained results from rearing chicks in isolation that suggested predator and other vocal signals are learned, cognitive behaviors. They had reported that chicks raised without parents and without predator experiences were unable to make appropriate alarm calls, an abnormal behavior that persisted into adulthood. "It seems clear that the development of normal antipredator behavior is influenced both by experience of predators and by the presence of parents," Marler and Evans had concluded, "although the precise mechanisms involved have still to be elucidated." On this basis they had argued in favor of a learned, cognitive dimension to such behaviors: "At this stage, it seems conceivable that a cognitive interpretation may be appropriate after all . . . Normal development seems to have been quite precluded by these rather drastic ontogenetic manipulations."[43]

But Griffiths ignores this aspect of Marler and Evans's report. He rejects their cognitive interpretation in favor of the view that strategic or transactional behaviors are not intentional or learned responses:

> [T]he evolutionary rationale for the emotions view, and the existence of audience effects in non-human animals, warn against any facile identification of the view that emotions are social transactions with the view that they are learnt or highly variable across cultures. *Indeed, the transactional view may seem less paradoxical to many people once the idea that emotions are strategic, social behaviors is separated from the idea that they are learnt behaviors or that they are intentional actions.* (my emphasis)[44]

And he has also made this point:

> The existence of audience effects in animals is in tension with some basic stereotypes about emotion. Emotions are the paradigm of something that happens without regard for the consequences. Emotions are also stereotypically "biological." Something sounds right in Konrad Lorenz's epigram that animals are highly emotional people

43. Peter Marler and Christopher Evans, "Animal Sounds and Human Faces: Do They Have Anything in Common?," in *The Psychology of Facial Expression*, 151.

44. Griffiths, "Emotions," 299.

> of limited intelligence: emotions are part of our "animal nature." Producing or suppressing behaviours so as to take account of social relationships, however, seems like a complex, cognitive achievement. It suggests processes that involve deliberation, perhaps even conscious deliberation. So the idea that the emotion system implements strategies of social interaction naturally suggests the idea that these aspects of emotion are learnt, and perhaps culture-specific rather than being part of our evolutionary heritage. But this inference may well be entirely spurious. The existence of sophisticated audience effects in animals suggests that the social, manipulative aspects of emotion may be as evolutionarily ancient as any others. The appraisal process that sets off a transparently Machiavellian response like sulking may very well resemble the ancient, "low road" to fear uncovered by LeDoux.[45]

It seems to me naïve to suggest that something as sensitive to context and meaning as an episode of human sulking can best be understood as a reflex response to a set of evolved "triggers" or stimuli. But this appears to be Griffiths's idea. On this basis he rejects as seriously inadequate or as "just too simple" views about the cognitive content of appraisal and the intentionality of the emotions associated with the work of various philosophers.[46]

Griffiths's claim is not only that all low-level "appraisals" are informationally encapsulated and hence inferentially impoverished, but that such appraisals lack conceptual content of any kind:

> There is no scientific puzzle about the nature of the information processing in affective computing. The forms of inferential impoverishment I have described all make good psychological and evolutionary sense. But there is a considerable philosophical puzzle about how to ascribe conceptual content to representational states in an isolated cognitive sub-system of this kind. If the concepts that figure in the content ascribed to a representation do not have their usual inferential role, then what is meant by attributing that content? The very idea that the [emotional] state has conceptual content is thereby called into question. The actual role of the representations involved in low-level appraisal and the inferential role of the content-sentences with which we describe those appraisals . . .

45. Griffiths, "Basic Emotions, Complex Emotions, Machiavellian Emotions," 56.

46. Paul E. Griffiths, "Toward a 'Machiavellian' Theory of Emotional Appraisal," in *Emotion, Evolution, and Rationality*, eds. P. Cruse and D. Evans (Oxford, 2004), 91.

strongly suggests [*sic*] that, in this role at least, appraisal theories simply are not theories of cognitive content.[47]

Griffiths supports this statement by citing the work of the philosopher Gareth Evans, who famously argued for the idea of the nonconceptual content of perceptual states. He also cites writings by José Luis Bermudez, Adrian Cussins, and others who have made similar claims.

In citing the work of those philosophers and extending their arguments to emotional phenomena, Griffiths is certainly aware that he is taking sides in a highly charged debate over the conceptual content of perceptual experience. In the past twenty or more years, Evans's arguments have been the target of criticisms by John McDowell, Bill Brewer, Matthew Boyle, and others, who argue on neo-Kantian grounds that mentation or mindedness is conceptual through and through, or "all the way out," as McDowell has put it.[48] Among many other issues, the debate concerns how to characterize the kinds of embodied copings that nonhuman and human animals exhibit when they negotiate their relations with the world and others in a highly skilled and apparently "automatic" fashion. For Griffiths and the other nonconceptual content theorists, such as Hubert Dreyfus in his recent debate with McDowell over precisely this issue, these are the kinds of skilled copings, such as playing a sport or the piano, but also emotional negotiations, that can and do occur without involving any form of conceptual rationality.[49]

This is not the place to adjudicate a controversy that concerns some of the deepest current questions in philosophy, including centrally the place of naturalism in the philosophy of mind. I will only observe that it is not as obvious as Griffiths seems to think it is that he is on the right side of the debate over the existence of nonconceptual mental content. On the contrary, it seems to me that those who argue the neo-Kantian position have the better of the argument. I would also point out that Griffiths proceeds as if he believes that the issues at stake in this dispute are largely empirical and hence can be decided by scientific consensus, as when he

47. Ibid., 95–96.

48. For contributions by Evans, McDowell, Cussins, Bermudez, and others on this topic, see *Essays on Nonconceptual Content*, ed. York H. Guenther (Cambridge, MA, and London, 2003). See also José Bermudez and Arnon Cohen, "Nonconceptual Mental Content," in *The Stanford Encyclopedia of Philosophy*, first published in 2003, revised January 14, 2008.

49. For the debate between McDowell and Dreyfus over the reach of conceptuality in skilled coping, see *Mind, Reason, and Being-in-the-World: The McDowell–Dreyfus Debate*, ed. Joseph K. Schear (London and New York, 2013).

advocates the idea of a level of appraisal that excludes the involvement of concepts on the grounds that many scientists support such a notion. But in fact, those issues are not empirical but philosophical—a point captured by the philosopher John Doris when some years earlier he had critically observed of Griffiths's undertaking that "there is some question about its implications: how much philosophical hay can be made with an empirical scythe?"[50]

"DON'T GIVE UP ON BASIC EMOTIONS"

In any case, for Griffiths and Scarantino the real battle lies elsewhere, not with philosophers who oppose the idea of nonconceptual mental content but with those psychologists who, since Fridlund, have again raised questions about the viability of the basic emotions model itself. In recent years especially, ever since in the name of the movement labeled "Psychological Constructionism," Lisa Feldman Barrett and James Russell have launched a vigorous challenge to the assumptions of the basic emotions model, Griffiths and Scarantino have been scrambling to defend the discrete emotion approach.

The fundamental claim of the psychological constructionists is that emotions do not form an "essence" or a set of fixed, hardwired categories in nature. Citing the absence of adequate empirical support for the claims of Ekman's school, Russell, Feldman Barrett, and other like-minded theorists have substituted for the idea that the basic emotions form a "natural kind" the notion that the mind constructs or produces our familiar categories of emotion out of a packet of components. Rather like Griffiths and Scarantino, the psychological constructionists argue that traditional categories of emotion, such as fear, are too broad and heterogeneous to serve as the basis for scientific induction and progress. But unlike them, they deny that the Basic Emotion Theory can be rescued by narrowing its claims or scope. Instead, they abandon the very idea of basic emotions in order to suggest that each instance of emotion "is not exclusively realized in the body of the emoter, or in the brain regions that regulate the body. The very existence of an emotional episode (either the self, in emotional experience, or another person, in emotion perception) also requires par-

50. John M. Doris, "Review of Paul E. Griffiths, *What Emotions Really Are: The Problem of Psychological Categories*," *Ethics* 110 (3) (2000): 617. But for yet another recent critique of the cognitivist position along lines close to that of Griffiths, see Andrea Scarantino, "Insights and Blindspots of the Cognitivist Theory of the Emotions," *British Journal for the Philosophy of Science* 61 (2010): 729–68.

ticipation from other parts of the brain that are involved in storing prior experience and knowledge within a perceiver."[51] It follows that for the psychological constructionists, particular emotion episodes are constituted by a range of components, no single one of which is an ingredient so crucial that its presence or absence is absolutely determining for the episode to occur. On this basis they vigorously critique claims that there are discrete emotions that can be correlated with distinct patterns of localized brain activity, physiological changes, and facial expressions.

More than at any other time, the criticisms of the Basic Emotion Theory made by the psychological constructionists have captured the attention of the emotion field. The line of defense adopted by Griffiths and Scarantino is to accept that there are flaws in current formulations of the "traditional" Basic Emotion Theory and then to offer a revised version in its place.[52] Following Griffiths's previous suggestions, they argue that the main shortcoming of traditional approaches to the emotions is the tendency of theorists to conflate two very different projects. On the one hand there is what Griffiths and Scarantino call the "Folk Emotion Project," an anthropological undertaking that aims to "describe the characteristics of ordinary emotion concepts," or to "unveil what makes something a member of a folk emotion category." On the other hand, there is what the authors call the "Scientific Emotion Project," an endeavor that aspires to discover "natural kinds of emotions and the reliable scientific generalizations that are true of them." Griffiths and Scarantino concede the validity of the by-now-widespread criticisms of Ekman's Basic Emotion Theory, such as the absence of a one-to-one correspondence between the basic emotions and any physiological, neurobiological, expressive, behavioral, or phenomenological responses, insofar as those emotions are considered folk categories.[53]

But the authors do not accept Fridlund's skeptical conclusions about the value of the concept of "emotion" itself. Rather, they continue to assume the existence of endogenous emotion states in the form of biologically basic emotions. They therefore propose that once we accept the distinction between folk emotion categories and scientific emotion cate-

51. Lisa Feldman Barrett, "Ten Common Misconceptions About Psychological Construction Theories of Emotion," in *The Psychological Construction of Emotion*, eds. Lisa Feldman Barrett and James A. Russell (New York and London, 2015), 48.

52. Andrea Scarantino, "Basic Emotions, Psychological Construction, and the Problem of Variability," in *The Psychological Construction of Emotion*, 334.

53. Andrea Scarantino and Paul E. Griffiths, "Don't Give Up on Basic Emotions," *Emotion Review* 3 (4) (2011): 444–54; hereafter abbreviated as "DGU."

gories, we can see that the criticisms of the basic emotions concept apply to the folk emotion categories, which are sufficiently heterogeneous in their manifestations that they are unlikely to share anything more than family resemblances or to have distinct biological signatures. Nevertheless, Griffiths and Scarantino argue—this is their key move—the fact that folk emotion categories exhibit such heterogeneous properties does not preclude the idea that some members of the "anger," "fear," "disgust," "happiness," and "sadness" categories do meet Ekman's criteria for being basic emotions. Hence they suggest that to avoid confusion, the biologically basic emotions should not be designated by folk emotion terms but should be replaced by neologisms or, preferably, by modified versions of folk categories, such as "fearB" (or "fearBASIC" or "fear*"), in order to signal that what is being referred to is not the whole of the folk category, but just a part of it ("DGU," 449–50). They cite as one example of a biologically basic emotion the fear produced in an individual by the sudden loss of support, a kind of fear that, they claim, has "a distinctive and possibly universal signal, a distinctive physiology, an automatic appraisal tuned to an antecedent universally present in all cultures, a distinctive developmental appearance, presence in other primates, quick onset, brief duration, unbidden occurrence, distinctive thoughts, memories and images, and distinctive subjective experience" ("DGU," 449).[54] On the basis of this example, they affirm that "there exist emotions in the world that are basicB in just the sense indicated by Ekman" ("DGU," 449). Or, as they state somewhat more hypothetically, "[T]he items that satisfy Ekman's markers of basic-ness are likely to constitute a homogenous domain for purposes of scientific extrapolation" ("DGU," 449).

An important feature of Griffiths and Scarantino's defense of the "new" Basic Emotion Theory is their acceptance of the fact that no single marker can be regarded as individually necessary or fundamental to any specific emotion. According to them, this is because natural kinds are defined by "property clusters" plus one or more mechanisms that cause the properties to co-occur. On this view, no single property is necessary for category membership, so that, depending on the circumstances, the causal mechanism or mechanisms involved bring about different subsets of the property cluster. In short, theirs is an approach to emotion that helps ex-

54. Scarantino and Russell here cite a paper by A. Öhman and S. Mineka, "Fears, Phobias, and Preparedness: Toward and Evolved Module of Fear and Fear Learning," *Psychological Review* 108 (2001): 483–522, in which the authors propose the existence of an evolved module for fear elicitation that is relatively impenetrable to cognitive control and originates in a dedicated neural circuitry.

plain why strict correlations between the behavioral, neuronal, and other components of an emotional response are usually lacking. But it is also an approach that risks making predictions and scientific generalizations difficult if not impossible to prove.

One of Griffiths and Scarantino's solutions to this dilemma is to suggest that prototypical cascades of behavioral, expressive, and physiological components held to be characteristic of discrete emotions do sometimes occur, but very rarely. Take, for example, the following statement by Scarantino:

> Whenever we expose Sally, and a great many other animals for that matter, to bears, a very similar set of components will co-occur, over and over again. This suggests the presence of an internal causal mechanism that *couples* dangerous stimuli with componential changes—the fear* mechanism. This is to say that the "very occasional occurrence" of prototypical fear . . . is not due to the laws of statistics, but rather to the fact that the fear* mechanism rarely encounters stimuli dangerous enough to demand a prototypical fear response. (his emphasis)[55]

On the one hand, this statement makes a claim for the existence of a prototypical fear response characterized by a rigid suite of associated expressions and physiological changes, which is precisely the claim that Fridlund and the psychological constructionists have been contesting all along. So, from the point of view of his critics, Scarantino appears to beg the question at stake in the controversy. On the other hand, the statement also acknowledges that such prototypical emotions occur very seldom in the scheme of things because animals and people seldom confront stimuli extreme enough to arouse such responses.

This is a surprising conclusion to the long-standing debate over the basic emotions, if for no other reason than that the Basic Emotion Theory was originally premised on the idea that prototypical emotional episodes were so common that people in very different cultures all across the world experienced them all the time and could readily recognize the facial and other characteristic signs of these emotional responses in others. Yet Scarantino now suggests that prototypical reactions only occur in exceptional circumstances. One wonders what kinds of experiments he thinks would be required to prove the effects of extreme stimuli: actually

55. Andrea Scarantino, "Discrete Emotions: From Folk Psychology to Causal Mechanisms," in *Categorical Versus Dimensional Models of Affect: A Seminar on the Theories of Panksepp and Russell*, 150–51; hereafter abbreviated as "DE."

exposing people, say, to oncoming bears or other wild animals while recording their facial expressions, heart rates, respiratory responses, and related changes? (In fact, this is precisely his suggestion!)[56]

Scarantino's further move is to assume the central importance of causal mechanisms subserving such basic emotional responses. He does not assert the existence of a unique causal mechanism common to all cases of specific emotions, such as fear, because according to him the folk psychology account of fear includes phenomena so heterogeneous that no single mechanism could serve them all. Rather, his claim is that there exist multiple causal mechanisms corresponding to different instances of the folk concept, such that, in his words, there is "an emotion mechanism—call it the *fear* mechanism* for now—that made Sally's heart pound, her palms sweat, her face broadcast danger and her legs freeze and then run" ("DE," 148). As he puts it:

> [P]ositing an internal emotion mechanism—what I have called . . . the fear* mechanism—provides a clear functional explanation of *why* the correlations among components take the form they do. On this proposal, the reason why the novelty constituted by a bear leads to the co-instantiation of heart rate acceleration and a face expression and sweaty palms and freezing and fleeing behaviors—possibly by means of causal relations among the components themselves and attention shifts—is one and the same: all such changes are functional to the avoidance of danger and brought about in a coordinated fashion by an internal mechanism for that very reason. ("DE," 150)

It follows that for Scarantino a properly scientific study of the emotions necessarily involves the need to identify the internal, dedicated neural

56. Scarantino writes: "Testing for this hypothesis requires exposing individuals to elicitors such as loud sounds or deadly predators for basic fear; physical restraint or sudden pokes in the back for basic anger; and dead insects in one's soup or feces for basic disgust; and so on. What the new BET predicts is that in such cases, and only in such cases, the basic emotions program will lead to a rigid cascade of responses that are specific to each basic emotion" (Scarantino, "Basic Emotions, Psychological Construction, and the Problem of Variability," 339). In a paper cited by Scarantino in support of this argument, Robert Levenson likewise states that when a basic emotion is activated by elicitors that are "focused, powerful, sudden, and closely match prototypical antecedent conditions," then a prototypical cascade of facial expressions, autonomic, behavioral, and subjective changes of the kind posited by traditional basic emotion theorists will occur (R. W. Levenson, "Basic Emotion Questions," *Emotion Review* 3 [4] [2011]: 382).

circuits or "programs" held to "couple" the eliciting stimuli with the componential changes—which is to say, to identify the neural "fear*" mechanism that causes the prototypical fear responses, or the "anger*" mechanism that causes the prototypical anger reaction, and so on. Scarantino associates the basic emotion program for fear with Panksepp's primary FEAR system, a system defined by the latter as a genetically based, subcortical brain network triggered by a limited number of sensory-perceptual inputs and connected to major life-challenging circumstances.[57] Scarantino admits that this fear mechanism is explanatorily insufficient in that it cannot account for all aspects of Sally's fear response, which involves perception, cognition, memory, and other processes. Nevertheless, over the objections of Russell, who thinks the suggestion is empirically false, Scarantino attributes to Panksepp's "fear*" mechanism a "*superordinate*" control function with respect to all the other control systems available to the organism, including perception, cognition, and memory ("DE," 152).[58] His idea is that the fear* mechanism coordinates the range of resources by "gating and modulating incoming sensory inputs relevant to behavioral and autonomic outputs and by engaging in reciprocal interaction with the brain networks involved in learning and higher thought" ("DE," 152), although how this control function is exercised is not specified.

As a research program, the new Basic Emotion Theory rests on the idea that evidence will be found for the existence of several hardwired neural circuits capable of explaining not only the distinctness of emotional responses to certain powerful elicitors but also the variability of affective reactions to less powerful stimuli. Scarantino regards LeDoux's and Panksepp's research projects on the neural basis of the fear response as the best models for the kind of research that needs to be carried out. Although these scientists' findings come from experiments on nonhuman animals such as rats, based on the argument from homology Scarantino is confident that such findings will be found in a similar form in humans.

Scarantino's new version of the Basic Emotion Theory has therefore been reduced to the suggestion that several yet-to-be-identified neural circuits will be discovered that have the explanatory scope claimed for them.

57. See for example J. Panksepp, "In Defense of Multiple Core Affects," in *Categorical Versus Dimensional Models of Affect: A Seminar on the Theories of Panksepp and Russell*, 31–78.

58. To which Russell objects: "This revision is an empirical claim and likely false. The FEAR circuit has known specific effects and influences other parts of the brain, but it does not function as an executive" (James A. Russell, "Final Remarks," *Categorical Versus Dimensional Models of Affect: A Seminar on the Theories of Panksepp and Russell*, 291).

Since LeDoux now states that his research concerns the study of neural circuits involved in defensive and survival functions in organisms but with minimal recourse to the terms "emotion" and "feeling," not much is left of the original Basic Emotion Theory.[59] This is especially the case when we remember that in Scarantino and Griffiths's approach, except in those rare instances when extreme situations cause a prototypical cascade of physiological responses and facial expressions, emotional reactions involve many different components whose coordination is so variable as to be virtually immune to systematic analysis and prediction. In sum, in the hands of Griffiths and Scarantino, Ekman's Basic Emotion Theory has been modified so drastically that it has been reduced to the bare hope that at some time in the future scientists will be able to identify the specific neural mechanisms assumed to subserve certain typical emotional events, although—as Russell has justifiably complained—just what those mechanisms are, and what the events are that they are supposed to account for, remain unspecified.

I will briefly return to aspects of these developments in the epilogue. Before doing so, in light of the arguments of this book, in the next and final chapter I offer an analysis of the recent "turn to affect" that has occurred in the humanities and social sciences.

59. Joseph LeDoux, "Rethinking the Emotional Brain," *Neuron* 73 (4) (2012): 653–76.

[CHAPTER SEVEN]

THE TURN TO AFFECT

A Critique

If you don't understand try to feel. According to Massumi it works.[1]

In this chapter, I plan to discuss the general turn to affect, particularly the turn to the neurosciences of emotion, that has recently taken place in the humanities and social sciences. The rise of interest in the emotions among historians has been well documented.[2] My concern is somewhat different. I want to consider the turn to the emotions that has been occurring in a broad range of fields, including history, political theory, human geography, economics, urban and environmental studies, architecture, literary studies, art history and criticism, media theory, and cultural studies. The work of the historians Daniel Lord Smail and Lynn Hunt, who have recently inaugurated the field of Neurohistory by arguing for the integration of history and the brain sciences, including the sciences of emotion, is a case in point.[3] But my inquiry will also consider the claims of

1. Elad Anlen, "Reflections on SCT 2009," in *Theory* (Fall 2009): 9. Anlen, a participant in the School of Criticism and Theory, was reporting on Brian Massumi's mini-seminar.

2. See especially William Reddy, *The Navigation of Feeling: Framework for the History of Emotions* (Cambridge, 2001); Barbara H. Rosenwein, "Worrying About Emotions in History," *American Historical Review* 107 (3) (June 2002): 821–45; Jan Plamper, "The History of the Emotions: An Interview with William Reddy, Barbara H. Rosenwein, and Peter Stearns," *History and Theory* 49 (2) (2010): 237–65; Nicole Eustace, Eugenia Lean, Julie Livingston, Jan Plamper, William M. Reddy, and Barbara H. Rosenwein, "AHR Conversation: The Historical Study of the Emotions," *American Historical Review* 117 (5) (2012): 1487–531; Jan Plamper, *The History of Emotions: An Introduction*, trans. Keith Tribe (Oxford, 2015) (Plamper's book first appeared under the title *Geschichte und Gefühl: Grundlagen der Emotionsgechichte* [Munich, Germany, 2013]); and Ute Frevert et al., *Emotional Lexicons: Continuity and Change in the Vocabulary of Feeling 1700–2000* (Oxford, 2014).

3. Daniel Lord Smail, *On Deep History and the Brain* (Berkeley, CA, 2008); hereafter abbreviated as *ODH*. See also Smail, "An Essay on Neurohistory," in *Emerging Disciplines: Shaping New Scholarly Fields of Scholarly Inquiry in and Beyond the Humanities*, ed. Melissa Bailer (Houston, 2010), 201–25; Andrew Shryock and Daniel Lord Smail, eds., *Deep History: The Architecture of Past and Present* (Berkeley, 2011); Smail, "Neuroscience and the Dialectics of History," *Análise Social* 47 (205) (2012): 894–909; Lynn

those cultural critics and others who, even before historians ventured into this terrain, in such newly designated fields as Neuropolitics, Neurogeography, Neuroaesthetics, and Neuroeconomics, have not only emphasized the importance of affect but have called for a renewal of their disciplines based on the findings of scientists working in the emotion field. In a brief compass, I cannot do justice to the full scope of the literature on affect or even to the range of issues that I find interesting. Instead, I shall focus on topics that seem to me to go most directly to the heart of what is at stake in the general turn to affect.

Let me begin by posing a simple question: Why are so many scholars today in the humanities and social sciences fascinated by the idea of affect? In an obvious sense, an answer is not difficult to find—one has only to attend to what those scholars say. "In this paper I want to think about affect in cities and about affective cities," geographer Nigel Thrift explains, "and, above all, about what the political consequences of thinking more explicitly about these topics might be—once it is accepted that the 'political decision is itself produced by a series of inhuman or presubjective forces and intensities.'"[4] Similarly, cultural critic Eric Shouse states that the importance of affect "rests upon the fact that in many cases the message consciously received may be of less import to the receiver of that message than his or her non-conscious affective resonances with the source of the message." He adds that the power of many forms of media lies "not so much in their ideological effects, but in their ability to create affective resonances independent of content or meaning."[5] In the same spirit, political philosopher and social theorist Brian Massumi, one of the most influential contemporary affect theorists in the humanities and social sciences, attributes Ronald Reagan's success as a politician to his ability to "produce ideological effects by non-ideological means . . . His means were affective." Characterizing Reagan as "brainless" and without content, Massumi asserts that "the statement that ideology—like every actual structure—is produced by operations that do not occur at its level

---

Hunt, *Inventing Human Rights: A History* (New York, 2007); idem, "The Experience of Revolution," *French Historical Studies* 32 (4) (2009): 671–78; and idem, "AHR Roundtable: The Self and Its History," *American Historical Review* 119 (5) (2014): 1576–86; hereafter abbreviated as "TSH."

4. Nigel Thrift, "Intensities of Feeling: Towards a Spatial Politics of Affect," *Geografiska Annaler* 86 (2004): 58; hereafter abbreviated as "IF." In this passage, Thrift is quoting Lee Spinks, "Thinking the Post-Human: Literature, Affect and the Politics of Style," *Textual Practice* 15 (1) (2001): 24.

5. Eric Shouse, "Feeling, Emotion, Affect," *M/C Journal* 8 (2005): 2–3; hereafter abbreviated as "FEA."

and do not follow its logic is simply a reminder that it is necessary to integrate infolding, or . . . 'implicate order,' into the account. This is necessary to avoid capture and closure on the plane of signification."[6] Likewise, political theorist William Connolly criticizes the "insufficiency of what might be called intellectualist and deliberationist models of thinking," asserting that "culture involves practices in which the porosity of argument is inhabited by more noise, unstated habit, and differential intensities of affect than adamant rationalists acknowledge."[7]

It is clear from such remarks—many others could be cited—that what motivates these scholars is the desire to contest a certain account of how, in their view, political argument and rationality have been thought to operate. These theorists are gripped by the notion that most philosophers and critics in the past (Kantians, neo-Kantians, Habermasians) have overvalued the role of reason and rationality in politics, ethics, and aesthetics, with the result that they have given too flat or "unlayered" or disembodied an account of the ways in which people actually form their political opinions and judgments. The claim is that we human beings are corporeal creatures imbued with subliminal affective intensities and resonances that so decisively influence or condition our political and other beliefs that we ignore those affective intensities and resonances at our peril—not only because doing so leads us to underestimate the political harm that the deliberate manipulation of our affective lives can do, but also because we will otherwise miss the potential for ethical creativity and transformation that "technologies of the self" designed to work on our embodied being can help bring about. As Thrift has put it in still another statement of the position: "[T]he envelope of what we call the political must increasingly expand to take note of the 'way that political attitudes and statements are partly conditioned by intense autonomic bodily reactions that do not simply reproduce the trace of a political intention and cannot be wholly recuperated within an ideological regime of truth'" ("IF," 64).[8]

Now, if it is true, as the authors I have just quoted affirm, that philosophers and critics have largely neglected the important role our corporeal-

6. Brian Massumi, *Parables for the Virtual: Movement, Affect, Sensation* (Durham, NC, and London, 2002), 39, 40, 41, 263; hereafter abbreviated as *PV*. Massumi continues in the passage I have just quoted: "Ideology is construed here in both the commonsense meaning as a structure of belief, and in the cultural-theoretical sense of an interpellative subject positioning" (*PV*, 263).

7. William E. Connolly, *Neuropolitics: Thinking, Culture, Speed* (Minneapolis and London, 2002), 10, 44; hereafter abbreviated as *N*.

8. Thrift is again citing Spinks, "Thinking the Post-Human: Literature, Affect and the Politics of Style," 23.

affective dispositions play in thinking, reasoning, and reflection, then it seems to follow that an account of affect and its place in our lives and institutions is called for. The passages I have cited give a preliminary glimpse of what that account will look like. They suggest that the affects must be viewed as independent of, and in an important sense prior to, "ideology"—that is, prior to intentions, meanings, reasons, and beliefs—because they are non-signifying, autonomic processes that take place below the threshold of conscious awareness and meaning. For the theorists in question, affects are "inhuman," "pre-subjective," "visceral" forces and intensities that influence our thinking and judgments but are separate from these. Whatever else may be meant by the terms *affect* and *emotion*—more on this in a moment—it seems from the remarks quoted above that the affects must be noncognitive, subpersonal, or corporeal processes or states. For such theorists, affect is, as Massumi asserts, "irreducibly bodily and autonomic" (*PV*, 28).

This is an interesting claim, not least because in certain obvious ways it matches the way in which today's psychologists and neuroscientists tend to conceptualize the emotions. As we have seen in previous chapters, for the past thirty years the dominant paradigm in the field of emotions, stemming from the work of the American psychologist Silvan S. Tomkins and his followers, assumes that affective processes occur independently of intention or meaning. According to that paradigm, our affects do not involve cognitions or beliefs about the objects in our world. Rather, they are rapid, phylogenetically old, automatic responses of the organism that have evolved for survival purposes and lack the cognitive characteristics of the higher-order mental processes. On this view, the affects can and do combine with the cognitive processing systems of the brain, but they are essentially separate from those. Tomkins and his followers thus posit a constitutive disjunction between our emotions on the one hand and our knowledge of what causes and maintains them on the other, because according to them affect and cognition are two separate systems. In Tomkins's words, there is a gap or "radical dichotomy between the 'real' causes of affect and the individual's own interpretation of these causes."[9]

In this book, I have given my reasons for questioning the validity of the basic emotions view of the affects. Specifically, I have argued that the experimental evidence for the existence of six or seven (or is it eight or nine

9. Silvan S. Tomkins, *Affect Imagery Consciousness: The Complete Edition* (New York, 2008), vol. 1–2, 137.

or even fifteen?) discrete emotions or "affect programs" located subcortically in the brain and characterized by distinct, universal facial expressions is seriously flawed, and that the theory underlying the paradigm is incoherent. Nor, as has emerged, am I alone in my criticisms. Nevertheless, the Basic Emotion Theory continues to dominate the research field.

Now, at first sight, the Tomkins–Ekman account of the emotions would appear to be too reductive for the purposes of the theorists in the humanities and social sciences whose turn to affect I have been considering so far. It is true that, like Tomkins and Ekman, many of them are committed to understanding the affects in biological terms. As Papoulias and Callard have helpfully observed, fifteen years ago cultural theorists influenced by social constructionism, psychoanalysis, and especially deconstruction tended to exclude the findings of biology from their models of subjectivity and culture for fear of falling into an essentialism they deemed hostile to the possibilities of cultural transformation.[10] But during the past several years, there has been a widespread reaction against what has come to be seen as the straitjacket imposed by the poststructuralist emphasis on language and psychoanalysis, a reaction also motivated by the view that the body in its lived materiality has been neglected in the humanities and social sciences. Within the field of literary studies, Eve Kosofsky Sedgwick, a brilliant critic who died in 2009, has been especially influential in emphasizing the value of Tomkins's approach to the affects for understanding the role of embodiment in (queer) identity formation and change. The general result of these developments has been that, as Thrift has put it, "distance from biology is no longer seen as a prime marker of social and cultural theory. It has become increasingly evident that the biological constitution of being . . . has to be taken into account if performative force

10. Constantina Papoulias and Felicity Callard, "Biology's Gift: Interrogating the Turn to Affect," *Body and Society, Special Issue on Affect*, 16 (1) (2010): 30; hereafter abbreviated as "BG." In this impressive article, the authors criticize the selective ways in which cultural theorists, such as Massumi and Connolly, have used the work of Damasio and other scientists to theorize affect. For another skeptical response to the work of the new affect theorists, especially Sedgwick and Massumi, see Claire Hemmings, "Invoking Affect: Cultural Theory and the Ontological Turn," *Cultural Studies* 19 (5) (2005): 548–67. Recent scholars, who on various grounds oppose the tendency to separate affect from meaning that is the focus of my critique, include Daniel Gross, *The Secret History of Emotion: From Aristotle's Rhetoric to Modern Brain Science* (Chicago, 2006); Martha Nussbaum, *Upheavals of Thought: The Intelligence of the Emotions* (Cambridge, 2001); and Barbara Rosenwein, in "The History of Emotions: An Interview with William Reddy, Barbara Rosenwein, and Peter Stearns."

is ever to be understood, and in particular, the dynamics of birth (and creativity) rather than death" (quoted in "BG," 31).[11]

Thrift's reference to the dynamics of birth and creativity suggests that, in embracing biology, many of today's affect theorists hope to avoid the charge of falling into a crude reductionism by positioning themselves at a distance from the geneticism and determinism that were a target of the previous phase of cultural theory. Instead, they seek to recast biology in dynamic, energistic, non-deterministic terms that emphasize its unpredictable and potentially emancipatory qualities.[12] Moreover, drawing on writings by Lucretius, Spinoza, Bergson, William James, Whitehead, and other dissenting philosophers of nature, especially two recent figures, Gilles Deleuze and Félix Guattari, many of these theorists make a distinction between "affect" and "emotion" in terms that, again at first sight, seem different from those of the Basic Emotion Theory.[13] Massumi, widely credited with emphasizing that distinction, defines affect as a nonsignifying, nonconscious "intensity" disconnected from the subjective, signifying, functional-meaning axis to which the more familiar psychological categories of emotion belong. "In the absence of an asignifying philosophy of affect," Massumi writes, "it is all too easy for received psychological categories to slip back in, undoing the considerable deconstructive work that has been effectively carried out by poststructuralism.

11. For further examples of the shift from deconstruction, language, and psychoanalysis to affect and embodiment, see also *The Affective Turn: Theorizing the Social*, ed. Patricia Ticineto Clough, with Jean Halley, foreword by Michael Hardt (Durham, NC, 2007); Maria Angel, "Brainfood: Rationality, Aesthetics, and Economies of Affect," *Textual Practice* 19 (2) (2005): 323–48; Elizabeth Wilson, *Psychosomatic: Feminism and the Neurological Body* (Durham, NC, and London, 2004); Teresa Brennan, *The Transmission of Affect* (Ithaca, NY, 2004); Derek P. McCormack, "Molecular Affects in Human Geographies," *Environment and Planning 1* (39) (2007): 359–77; Elizabeth Wilson, *Affect and Artificial Intelligence* (Seattle, 2010); Adam Frank, *Transferential Poetics: From Poe to Warhol* (New York, 2014). For an exchange over affect between Frank, Wilson, and Charles Altieri and myself, see Adam Frank and Elizabeth E. Wilson, "Like-Minded," *Critical Inquiry* 38 (Summer 2012): 870–77; Charles Altieri, "Affect, Intentionality, and Cognition: A Response to Ruth Leys," in ibid., 878–81; and Leys, "Facts and Moods: Reply to My Critics," in ibid., 882–91.

12. For a discussion of the influence of ideas about chaos and complexity, associated with the work of Ilya Prigogine and Isabelle Stengers, on the theorization of affect see Clough, introduction, *The Affective Turn: Theorizing the Social,* 1–33.

13. Probably the most influential figure in the rise of the new affect theory is Deleuze, but it is invariably an open question as to the accuracy with which one or another affect theory represents his views. In this chapter, I shall leave this question to the side in order to focus on the claims made by the theorists under consideration here.

Affect is most often used loosely as a synonym for emotion. But . . . emotion and affect—if affect is intensity—follow different logics and pertain to different orders" (*PV*, 27).[14]

Similarly, Thrift rejects or sets aside approaches that "tend to work with a notion of individualised emotions (such as are often found in certain forms of empirical sociology and psychology)" in favor of approaches that posit "broad tendencies and lines of force" and in which, adhering to an "'inhuman'" or "'transhuman'" framework, "individuals are generally understood as effects of the events to which their body parts (broadly understood) respond and in which they participate" ("IF," 60). Likewise, Shouse follows Massumi by remarking that "it is important not to confuse affect with feelings and emotions . . . [A]ffect is not a personal feeling. Feelings are *personal and biographical*, emotions are *social*, and affects are *prepersonal* . . . An affect is a non-conscious experience of intensity; it is a moment of unformed and unstructured potential . . . Affect cannot be fully realized in language . . . because affect is always prior to and/or outside consciousness . . . Affect is the body's way of preparing itself for action in a given circumstance by adding a quantitative dimension of intensity to the quality of an experience. The body has a grammar of its own that cannot be fully captured in language" ("FEA," para. 1, 5; his emphasis).[15]

The claim that affect is a formless, unstructured, non-signifying force or "intensity" that escapes the categories of the psychologists suggests that Tomkins's or Ekman's or Damasio's talk about the existence of six or seven or eight or nine structured, evolved categories of innate emotions is at odds with the views of writers such as Massumi who espouse Spinozist–Deleuzean ideas about affect. Yet it is striking how compatible Deleuze-inspired definitions of affect as a nonlinguistic, bodily "intensity" turn out to be with the Tomkins–Ekman paradigm. Thus, Thrift states that he wants to avoid the emotion categories of the empirical psychologists and social scientists. But he then proceeds to draw on four "translations"

14. Massumi continues: "An emotion is a subjective content, the sociolinguistic fixing of the quality of an experience which is from that point onward defined as personal. Emotion is qualified intensity, the conventional, consensual point of insertion of intensity into semantically and semiotically formed progressions, into narrativizable action-reaction circuits, into function and meaning. It is intensity owned and recognized" (*PV*, 28).

15. In many texts, the concept of affect is tied to a "nonrepresentationalist" ontology that defines affect in terms derived from Spinoza as *the capacity to affect and be affected*. Characterized in this way, affect is seen to function as a layer of preconscious "priming to act" such that embodied action is a matter of being attuned to and coping with the world, without the input of rational content.

of affect that include references to the ideas of Tomkins, Ekman, and Damasio—the last of whom, in spite of a declared Spinozism and anti-dualism that makes his work especially attractive to many cultural critics, follows the Tomkins–Ekman paradigm in his approach to the study of the basic emotions ("IF," 61–64).

Similarly, in a discussion of "berserker rage," cultural theorist John Protevi combines appeals to Deleuzean ideas about emergent assemblages and the Tomkins–Ekman theory of the basic emotions to explain why the agent of killing during military combat is not the individual soldier or person but the precognitive, affective processes and forces that overtake him in the moment of battle. "[I]n some cases the military unit and non-subjective reflexes and basic emotions are intertwined in such a way as to bypass the soldiers' subjectivity controlled intentional action," Protevi observes. "In these cases the practical agent of the act of killing is not the individual person or subject, but the emergent assemblage of military unit and non-subjective reflex or equally non-subjective 'affect program.'"[16] In other words, Protevi links Deleuze and Guattari's ideas about "emergent assemblages" to Ekman's theory of "affect programs" in order to suggest that in the act of killing a form of subpersonal "agency" short-circuits the individual's capacity for intentional action.

The regularity with which Deleuze-inspired affect theorists find a use for such Ekman-style scientific approaches to the emotions suggests that however complex the negotiations between such theorists and neuroscientists are said to be, and however those negotiations are described—as involving a renewed "conversation" between the humanities and neurosciences, or as involving a more inventive and shameless form of borrowing by the humanities from the sciences[17]—what fundamentally binds together the new affect theorists and the neuroscientists is their *shared anti-intentionalism.* My claim is that whatever differences of philosophical-intellectual orientation there may be among the new affect theorists themselves, and between them and the neuroscientists whose findings they wish to appropriate (differences do of course exist), the important point

16. John Protevi, "Affect, Agency and Responsibility: The Act of Killing in the Age of Cyborgs," *Phenomenology and Cognitive Sciences* 7 (2008): 408.

17. Connolly states that his aim is not to "derive the logic of cultural activity" from the neurosciences, but to "pursue *conversations* between cultural theory and neuroscience" (*N*, 8); Massumi declares that the point is to "borrow from science in order to make a difference in the humanities," a process he also characterizes as a kind of "piracy" or "poaching" (*PV*, 20, 21).

to recognize is that they all share a single belief: the belief that affect is independent of signification and meaning. In short, I propose that although at first sight the work of Tomkins—or Ekman, or Damasio—might appear to be too reductive for the purposes of those cultural theorists indebted to Deleuzean ideas about affect, there is in fact a deep coherence between the ideas of both groups.

That coherence concerns precisely the separation presumed to obtain between the affect system on the one hand and intention or meaning or cognition on the other. For both the new affect theorists and the neuroscientists from whom they variously borrow—and transcending differences of philosophical background, approach, and orientation—affect is a matter of subpersonal, autonomic responses that are held to occur below the threshold of consciousness and cognition and to be rooted in the body. What the new affect theorists and the neuroscientists share is a commitment to the idea that there is a gap between the subject's affects and its cognition or appraisal of the affective situation or object, such that cognition or thinking comes "too late" for reasons, beliefs, intentions, and meanings to play the role in action and behavior usually accorded to them. The result is that action and behavior are held to be determined by affective dispositions that are independent of mindedness.

This is the thesis I wish to test in the remainder of this chapter. What I propose to do is to examine the interface between the new affect theory and the neurosciences by examining some experimental studies that play strategic roles in recent arguments about affect. Much of the time, Massumi, whose use of experiments I shall be examining, engages in rather opaque philosophical-speculative reflections in which the neurosciences make only fleeting appearances. In texts by Connolly, whose employment of experiment I shall also briefly consider, the neurosciences play a more prominent role. But not only do both scholars argue for the importance of the neurosciences in the study of affect, they also appeal to particular neuroscientific experiments in order to justify their views, and it is this that interests me. In selecting for analysis and discussion three such experiments and the uses that have been made of them, I shall be following the method of working through examples advocated by Massumi, a method whose success, he observes, "hinges on the details" (*PV*, 18).

THE SNOWMAN EXPERIMENT

My first example comes from Massumi's influential essay "The Autonomy of Affect" (1995), which from the outset plunges us into the minutiae of a little-known 1980 German study of the emotional effects of the media.[18] The study in question was undertaken when a short film with sound but no words was shown on Munich TV as a filler between programs. The basic plot was simple. "A man builds a snowman on his roof garden. It starts to melt in the afternoon sun. He watches. After a time, he takes the snowman to the cool of the mountains where it stops melting. He bids it good-bye and leaves" (*PV*, 23). The film drew protests from parents complaining that it had frightened their children. A team of investigators, headed by media researcher Hertha Sturm, decided to assess the film's emotional impact by conducting several experiments. The team used three versions of the film: the original version, a version with a "factual" soundtrack commenting on the various situations and actions, and a version in which the factual text was further (slightly) supplemented with emotional attributions. The verbal material added to the original film consisted of fifteen short sentences of fifty seconds each; each version of the film was twenty-eight minutes long.[19]

The emotional reactions of nine-year-old children from an elementary school in Vienna were tested on three levels: the physiological, the verbal-cognitive, and the motor. On the *physiological level*, the variables of heart rate, respiration, and skin conductance were measured by using peripheral recording devices: a clip on the middle finger monitored the children's heart rates; a belt measured respiration frequency; and electrodes on the children's hands recorded the galvanic skin responses. On the *verbal-cognitive level*, three variables were selected for testing. Inquiries were made during the presentation of the film versions about whether a given scene was "pleasant" or "unpleasant." A disc showing laughing and weeping faces on the ends of a scale allowed the children to choose between degrees of "happy" or "sad." Also, the children's recollections of what they had seen were registered by asking them to "reproduce" as many of the film scenes as possible (meaning what exactly? It is not clear). Finally, the *motor level* of response was measured by making videorecordings of the children's "mimic" reactions while they were watching one

18. Massumi's essay "The Autonomy of Affect" was first published in 1995 and was reprinted in a slightly revised form as chapter 1 in *Parables for the Virtual.*

19. See Hertha Sturm and Marianne Grewe-Partsch, "Television—The Emotional Medium: Results from Three Studies," in *Emotional Effects of Media: The Work of Hertha Sturm*, ed. Gertrude Joch Robinson (Montreal, 1987), 25–44; hereafter abbreviated as "TEM." In fact, Massumi gives a somewhat simplified account of the film's content. Without specifying further, Sturm and Grewe-Partsch note that "after some thought and complications" the man takes the snowman into the high mountains where the snowman will not melt ("TEM," 30).

of the film versions (I take this to mean that the children's facial-bodily movements were [secretly?] videotaped during the screenings). In addition, the investigators collected various kinds of personal data about the children and their television viewing habits through personality tests and parent interviews. Different groups of the children were exposed to the three different versions of the film, and the results were compared. The experiment was also designed to assess the impact of repetition by reshowing the films after an interval of three weeks ("TEM," 31).

In their overview of the experiments, Sturm and coauthor Marianne Grewe-Partsch remark that the findings were "extremely complex" ("TEM," 30).[20] The summary of the physiological data obtained from the first film presentations reports that the children who saw the "factual" version of the film had a higher heart rate than the children who saw the other versions. According to the authors, the higher heart rate indicated a higher activation level during the presentation of the film, suggesting that the children were more aroused by the factual version than by the other two versions. This result was reinforced by the finding that skin resistance decreased during the presentation of the factual version. Sturm and Grewe-Partsch observe that such a decrease in skin resistance (or increase in skin conductance) is usually linked to an increase in the general activation level. No significant differences in respiration frequency between the three film versions were found. The verbal-cognitive data showed that the children judged the factual version of the film to be significantly less pleasant than the other versions, whereas the original, wordless version was considered the most agreeable. The authors report that the highest level of galvanic skin response accompanied the original version (I take this to mean that skin resistance was high, indicating lower conductance owing to the lower level of arousal or activation). No differences in the "happy-sad" evaluation of the three versions were found. As for recall, the emotional version of the film was clearly the most easily remembered.

In order to explore further the effects of dramaturgy, such as cuts, zooms, and lighting, the investigators divided the three film versions into ten segments. No differences in physiological reactions to these segments were found, except that, as expected, respiration seemed to run parallel to

20. In researching the effects of television, Sturm and Grewe-Partsch made use of a concept of arousal that distinguished between two arousal systems: a "reticular" activating system, viewed as a primary apparatus for producing cortical arousal, and a "limbic" system responsible for vegetative processes, including the emotions. They suggested that there occurs an interplay and potential interference between the two systems ("TEM," 29). But Massumi makes no reference to this distinction.

dramaturgy. Decreases in skin resistance during the ten scenes were interpreted as an increase in activation owing to increased attention to and interest in the film segments being presented. On the cognitive level, however, the ten segments received different ratings of "pleasant-unpleasant" and "happy-sad," showing a similar trend in all three versions. Thus, the children scaled scenes 1–3 as "sad," scenes 4–8 as more "happy," and scenes 9–10 as "sad." The authors report an inverse relationship between the judgment "happy-sad" and the scaling of "pleasant-unpleasant," in that the sadder the scene was perceived to be, the more pleasant it was rated. Retention of the ten scenes was related to these ratings in that the more pleasantly experienced scenes were the ones that were better remembered ("TEM," 32).

The authors further observe that the repetition findings produced some unanticipated and interesting results. Whereas the second viewing of the nonverbal and emotional versions of the films decreased heart rate, it increased heart rate in the factual version, even reaching a higher level than in the first presentation, while respiration and skin responses showed no significant differences. Moreover, all three versions were rated more pleasant the second time around, though the factual one was again rated as least pleasant.

In their report, the authors express surprise that the factual film version produced the highest level of activation in the viewers, in spite of the fact that it differed from the emotional version by only a minimal substitution of words. They observe that all these findings were congruent with physiological theories according to which moderate increases in arousal are perceived as pleasant, whereas extreme increases in arousal are perceived as unpleasant. They note in this regard that these findings had been corroborated by D. E. Berlyne, who had shown that a sequential decrease of arousal after short activation is experienced as "pleasant," while an increase in activation that lasts longer is experienced as "unpleasant." "This explains why the factual version was perceived as more 'unpleasant' than the two other versions. The viewers remained on a higher activation level for a long period of time" ("TEM," 32). The authors suggest that the same high activation level appeared to be the cause of the poorer recall, because it inhibited memory.

On the basis of these findings, the investigators conclude that very slight verbal changes in the form of the presentation resulted in quite different viewer experiences. The results demonstrated that viewers reacted to minute changes in word-picture relationships and that narrative connections were extremely important—a finding of interest for those wishing to understand the effects of the media. The authors especially note the clear results obtained in the case of the factual version because viewers of that version experienced a distinctly higher physiological arousal than that of the other groups of children. The authors suggest that the latter effect might be the result of a discrepancy between the emotional-visual presentation and the factual text, a discrepancy apparently experienced as "unpleasant," since the viewers rated this version as such. This result confirmed Berlyne's hypothesis: a strong increase of activation is experienced as unpleasant. On the other hand, viewers experienced presentations as pleasant when pictures and language coincided. The investigators propose that their results also might explain the well-documented finding that viewers recall TV news programs poorly, suggesting that the discrepancy between emotional pictures and the factual language employed in such programs interferes with comprehension and recall. Nevertheless, Sturm and Grewe Partsch state that it was a "total surprise" to realize that the children judged the "sad" scenes in the film segments as "pleasant": "The sadder the segment the higher was the positive estimation and the deeper was the respiration" ("TEM," 33).

These are the findings that interest Massumi and that he uses to help establish his views on affect. What immediately attracts Massumi's attention is the apparent discrepancy between the children's happy-sad and pleasant-unpleasant responses to the snowman film segments—the fact that the saddest film segments were also rated the most pleasant. Rejecting the possibility of any ambiguity or vagueness in the scales or tests used to test the children's responses (how would one quantify the scale "pleasant-unpleasant"? would a "more pleasant" rating mean that the child experienced more affect?), and evidently regarding as inherently contradictory the idea that someone could simultaneously experience something as both sad and pleasant (but are not sad films sometimes also pleasurable or enjoyable?), Massumi locates affect in the alleged incongruity between the children's responses on these two scales. According to him, the experimental data suggest the existence of a bifurcation between two different responses or systems, or what he characterizes as a gap between the *content* of the image and its *effect*. He writes, "the primacy of the affective is marked by a gap between *content* and *effect*: it would appear that the strength or duration of an image's effect is not logically connected to the content in any straightforward way" (*PV*, 24).

Taking the research findings in a direction not envisaged by Sturm and her team, Massumi therefore claims that the content of the image is "its indexing to conventional meanings in an intersubjective context, its sociolinguistic qualification. This indexing fixes the determinate *qualities* of the image" (*PV*, 24). (It's unclear to me, though, what he thinks the "conventional meanings" of such images might be—the man leaving the snowman in the mountains, for example.) But the "strength or duration of the film image's effect," or what he calls its "*intensity*," is characterized by him as involving an asignifying logic or "crossing of semantic wires: on it, sadness is pleasant" (*PV*, 24). Massumi argues that there is an immediate "bifurcation in response" into two systems:

> The level of intensity is organized according to a logic that does not admit the excluded middle. This is to say that it is not semantically or semiotically ordered. It does not fix distinctions. Instead, it vaguely but insistently connects what is normally indexed as separate. When asked to signify itself, it can only do so in a paradox. There is disconnection of signifying order from intensity—which constitutes a different order of connection operating in parallel. The gap . . . is not only between content and effect. It is also between the form of content—signification as a conventional system of distinctive difference—and intensity. (*PV*, 24–25)

Furthermore, Massumi states that this gap between content (in the form of signification) and effect (which appears as intensity) is matched by a comparable gap between two different kinds of embodiment. In a move seemingly unwarranted by the experimental results, Massumi claims that the children in the experiment were "physiologically split" because "factuality made their heart beat faster and deepened their breathing, but it also made their skin resistance fall" (*PV*, 24). Sturm and her group of researchers did not view these results as incoherent or contradictory because, they suggested, both decreased skin resistance—that is, increased skin conductance—and increased heart and lung responses are usually linked to an increase in emotional-cognitive activation (see "TEM," 30). But Massumi takes these findings to demonstrate the existence of a bifurcation or distinction between the depth of the heart and lungs and the surface of the skin to match the distinction between "qualification" (or meaning) on the one hand and "intensity" on the other. He asserts that although both qualification (or meaning) and intensity (or non-signifying affective intensity) are embodied, depth reactions "belong more to the form/content (qualification) level, even though they also involve autonomic functions such as heartbeat and breathing," whereas intensity is "embodied in purely autonomic reactions most directly manifested in the skin—at the surface of the body, at its interface with things" (*PV*, 25).

Massumi then uses this distinction between depth and surface to buttress the idea of the existence of two different systems: the one, a "conscious-automatic" system functioning on a depth or vertical axis and linked by him with signification, expectation, common sense, and narrative continuity and associated with modulations of heart and breathing; the other, an intensity system functioning on a horizontal or surface axis "spreading over the generalized body surface like a lateral backwash from the function-meaning interloops that travel the vertical path between head and heart" and defined by him as an autonomic reaction system that is separate from meaning and signification. "Intensity is . . . a non-conscious, never-to-be- conscious autonomic remainder," he writes. "It is outside expectation and adaptation, as disconnected from meaningful sequencing, from narration, as it is from vital function" (*PV*, 25).

The "meaning" and "intensity" (or affect) systems are said by Massumi to resonate or interfere with one another in various ways, but to the system of intensity belong all the attributes so prized by today's self-professed Deleuzean affect theorists—the attributes of the non-semantic, the nonlinear, the autonomous, the vital, the singular, the new, the anomalous, the indeterminate, the unpredictable, and the disruption of fixed or "conventional" meanings.[21] For Massumi, the system of intensity is the "system of the inexplicable: emergence, into and against regeneration (the reproduction of a structure). In the case of the snowman, the unexpected and inexplicable that emerged along with the generated responses had to do with the differences between happiness and sadness, children and adults, not being all they're cracked up to be, much to our scientific chagrin: a change in the rules. Intensity is the unassimilable" (*PV*, 27). And intensity is another word for affect defined in these asignifying terms. "For present purposes," Massumi writes, "intensity will be equated with affect" (*PV*, 27). Further on: "Affect is autonomous to the degree to which it escapes confinement in the particular body whose vitality, or potential for interaction, it is. Formed, qualified, situated perceptions and cogni-

21. Thus, Massumi says that the factual version of the film "dampens" intensity: "Matter-of-factness dampens intensity. In this case, matter-of-factness was a doubling of the sequence of images with narration expressing in as objective a manner as possible the commonsense function and consensual meaning of the movements perceived on screen. This interfered with the 'images' effect" (*PV*, 25). Massumi suggests that the phrases or textual "qualifications" added to the emotional film version "enhanced the images' effect, as if they resonated with the level of intensity rather than interfering with it. An emotional qualification breaks narrative continuity for a moment to register a state—actually to re-register an already felt state, for the skin is faster than the word" (*PV*, 25). The impression conveyed by these less than perspicuous remarks is that Massumi thinks the lowered skin resistance in the viewers of the factual film version was a measure of the dampening of intensity. But since lowered skin resistance is a sign of increased skin conductance associated with higher arousal, it is unclear from the data what if anything was dampened in this version.

tions fulfilling functions of actual connection or blockage are the capture and closure of affect. Emotion is the most intense (most contracted) expression of that *capture*—and of the fact that something has always and again escaped" (*PV*, 35; his emphasis). *Affect* is the name for what eludes cognition and meaning.

It is important to notice that Massumi imposes on Sturm's experimental findings an interpretation motivated by a set of assumptions about the asignifying nature of affect. These assumptions drive his analysis of Sturm's data in order to produce a distinction between, on the one hand, the conscious, signifying ("emotional" and intellectual) processes held to be captive to the fixity of received meanings and categories, and on the other hand, the nonconscious affective processes of intensity held to be autonomous from signification. *Differently from Tomkins and Ekman but to the same end*, Massumi conceptualizes affect as inherently independent of meaning and intention. What he and other Deleuzean affect theorists share with Tomkins and Ekman—hence also with Sedgwick, Smail, and Hunt—is a commitment to the idea that there is a disjunction or gap between the subject's affective processes and his or her cognition or knowledge of the objects that caused them. The result is that the body not only "senses" and performs a kind of "thinking" below the threshold of conscious recognition and meaning, but—as we shall see in a moment—because of the *speed* with which the autonomic, affective processes are said to occur, it does all this before the mind has time to intervene.

And now the larger stakes of Massumi's effort to distinguish "affect" from signification begin to become clear. He is not interested in the cognitive content or meaning that film or fictional or artistic or political representations may have for the audience or viewer, but rather in the effects of those representations on the subject regardless of signification. The whole point of the turn to affect by Massumi and like-minded cultural critics is to shift attention away from considerations of meaning or "ideology" or indeed representation to the subject's subpersonal material-affective responses, where, it is claimed, political and other influences do their real work. The disconnect between "ideology" and affect produces as one of its consequences a relative indifference to the role of ideas and beliefs in politics, culture, and art in favor of an "ontological" concern with different people's affective experiences and reactions. We find a similar disconnect between meaning and affect in Smail's neurohistory, where, for example, gossip is said to have nothing to do with meaning but is a "meaningless social chatter whose only function is the mutual stimulation of peace-and-contentment hormones." Gossip on this model remains important as a medium of communication, as Smail observes, but what get communicated are not primarily words and their meanings but "chemical messengers" (*ODH*, 176).

For both the affect theorists and Smail, then, political campaigns, advertising, literature, visual images, and the mass media are all mechanisms for producing such effects below the threshold of meaning and ideology. An entire aesthetic is involved here, one that emphasizes the reader's or viewer's affective experience of a text or image to the extent that that experience might be said to stand in for the text or image in question. An opposite position would insist that although a work of art might make us feel happy or sad or envious or ashamed, what matters is the meaning of the work itself, which is to say the structure of intentional relationships built into it by the artist. The fact that a novel or painting makes me feel or think a certain way may be a significant aspect of my response to the work, but simply as my response, it has no standing as an interpretation of it. But cultural theorists who have turned to affect convert questions about the meaning of works of art into ones concerning their traumatic-affective effect or influence on the reader or viewer.[22] In short, according to such theorists—and here we may recognize a resonance between Massumi's ideas about ideology and Tomkins's earlier claims about the psychology of knowledge—affect has the potential to transform individuals for good or ill without regard to the content of argument or debate. These are the reasons Massumi and the others are interested in scientific studies al-

22. See for example Mark Hansen, "The Time of Affect, or Bearing Witness to Life," *Critical Inquiry* 30 (Spring 2004): 584–26; Mark Hansen, *New Philosophy for New Media* (Cambridge, MA, 2006); Jill Bennett, *Empathic Vision: Affect, Trauma, and Contemporary Art* (Stanford, CA, 2005); and Marco Abel, *Violent Affect: Literature, Cinema, and Critique After Representation* (Lincoln, NE, 2007). The position adopted by these affect theorists is recognizable as a version of the "affective fallacy" defined by the critics W. K. Wimsatt and Monroe Beardsley as the error of judging the importance or success of a work of art in terms of its emotional effects on the reader. In art criticism, the same issue has been a focus of debate ever since Michael Fried, in "Art and Objecthood," defended high modernism against minimalism (or, as he also called it, literalism) on the grounds that the minimalist/literalist position made the viewer's subjective, present-tense experience stand in for—take the place of—the work itself (I am simplifying, of course). See Michael Fried, "Art and Objecthood" (1967), in Fried, *Art and Objecthood: Essays and Reviews* (Chicago and London, 1998), 148–72. See also Fried, "An Introduction to My Art Criticism," in ibid., 1–74. Fried's views are defended and generalized in Walter Benn Michaels, *The Shape of the Signifier: 1967 to the End of History* (Princeton, 2004). For an incisive critique of affective neuroaesthetics, see also Jennifer Ashton, "Two Problems with a Neuroaesthetic Theory of Interpretation," *nonsite.org*, no. 2 (2011). See also Leys, "Trauma and the Turn to Affect," in *Trauma, Memory, and Narrative in the Contemporary South African Novel: Essays*, eds. Ewald Mengele and Michela Borzaga (Amsterdam and New York, 2012), 3–27.

legedly showing that affective processes and even a kind of intelligence go on in the body independently of cognition or consciousness and that the mind operates too late to intervene.[23]

THE MISSING HALF SECOND

To be exact, according to Massumi and many others, the mind intervenes *half a second too late* to play the role usually attributed to it in human behavior. In proof of this claim, Massumi in "The Autonomy of Affect" makes use of a well-known experiment on consciousness and the body that has come to play a strategic role in his and other like-minded theorists' arguments about affect. The experiment in question concerns the relationship between conscious intention and brain activity and belongs to a group of studies performed over a stretch of years between the 1970s and the 1990s by the American scientist Benjamin Libet.

In the experiment briefly described by Massumi, subjects with their hands on a tabletop were asked to flex a finger at a moment of their choosing and to report when they were first aware of their decision or intention to perform that movement by noting the spatial position of a revolving dot on a large clock that measured fractions of a second. Libet found that the actual finger flexes occurred 0.2 seconds after the experimental subject clocked his decision, but that the EEG machine employed to monitor brain activity registered significant activity 0.3 seconds *before* the subject registered his awareness of his decision. In other words, there seemed to be *a half-second delay* between the start of the body-brain event and its

23. According to Thrift and others who share his views, affective responses involve a kind of "thinking" that takes place in a non-reflective, nonrepresentational manner in the form of embodied habits, that is, in the form of subpersonal bodily thinking that is said to precede cognition and intentionality. "Only the smallest part of thinking is explicitly cognitive," he states. "Where, then, does all the other thinking lie? It lies in body, understood not as a fixed residence for 'mind,' but as a 'dynamic trajectory by which we learn to register and become sensitive to what the world is made of'" (citing Bruno Latour) (Thrift, "Summoning Life," in *Envisioning Human Geographies*, eds. Paul Cloke, Philip Crang, and Mark Goodwin [London, 2004], 90). The word "explicitly" in this statement might suggest that the author believes bodily thinking is connected to "implicitly" cognitive capacities. But the rapidity with which Thrift turns to the body as the source of this "thinking" suggests that he imagines such modes of intelligence to be entirely corporeal in nature—as if bodily thinking, embodied habits, and skillful copings can be theorized in entirely nonconceptual, noncognitive terms. The emphasis thus falls on the role of affective neural and neurochemical networks considered to be capable of emergent, unpredictable activity creating possibilities for political and personal change.

completion in the form of the movement of the finger. Libet concluded that unconscious cerebral processes initiate voluntary actions before conscious intention appears, although, he proposed, the brain fools us into thinking that we consciously decide matters and that our actions are personal events.[24] As Massumi reports, when Libet was asked to speculate on the implications of his findings for the doctrine of free will, he suggested that "'we may exert free will not by initiating intentions but by vetoing, acceding or otherwise responding to them after they arise'" (*PV*, 29).[25]

On the basis of the "exemplary case" of Libet's experiment (*PV*, 206), Massumi concludes that the "half second is missed not because it is empty, but because it is overfull, in excess of the actually-performed action and of its ascribed meaning." As he puts it, during the mysterious half second

> what we think of as "free," "higher" functions, such as volition, are apparently being performed by autonomic, bodily reactions occurring in the brain but outside consciousness, and between brain and finger but prior to action and expression. The formation of a volition is necessarily accompanied and aided by cognitive functions. Perhaps the snowman researchers of our first story couldn't find cognition because they were looking for it in the wrong place—in the "mind," rather than in *the body* they were monitoring. (*PV*, 29; his emphasis)[26]

As in the case of his analysis of the experiments by Sturm and her group, so in the case of his interpretation of Libet's findings, Massumi's emphasis falls on the determining role in thinking of subpersonal affective

24. See Benjamin Libet, "Unconscious Cerebral Initiative and the Role of Conscious Will in Voluntary Action," *Behavioral and Brain Sciences* 8 (December 1985): 529–39.

25. Massumi is quoting from John Horgan's article on Libet in "Can Science Explain Consciousness?," *Scientific American* (July 1994): 76–77. In a note, he also gives the reference to Libet's original 1985 paper.

26. Massumi denies at this juncture that his ideas about affect or intensity involve an "appeal to a prereflexive, romantically raw domain of primitive experiential richness—the nature in our culture. It is not that." It is not that for him, first, "because something that is happening out of mind in a body directly absorbing its outside cannot exactly be said to be experienced" (*PV*, 29); and second, "because volition, cognition, and presumably other 'higher' functions usually presumed to be in the mind . . . are present and active in that now not-so-'raw' domain. Resonation assumes feedback. 'Higher functions' belonging to the realm of qualified form/content in which identified, self-expressive persons interact in conventionalized action-reaction circuits, following a linear time line, are fed back into the realm of intensity and recursive causality" (*PV*, 29–30). For further references to Libet's ideas by Massumi, see *PV*, 195, 206.

processes. "Thought lags behind itself," Massumi observes in another reference to Libet's experiment. "It can never catch up with its own beginnings. The half-second of thought-forming is forever lost in darkness. All awareness emerges from a nonconscious thought-o-genic lapse indistinguishable from movements of matter" (*PV*, 195). Simply put, he takes Libet's experiment to prove that the material processes of the body-brain generate our thoughts, and that conscious thought or intention arrives too late to do anything other than supervise the results.[27]

But is this interpretation of Libet's findings valid? There are good reasons to doubt it. I will pass over the technical criticisms that have been leveled at Libet's experiment, in order to focus on some of the more conceptual-philosophical problems it raises.[28] Massumi and many other cultural theorists present themselves as Spinozists who oppose dualism in all its guises. Yet a little reflection suffices to demonstrate that in fact a classical dualism of mind and body informs both Libet's and Massumi's shared interpretation of Libet's experimental findings. Indeed, it is only by adopting a highly idealized or metaphysical picture of the mind as completely separate from the body and brain to which it freely directs its intentions and decisions that they can reach the skeptical conclusions they do.

27. Similarly, Thrift cites Damasio, Libet, Ekman, and others in order to claim that "we are 'late for consciousness' [citing Damasio, *The Feeling of What Happens*, 1999] . . . That insight was subsequently formalized in the 1960s by Libet using new body recording technologies. He was able to show decisively that an action is set in motion before we decide to perform it . . . 'the brain makes us ready for action, then we have the experience of acting' . . . Thus we can now understand emotions as a kind of corporeal thinking [citing LeDoux, *The Emotional Brain,* and Damasio's texts again]" ("IF," 67). In support of similar ideas, Thrift cites Tor Norretranders's *The User Illusion*, whose discussion of Libet's experiments and general claim that our consciousness is a user illusion is also cited favorably by Massumi and many other affect theorists (Norretranders also blurbs Smail's *Of Deep History and the Brain*). As Norretranders puts it: "Even when we think we make a conscious decision to act, our brain starts a half second before we do so! Our consciousness is not the initiator—unconscious processes are! . . . Our consciousness dupes us!" (Tor Norretranders, *The User Illusion: Cutting Consciousness Down to Size* [New York, 1999], 220).

28. For a helpful collection of essays on Libet, see *Does Consciousness Cause Behavior?*, eds. Susan Pockett, William P. Banks, and Shaun Gallagher (Cambridge, MA, 2006). See also M. R. Bennett and P. M. S. Hacker, *Philosophical Foundations of Neuroscience* (Malden, MA, and Oxford, 2003), 228–31; and Daniel C. Dennett, *Freedom Evolves* (London, 2004), 221–57. For the recent debate in Germany between neurobiologists and philosophers over Libet's experiments and their implications for theories of agency and free will, see also Benedikt Korf, "A Neural Turn? On the Ontology of the Geographical Subject," *Environment and Planning A* 40 (2008): 715–32.

Already in 1985, in discussions Massumi ignores, several of the researchers invited to comment on Libet's results observed that the kinds of finger and wrist movements the subjects in the experiment were asked to perform were those that are normally carried out without one's awareness of an intention to act. They thus suggested that in his experiment Libet had imposed an artificial requirement when he asked his subjects to pay conscious attention to such movements. As these researchers pointed out, skilled pianists are not consciously aware of the innumerable movements their fingers must enact during a performance, but this does not make those movements unintentional or negate the fact that the pianists intended to play the music. Indeed, as Libet's critics also argued, the movements Libet's experimental subjects were asked to perform were part of an overall intentional structure or situation that included the subjects' willingness (that is, their intention) to participate in the experiment and to comply with the researcher's expectations.[29]

Furthermore, all the subjects went into the experiment knowing what actions they were expected to perform, and if they were uncertain they were allowed to practice them first. As science writer John McCrone has observed of criticisms by the well-known consciousness researcher Bernard Baars: "Baars could . . . see that there was a deceptive simplicity to Libet's finger-lifting task which was to blame for much of the controversy. People were taking the experiment to mean that lower-level brain processes generated our thoughts and the conscious-level mind arrives too late in the day to do more than supervise the results. Yet that was not the case at all . . . As Baars put it, there was always a conscious-level context in place, framing whatever occurred." McCrone adds: "Libet's subjects knew what action to produce. The point that students went into the freewill experiment with a consciously-held context was obvious to many commentators."[30] In short, it is a confusion on both Libet's and Massumi's part to think that because such actions usually go on automatically, below the threshold of consciousness, it is necessary to break with the whole idea of intentionality and to assume that they can be explained only in corporeal terms.

The problem here is not the idea that many bodily (and mental) pro-

29. For technical and conceptual criticisms of Libet's experiments by Bruno G. Brietmeyer, Arthur C. Danto, Richard Latto, Donald M. McKay, and others, and for Libet's responses to these, see "Open Peer Commentary," *Behavioral and Brain Sciences* 8 (Dec. 1985): 539–66.

30. J. McCrone, *Going Inside: A Tour Round a Single Moment of Consciousness* (London, 1999), 137–38, 334n.

cesses take place subliminally, below the threshold of awareness. Who would dream of doubting that they do? Rather, the problem concerns the implications Massumi appears to draw from this state of affairs. Philosopher Shaun Gallagher has recently argued that it is only when normal motor control mechanisms fail that people are put in the position of Libet's experimental subjects. So, for example, patients suffering from deafferention, or complete loss of proprioceptive feedback, lack the normally automatic processes governing motor behavior, which means they are compelled to think consciously every time they make a normally habitual movement. Such patients find it necessary to make a conscious mental decision for every simple motion, with the result that they can barely move properly at all. As Gallagher points out, such pathological cases require a picture of mental causation that is completely in line with the standard Cartesian account of the mind as a mental space separate from the body in which the subject freely controls his or her own thoughts and actions.[31] In other words, both Massumi and Libet seem to be in the grip of a false picture of how the mind relates to the body. Their error is to idealize the mind by defining it as a purely disembodied consciousness and then, when the artificial requirements of the experimental setup appear to indicate that consciousness of the willing or intention comes "too late" in the sequence of events to account for the movements under study, to conclude in dualist fashion that intentionality has no place in the initiation of such movements and that therefore it must be the brain that does all the thinking and feeling and moving for us. (All the "willing," so to speak.)

Gallagher offers his critique of Libet's experimental work from the perspective of a phenomenology of embodiment.[32] But Massumi rejects phenomenology on the grounds that its intentional structures remain stuck in repetition and prevent the emergence of the new. He complains:

> For phenomenology, the personal is prefigured or "pre-reflected" in the world, in a closed loop of "intentionality." The act of perception

31. Shaun Gallagher, "Where's the Action? Epiphenomenalism and the Problem of Free Will," in Pockett, Banks, and Gallagher, *Does Consciousness Cause Behavior?*, 109–24. Compare idem, "Body Schema and Body Image in a Deafferented Subject," *Journal of Mind and Behavior* 16 (Autumn 1995): 369–90; idem, *How the Body Shapes the Mind* (Oxford, 2005); and Gallagher and Anthony J. Marcel, "The Self in Contextualized Action," *Journal of Consciousness Studies* 6 (4) (1999): 4–30.

32. Gallagher, "Where's the Action?," 115–16. Gallagher draws a distinction in this regard between intentional actions, which are usually (but not always) conscious, and motor movements, which are usually unconscious, suggesting that Libet's experiments apply not to intentional actions as such, but to motor movements, the control of which we would normally expect to be unaware.

or cognition is a reflection of what is already "pre"-embedded in the world. It repeats the same structures, expressing where you already were . . . This is like the déjà vu without the portent of the new . . . Experience, normal or clinical, is never fully intentional. No matter how practiced the act, the result remains at least as involuntary as it is elicited. (*PV*, 191; see also 287, n. 4)

The words "fully intentional" in this passage—as in "experience, normal or clinical, is never fully intentional"—mark the moment when Massumi succumbs to a false dichotomy between mind and matter. They mark the moment when he commits himself to the (essentially metaphysical) idea that for something to be "elicited" or intended it must be "fully" conscious, and that, since not all experience can be described in those terms (but can *any* "experience" be so described?), the only alternative is to regard it as corporeal or material. Libet's experiments and interpretations appeal to Massumi precisely because they are formulated in terms of this false dichotomy and thus seem to provide scientific evidence for the priority of brain matter in the origin of thought.[33]

This last point can be generalized: even as the new affect theorists condemn the subject-object split, there is a constant tendency among them to adhere to this same false opposition between the mind and the body. Music is often cited by affect theorists as exemplifying the power of the affects. For example, Shouse suggests that music provides perhaps the clearest example of how the "intensity of the impingement of sensations of the body can 'mean' more to people than meaning itself." He observes in this regard that "'music has *physical effects*, which can be identified, described, and discussed but which are not the same thing as it having *meanings*, and any attempt to understand how music works in culture must . . . be able to say something about those effects without trying to collapse them into meanings" ("FEA," para. 13; his emphasis). Here, Shouse puts everything that is not a question of "meaning," defined in some highly limited sense,

33. In another reference to Libet's experiment, Massumi observes of his own position: "The perspective suggested here displays a tropism toward realist materialism. . . . At virtually every turn in the discussion, dynamics that seemed 'subjective' to the extreme made a literal end run back to impersonal matter. The end run of mindedness back to matter always somehow coincided with its emergence from it, the exemplary case being Libet's feedback loop between the dawning of perceptual awareness and the ever-present previousness of movements of brain matter capable of coloring experience without themselves becoming aware. Accepting this insistence of the material and impersonal (the 'involuntary') *in* bootstrapped personal experience distinguishes the current account most sharply from phenomenological approaches" (*PV*, 206; his emphasis).

over against the body or affect. What seems wrong or confused about this is the sharpness of the dichotomy, which operates at once with a highly intellectualist or rationalist concept of meaning and an unexamined assumption that everything that is not "meaning" in this limited sense belongs to the body. This too is a false dichotomy, one that—in spite of a professed hostility to dualism—threads its way throughout much of the new literature on affect.

OUT OF THE BLUE

The same half-second delay between the operations of the brain and the emergence of consciousness plays a role in Connolly's efforts to rethink the role of reason, argument, and decision in everyday life. In line with Massumi, to whom he acknowledges a debt,[34] Connolly advocates an "immanent naturalism" according to which the Kantian transcendental field can be "translated" into an immanent, material field that decisively alters the direction of our conscious thought. The emphasis falls on the "layered" character of thinking, and especially on the priority of fast-acting, subcortical, or "subliminal" perceptions, "thought-imbued affects," visceral intensities, and corporeal habits and sensibilities over intentional consciousness, reason, propositional knowledge, and explicit argument in political life. What all this means is not entirely clear, but among several neuroscientific studies that interest Connolly I will single out one in particular—my third and last example—as revelatory of the stakes involved in the new affect theorists' appropriations of the findings of neuroscience.

The study concerned the case of a sixteen-year-old girl suffering from epilepsy who, prior to surgery, was undergoing stimulation by intracranial electrodes in order to locate precisely the brain areas responsible for her seizures. During stimulation the patient was asked to perform a variety of tasks, including naming objects, reading paragraphs of texts, counting, and various movements of the arms, fingers, and feet. When her physicians began stimulating a region of the left frontal lobe, they discovered that an electrode touching a tiny patch in the "supplementary motor area" made the patient laugh. According to the physicians, not only was the girl's laughter accompanied by a "sensation . . . of mirth," but each time

34. For Connolly's praise of Massumi's "superb exploration of the 'missing half second'" which, he says, prompted some of his own thinking, see William E. Connolly, "Brain Waves, Transcendental Fields and Techniques of Thought," *Radical Philosophy* 29 (94) (1999): 28, n. 6; hereafter abbreviated as "BW"; and *N*, 209, n. 7.

the laughter was involuntarily produced by stimulation in this way, when asked to identify the cause of her laughter, the girl offered a different explanation for it, attributing it to "whatever external stimulus was present." As the physicians reported: "Thus, laughter was attributed to the particular object seen during naming ('the horse is funny'), to the particular content of a paragraph during reading, or to persons present in the room while the patient performed a finger apposition task ('you guys are just so funny . . . standing around')."[35]

What appeals to Connolly about this study is precisely the idea that the girl was obliged to offer reasons for her laughter after the fact—in other words, the idea that, as in the case of Libet's subjects, the behavior came first and only afterward could the girl come up with various reasons, or rather rationalizations, for it: "The young girl, following the time-honored principle of retrospective interpretation, decided that these researchers were extremely funny guys" (*N*, 83).[36] Connolly views the case as offering further evidence that a "lot of thinking and interpretation" goes on during the "'half-second delay' between the reception of sensory material and conscious interpretation of it" (*N*, 83).[37] "We move here . . . into the quick, crude reaction time of the amygdala that precedes feeling and consciousness," he observes, in one of several references in his book to the claim, associated with the experiments of Damasio, LeDoux, and others, that the amygdala is a crucial component in the generation of rapid emotional responses operating below the reach of conscious cognition and judgment

35. Itzhak Fried et al., "Electric Current Stimulates Laughter," *Nature* (12 Feb. 1998): 650.

36. Connolly does not cite the original scientific paper but an article on it in the *New York Times*. There, the author prefaces his discussion by observing that back in the 1930s, the humorist author Robert Benchley, who liked to slyly poke fun at scientists, had written a mock analysis of laughter. After asserting that "'all laughter is merely a compensatory reflex to take the place of sneezing,'" Benchley had added a footnote: "'Schwanzenleben, in his work 'Humor After Death,' hits on this point indirectly when he says, 'All laughter is a muscular rigidity spasmodically relieved by involuntary twitching. It can be induced by the application of electricity as well as by a so-called joke.'" But the science writer takes the claim seriously. "Little did Mr. Benchley imagine that the imaginary Schwanzenleben would turn out to be nearly right," he comments, concluding that "Science may yet prove more potent than Shakespeare or Monty Python" (Malcolm W. Browne, "Who Needs Jokes? Brain has a Ticklish Spot," *New York Times*, March 10, 1998, D1).

37. In another reference to the girl who was made to laugh "out of the blue," Connolly sees in the patient's after-the-fact attempts to explain her provoked laughter, attributing it to objects or situations that were to hand, a sign of the creative potential of the mind to come up with new possibilities of interpretation ("BW," 25).

(*N*, 208, n. 6).[38] On the basis of these and related studies, he contests "neo-Kantian" political theorists for overrating the importance of reason and underestimating the role of what he calls "technique" or "external tactics," such as drugs, in influencing our thinking and ethics ("BW," 23).

Connolly makes an interesting choice when he illustrates the same phenomenon of the half-second delay by citing the reflex movements we make when we recoil from the painful touch of a hot stove. "A half-second delay?" Connolly asks, and replies:

> It can be illustrated phenomenologically. When you place your hand over a hot stove, your hand recoils before you experience a feeling of pain, even though you tend to interpret the recoil as if it were caused by the feeling that followed it. The reflex action precedes the feeling commonly thought to cause it; in this case, at least, close attention to the order of action can verify the discrepancy between normal retrospective interpretation of temporal order and the actual order. It seems that "incomprehensible quantities of unconscious calculation" take place during the half-second delay between the reception of sensory material and the consolidation of perceptions, feelings, and judgments. (*N*, 82)[39]

It is hard to know what to make of this. Is Connolly implying that by analogy with the pain reflex, laughter can also be understood in reflex terms?[40] If so, he is implicitly arguing that far from being a complex, social-cognitive phenomenon, laughter as an expression of amusement can be conceptualized as an automatic response to stimuli without regard to the meaning those stimuli might have for us, since they are intrinsically capable of triggering a laugh reflex.

In fact, this is just how Damasio interprets laughter. In his discussion

38. Connolly refers to LeDoux's ideas about the role of amygdala in affect in *N*, 76, 90–91, 206, n. 27, and 211. In a more recent publication, Connolly criticizes LeDoux's work as reductive, and pays more attention to the work and ideas of Damasio, V. S. Ramachandran, Francesco Varela, and others (see Connolly, "An Interview with William Connolly," in *The New Pluralism: William Connolly and the Contemporary Global Condition*, eds. David Campbell and Morton Schoolman [Durham, NC, 2008], 327).

39. In this passage, Connolly is citing from Norretranders, *The User Illusion*, 164.

40. Unless he is speaking merely figuratively, this is how Massumi theorizes the way affect exerts its influence in political life today. See Brian Massumi, "Requiem for Our Prospective Dead (Toward a Participatory Critique of Capitalist Power)," in *Deleuze and Guattari: New Mappings in Politics, Philosophy, and Culture*, eds. Eleanor Kauffman and Kevin John Heller (Minneapolis, 1998), 58.

of the case just referred to, Damasio emphasizes the fact that the girl's laughter came "out of the blue" and was "entirely unmotivated."[41] What intrigues him about this and similar cases of electrode-induced emotional reactions is the idea that although such responses seem to manifest the presence of thoughts capable of causing the emotion, the thoughts come only after the emotional behavior has been triggered. "[T]he effect appeared to manifest, for all intents and purposes, the presence of thoughts capable of causing sadness," he writes of a case of sudden sobbing unexpectedly induced by an electrode probe in a woman suffering from Parkinsonian motor symptoms. "Except, of course, that no such thoughts had been present prior to the unexpected incident, nor was the patient even prone to having such thoughts spontaneously. Emotion-related thoughts only came *after* the emotion began." He concludes that the evidence speaks to the "relative autonomy of the neural triggering mechanism of emotion." In the case of the patient who laughed "out of the blue," he proposes that the electrical stimulus mimicked the neural results that the "laughter-competent" stimulus would have normally produced (*LS*, 69–70).[42] (Presumably because laughing at such a stimulus has evolutionary value: it would so startle a crocodile that he would not want to eat you.) The point for Damasio is not to define laughter or sadness in terms of cognitively defined objects or beliefs about the world but as intentionless states, such that my ability to give a reason for my feeling something must

41. Antonio Damasio, *Looking for Spinoza: Joy, Sorrow, and the Feeling Brain* (Orlando, FL, 2003), 75; hereafter abbreviated as *LS*.

42. In keeping with such a view of laughter, which supposes that affects are independent of context and meaning, Damasio accepts Ekman's questionable claim that the difference between genuine and simulated laughter can be detected on the face because only in a person who authentically feels the emotion do the relevant facial muscles involuntarily contract. As we saw in chapter 1, the distinction goes back to the nineteenth-century scientist Duchenne de Boulogne, who proposed that an authentic laugh or smile cannot be feigned because it requires the contraction of muscles not under voluntary control. On this view, a genuine laugh can be produced only by someone really feeling the emotion involved; by the same token, actors cannot convincingly portray the emotion they are trying to represent unless they experience the emotion themselves—if they do not, they can only simulate it and the simulation will show. The neurological data are said to confirm this. Thus, Damasio reports that patients with damage to the same brain areas that were stimulated in the case of the girl who laughed "out of the blue" have difficulty smiling a "natural" smile—a smile spontaneously induced by getting a joke—and are limited instead to a fake sort of "say cheese" smile (*LS*, 76). But as we have seen, the validity of the distinction between authentic and fake smiles has been challenged by Fridlund and many other researchers.

be based on an illusion, in that what I feel is just a matter of my physiological condition.[43] The significance of the case of the girl whose laughter was caused by electrode stimulation for Damasio is thus that it exemplifies the way all the basic affects are supposed to work.

Connolly has suggested that on reviewing the study of her case, the patient who was made to laugh when her brain was stimulated enriched her interpretation of laughter: "As time unfolded she appreciated even more how spontaneous laughter is both a joy in itself and one sign of the excess of affect over epistemically available belief and perception."[44] But what is "joy in itself" when Connolly has stripped it of all conceptual meaning and indeed from intentional life? When the "meaning" we ascribe to "laughter," here defined in terms of the movements of the person's body, is cut off from the reasons we might have for it, when laughter is seen as a purely neurological event, then we have totally lost our grip on the meaning of the word "joy." The girl's brain might be stimulated, for example, when she was given what counted for her as bad news, and she would then be making all the physical movements that could be described as laughter while feeling sad that her dog had died. So, either joy has its usual meaning of happiness and the girl's simulated laughter has nothing to do with it, or joy has no meaning beyond the designation of the phys-

43. Like many theorists of affect, Damasio has been influenced by William James's well-known theory of emotion, according to which we are frightened because we run, and sad because we cry (see for example *LS*, 57). So has Massumi (Massumi, "Fear [The Spectrum Said]," 36–37). This is not the place for a detailed discussion of the role of James's theory of emotion in the recent turn to affect. But for a valuable discussion of the history of that theory, its critical reception, and James's subsequent revisions, see Thomas Dixon, *From Passions to Emotions: The Creation of a Secular Psychological Category* (Cambridge, 2003), 204–30. I have also benefited from the valuable discussions of Wittgenstein's critiques of James's theory of emotion by Michel Ter Hark's *Beyond the Inner and the Outer: Wittgenstein's Philosophy of Psychology* (Dordrecht, 1990), 213–21; and Russell B. Goodman, *Wittgenstein and William James* (Cambridge, 2002).

44. William E. Connolly, "The Complexity of Intention," *Critical Inquiry* 37 (Summer 2011): 797; hereafter abbreviated as "CI." A similar contradiction comes up when in the same paper Connolly cites his previous article, "Experience and Experiment" (2006), in which he makes use of Damasio's analysis of the patient with bilateral amygdala damage who, it was held, could not experience fear ("CI," 796, n. 7). Connolly takes this case, among others, to demonstrate the insufficiency of "intellectualist" and "deliberationist" models of thinking and the importance of granting the priority of subconscious processes in perception and judgment. But Damasio's investigations of that patient have depended on the same "restricted" assumptions of emotion that Connolly now says he rejects. Moreover, as I have shown in chapter 2, Damasio's interpretation of that case has been called into question by subsequent findings.

ical movements associated with laughing. Connolly's inability to choose between these options or to recognize that there is a choice to be made is symptomatic of the reductionism intrinsic to his project.

More recently, Connolly has begun to take his distance from Damasio, whose work he has supported in the past. He now criticizes him for adopting "too functionalist a reading of evolution and too restricted a notion of emotion" ("CI," 797). It appears that Connolly has come to recognize that Damasio accepts a "restricted" notion according to which there is a limited number of evolved, built-in, basic emotions that function independently of cognition or signification. But if Connolly now repudiates this central aspect of Damasio's position, it is hard to see how he can simultaneously and without embarrassment continue to find "valuable" ("CI," 798) Damasio's Somatic Marker Hypothesis, which depends on precisely that same "restricted" notion. (For a further discussion of Damasio's Somatic Marker Hypothesis, see appendix 2.)

SOME CONCLUSIONS

It is time to take stock. I will bring my discussion to a close by offering a few summary comments about the ways in which the new affect theorists are making use of the neurosciences to forward their views and about the general implications of their work.

1. In the case of certain scholars in the humanities and social sciences, the situation is relatively straightforward. In their turn to affect, they have drawn on the dominant research paradigm in emotion research, the "affect program" or Basic Emotion Theory associated especially with the work of Tomkins and Ekman, and adopted by Damasio. For them, the appeal of the Basic Emotion Theory is that it provides a picture of the emotions as inherently independent of intentions in that the affects are held to be a set of innate, automatically triggered brain-body behaviors and expressions operating outside the domain of consciousness and intentional action.

For Sedgwick, the paradigm serves her theoretical and political interests in several ways, not least in its emphasis on the role of contingency and error in emotional life. Following Tomkins, she holds that it is because the affects can be triggered by virtually any object without our cognitive system's knowledge of the object or "stimulus" that elicits it that we are so liable to be wrong about ourselves. The alleged disjunction between emotion and cognition is attractive to Sedgwick precisely because of what she describes as "the unexpected fault lines between regions of the calculable and the incalculable."[45] In other words, for Sedgwick the shift away from questions of meaning and intention in Tomkins's approach to the emotions produces as one of its important consequences an emphasis on the attributes of a subject who can incidentally attach itself to objects but who has no essential relation to them. The effect is to replace the idea of one's intentions toward objects and situations or the meanings those objects and situations might have for one with the idea of the singularity of one's affective experiences, which is to say with the idea of one's difference from all other subjects.[46]

In the case of the medieval historian Smail, the work of Damasio (and of LeDoux) likewise serves his purposes because he believes that the study of intentionality and meaning is an inadequate basis for history. Treating the search for meaning in texts based on the interpretation of authorial intentions as inherently untrustworthy because authors may lie, Smail advocates the study of the traces of the past that are unintentionally sedimented in documents—material marks such as those provided by archeological remains, stone tools, fossils, and sedimentary deposits—that not only can be considered more reliable records of what has happened in the past but that, as a matter of "information" rather than intended meaning, can be interpreted in much the same way a population geneticist reads or decodes a strand of DNA.[47]

Smail's critique of authorial intention is aimed at dissolving the prejudice against studying prehistory, which lacks written records. The result is a "neurohistorical" approach that by collapsing the distinction between "pre-history" and written history, brings the study of the past and neurobiology together in the study of all those allegedly non-intentional processes that have influenced human behavior from the "deep time" of

45. Sedgwick, *Touching Feeling*, 106.

46. For a detailed discussion of Sedgwick's views, see Leys, *From Guilt to Shame: Auschwitz and After* (Princeton, 2007), chapter 4.

47. Commenting on the "information" that drifted into medieval notarial documents without the notary or client "really being aware of it," Smail argues: "This absence of intention or even awareness means that we can trust the facts that emerge . . . in just the same way that we can never really trust the facts intentionally conveyed by notaries and their clients. The unintended meanings found in all documents are like sediments that have precipitated out of solution . . . We search not for the meaning that an author chose to leave behind but rather for the information that was accidentally or unintentionally preserved inside that little trace of the past . . . [D]ocuments bearing intended meanings cannot be seen as qualitatively superior to nondocumentary traces" (*ODH*, 65).

the prehistoric ages to the present.[48] It is not surprising, then, that Smail endorses an account of the affects as basic automated body states or that, in an echo of Libet's ideas, he claims that "the brain often likes to do its communicating all by itself, and it only grudgingly allows the mind a say in the process" (*ODH*, 165).[49]

Among the non-intentional, nonconscious motors of change that interest Smail are "autotropic" devices, defined by him as stimulants such as tea, coffee, alcohol, opium, and psychotropic drugs, but also relatively new addictive cultural practices of novel-reading, going to the theater, and shopping, which in modernity have worked primarily to activate the basic emotional predispositions, or reward and stress systems, that humans have inherited from their ancestral past. Focusing especially on the rise of such devices in the eighteenth century, he suggests that their widespread use transformed the neurochemistry of the aggregate brain, creating generalized states or conditions in entire groups or populations—drug-like states or conditions that, as he puts it, have the appearance of being hardwired without being genetic.[50] In effect, Smail comes close to suggesting that the political changes and ideas associated with the Enlightenment should be conceptualized less as the product of individual human

48. With reference to Damasio's and LeDoux's theories, Smail observes: "Fear, in their model, is automated . . . Emotional expressions, for example, normally lie outside voluntary control. In this way, the integrity of the emotion is guaranteed by the body. Would a person respond so readily to someone's angry demeanor if anger was easily faked? The response to an angry demeanor, typically, is equally automated for much the same reason" (*ODH*, 150–51). Smail goes on to accommodate the fact that, as he concedes, we are ordinarily capable of "faking" expressions reasonably well and that actors do this all the time, by emphasizing the plasticity of the nervous system and by proposing an interaction between biologically given body states and cultural norms and conventions in terms not unlike Ekman's neurocultural theory of the emotions (*ODH*, 151–54).

49. Smail's note at this juncture refers to Timothy D. Wilson's *Strangers to Ourselves: Discovering the Adaptive Unconscious* (Cambridge, MA, 2002), a book blurbed by the same Norretranders who also blurbs Smail's book and whose *The User Illusion: Cutting Consciousness Down to Size* is also cited by Thrift (see my n. 27, this chapter). In his book, Wilson cites the work of Daniel Wegner, *The Illusion of Conscious Will* (2002), who in turn cites Libet, in order to claim that the brain thinks for us. This is the same Timothy D. Wilson who, with Richard Nisbett, published the influential paper "Telling More Than We Can Know: Verbal Reports on Mental Processes" (1977), which I discussed earlier in chapter 4.

50. Daniel Lord Smail, "An Essay on Neurohistory," in *Emerging Disciplines: Shaping New Fields of Scholarly Inquiry In and Beyond the Humanities*, ed. Melissa Bailar (Houston, 2010), 201–2.

thought and agency than as the unintended consequence of large-scale, impersonal shifts in cultural practices interacting with subpersonal, unconscious brain processes. As Smail puts it, neurobiology "confirms on a chemical level that there is no meaningful distinction between cultural practices and psychoactive chemicals. Both kinds of input are translated into the language of the nervous system . . . Culture, in this sense, is like a drug."[51]

The idea that politics and culture can be understood in neurobiological terms is precisely the recent proposition of historian Lynn Hunt who, acknowledging the influence of Smail, advocates neurohistory as a radical solution to what she sees as impasses in the interpretation of the French Revolution. Hunt's goal appears to be to put a stop to what seem to her the endlessness and fruitlessness of recurrent debates over the origins and meaning of that revolution. It's as if she thinks that by now those debates ought to have come to a stop in definitive answers and that, since this has not happened—but is it likely that debates among historians over major events such as the French Revolution will ever issue in complete agreement? Has such an outcome ever happened?—it is because scholars have looked for answers in the wrong place. Her proposal is that historians of the revolution should seek consensus, however provisional, in the hitherto "unrecognized common ground on which all these debates have taken place."[52]

51. Smail, "An Essay on Neurohistory," 210. Smail regards it as a strength of his approach that, unlike those who offer a psycho-evolutionary analysis of the past and assume the existence of a fixed human nature, he can provide a genuinely historical account of change that is attentive to issues of natural selection while emphasizing the nature of the contingent-dynamic interactions between culture and the brain. For incisive criticisms of his approach, see Reddy, "Neuroscience and the Fallacies of Functionalism," *History and Theory* 49 (3) (2010): 412–35; and Plamper, *The History of Emotions: An Introduction*, 270–76.

52. Lynn Hunt, "The World We Have Gained: The Future of the French Revolution," *American Historical Review* 108 (1) (February 2003): 3. "The rat-a-tat of scholarly and political cross-fire threatens to obliterate the real accomplishments made in historical understanding over the centuries," Hunt writes in this regard:

> I do not envisage myself as the circus performer at the top of the human pyramid, with Edmund [Burke], Tom [Paine], and Mary [Wollstonecraft] at the bottom, and Karl [Marx] and Alexis [de Tocqueville] in the next row, and myself at the top straining to juggle several different interpretations in the air at once. The process resembles more a rambunctious history department meeting in which out of the cacophony of discordant voices finally issues, in part out of exhaustion, a partial, provisional, and always revocable agreement on what needs to be explained, if not how to explain it. (3)

Hunt therefore suggests that they should turn their attention to the subjective experiences of the individual self and specifically to those changes in the brain that, she suggests, in the late eighteenth century caused the emergence, through the experience specifically of reading novels and other texts and the consequent activation of the predisposition or hardwired tendency to identification, or what she calls empathy or sympathy, of new attitudes or views about the rights of others. "My argument depends," she writes, "on the notion that reading accounts of torture or epistolary novels had physical effects that translated into brain changes and came back out as new concepts about the organization of social and political life. New kinds of reading (and viewing and listening) created new individual experiences (empathy), which in turn made possible new social and political concepts (human rights)." Or, as she also states: "I am insisting that any account of historical change must in the end account for the alteration of individual minds. For human rights to be become self-evident, ordinary people had to have new understandings that came from new kinds of feelings."[53]

Like many other humanists drawn to the neurosciences today, Hunt turns especially to the work of Damasio, whose hypotheses about feelings and the formation of identity she embraces as a way of getting at what goes on inside the "black box" of the self and subjectivity. Shifting the focus of attention away from arguments about meaning, in this case the meaning of the French Revolution, Hunt turns to a brain-based historiography founded on the recent neurosciences of emotion and the self, presented especially by Damasio, as resources for writing such a "history from within."[54] "The French Revolution, like all revolutions, was first and foremost an experience," Hunt writes.

> I use the word advisedly . . . in order to signal that attention must be paid to the way in which events were subjectively viewed; these subjective views had everything to do with how events developed. One *Oxford English Dictionary* definition of experience is "an event by which one is affected." I want to get at what it means for an event such as the revolution to alter the mental state of millions of people.[55]

53. Lynn Hunt, *Inventing Human Rights: A History* (New York and London, 2007), 33–34.

54. The phrase is Jeremy Trevelyan Burman's in "History from Within? Contextualizing the New NeuroHistory and Seeking Its Methods," *History and Psychology* 15 (1) (2012): 84–99, a favorable analysis of Hunt's neurohistorical approach, cited by Hunt, *Writing History in the Global Era* (New York, 2014), 179, n. 17.

55. Lynn Hunt, "The World We Have Gained," 3.

Like Sedgwick and the Deleuzean affect theorists in this regard, she advocates studying the bodily emotions and processes that are held to occur independently of beliefs and ideology. In short, Hunt's arguments are consistent with the general turn to affect I have been examining in this chapter in the latter's depreciation of meaning and signification and its emphasis on scientific claims about the personal impact of bodily processes and feelings operating independently of beliefs and ideas.[56]

2. When we examine the work of affect theorists such as Massumi, Shouse, Thrift, and Connolly, who claim to be influenced by Spinoza, James, Bergson, Deleuze, Guattari, and others, their relationship to the sciences at first appears to be more complicated than that of Sedgwick, Smail, and Hunt. It might even seem that the neurosciences would have little to offer these theorists because they define affect in terms that appear to be inimical to scientific analysis. As literary critic Sianne Ngai has recently observed in this connection, Massumi's characterization of affect as an "asignifying intensity" that is prior to or apart from any qualification or quantification "creates difficulties for more positivistic kinds of materialist analysis."[57]

In fact, Massumi has some rather harsh things to say about the sciences, accusing them of seeking to tame, instrumentalize, and render

56. On the basis of Damasio's latest book, *Self Comes to Mind: Constructing the Conscious Brain* (New York, 2010), Hunt denies that he is an anti-intentionalist (Hunt, "TSH," 1580, n. 23). But from my perspective it is not enough for her to simply declare this to be the case; she would have to demonstrate it, something I believe she would find hard if not indeed impossible to do. Hunt also faces difficulties when she attempts to refute John Searle's criticism of Damasio's book. Searle argues that, far from showing how consciousness emerges out of the body, Damasio presumes the very consciousness whose origin he is trying to explain, so that his arguments are circular. In response, Hunt points to the claim of Shaun Gallagher that newborns are born with an innate body schema, as is demonstrated by their ability to imitate other people at birth. This suggests to her that the elements of a rudimentary self exist at birth and hence that a proto-self precedes consciousness, as Damasio maintains, and not the other way around. But Hunt is on shaky ground here, because the evidence in favor of neonatal mimicry has been found wanting in a recent impressive reanalysis by Susan Jones, an expert on infant behavior, suggesting that the argument for an innate capacity for imitation or empathy in newborns, based on the existence of mirror neurons or some other mechanism, is unjustified. On these points, see John Searle, "The Mystery of Consciousness Continues," *New York Review of Books*, June 9, 2011; Shaun Gallagher, *How the Body Shapes the Mind*, 65–78; Susan Jones, "The Development of Imitation in Infancy," *Philosophical Transactions of the Royal Society of London, Series B Biological Sciences* 364 (2009): 2325–35; and Greg Hickok, *The Myth of Mirror Neurons: The Real Neuroscience of Communication and Cognition* (New York and London, 2014), 205.

57. Sianne Ngai, *Ugly Feelings* (Cambridge, 2005), 26.

profitable the singularity, unpredictability, immanence, and liveliness of a world in flux. "Scientific method is the institutionalized maintenance of sangfroid in the face of surprise," he writes. "Properly scientific activity starts from a preconversion of surprise into cognitive confidence" (*PV*, 233). But this is not the whole story, since he and like-minded affect theorists are also keen to enter into some sort of relationship with the sciences. Indeed, it is an interesting feature of the present situation that according to Massumi the humanities need the sciences "for their own conceptual health a lot more than the sciences need the humanities" (*PV*, 21, quoted in "BG," 39). What he hopes is that the humanities can borrow or pilfer from the sciences in such a way as to stir things up and, ideally, change the terms of the encounter between the two fields.

In a generous mood, one might be willing to concede that this is what Massumi is doing when he makes use of the snowman experiments of Sturm and her team. He could be said to be catching those scientists in the very process of trying to tame the complexities and apparent paradoxes inhering in their findings about the way children respond emotionally to the media—once those findings are reinterpreted along the lines he proposes. In a less generous mood, however, one could argue that not only is Massumi unfair to Sturm and her group, who acknowledge the complexity of their results and the difficulty of interpreting them, but that he willfully or otherwise misreads the data in order to create paradoxes where none exist.

In any case, creative misreading can hardly be said to characterize Massumi's and Connolly's appropriations of Libet's experiments, or the uses Connolly makes of experiments and pathological case histories described by Damasio and others, appropriations and uses that amount to straightforward endorsements.[58] In Massumi's case, in spite of his claim to embrace a form of "radical empiricism" inspired in part by William James, he comes across as a materialist who invariably privileges the "body" and its affects over the "mind" in straightforwardly dualist terms, forgetting that for James the "body" is not a pure state of being but rather a pragmatic classification of the operations of "pure experience," just as is the

58. Papoulias and Callard observe in this regard that the language in which the new affect theorists invoke the neurosciences is often the language of evidence and verification ("BG," 37), citing Massumi for stating that the "'time-loop of experience has been *experimentally verified*'" (Massumi, "Fear [The Spectrum Said]," 195; their emphasis), and Connolly for stating that one of Damasio's experiments in the case of the patient with defective amygdalae "'*reveals* how much of perception and judgment is prior to consciousness'" (Connolly, "Experience and Experiment," 73; their emphasis).

"mind."[59] That is why Massumi finds the work of scientists such as Libet so congenial, because as we have seen, Libet privileges the body in such a way as to claim that the mind always functions "too late" for intention and reason to play a decisive role in action and behavior. In this regard, Massumi's attitude toward the sciences is scarcely to be differentiated from that of non-Deleuzean affect scholars, such as Sedgwick, Smail, and Hunt.[60]

3. My critique of the new affect theorists goes well beyond the suggestion that they are poorly informed about the neurosciences on which they lean, or that they are determined to find in certain neuroscientific claims precisely what they are looking for even if those claims are poorly supported empirically, although these are certainly among my criticisms. In the course of this chapter, I have already hinted at some of the problems with the political, aesthetic, and philosophical implications of the new affect theorists' views. At the most general level, the theories of affect that I have been analyzing can be seen to belong to a body of thought that has been criticized in devastating terms by Walter Benn Michaels, in his book *The Shape of the Signifier: From 1967 to the End of History* (2003). Very briefly, Michaels argues that the "irrelevance of political beliefs or ideas," and their

59. "Subjectivity and objectivity are affairs not of what an experience is aboriginally made of, but of its classification." Massumi cites this passage from James's essay, "The Place of Affectional Facts in a World of Pure Experience," first published in the latter's *Essays in Radical Empiricism* in 1912 (*PV*, 296). In his essay, James suggests that the categorization we choose is a pragmatic question and not a constitutive one, whereas Massumi invariably privileges the body over the mind. See William James, *Essays in Radical Empiricism*, gen. ed. Frederick Burkhardt (Cambridge, MA, 1976), 71, 73.

60. Apropos of the title of Massumi's book, *Parables for the Virtual*, the notion of the "virtual" deserves a further comment. If, according to Massumi, affect is "virtual" because it is something that "happens too quickly to have happened" (*PV*, 30); if affect *as* the virtual is the "incorporeality of the body" (*PV*, 21); if affect is "*incipience*, incipient action and expression" (*PV*, 30); if affect is the realm of "*potential*" (*PV*, 30); if affect is the "unclassifiable" or the "never-yet-felt" (*PV*, 33); if affect is "undecidability fed forward into thought" (*PV*, 37); if affect or the virtual "as such" is "inaccessible to the senses" (*PV*, 133); if affect can only be grasped topologically, which is to say, unempirically (see *PV*, 134); if affect is "prior to or apart from the qualitative" and is not a matter of quantitative investment (*PV*, 260); then it is not at all clear that it makes sense for Massumi to cite Libet's experiment in support of his views, because by doing so he makes it seem as if the "virtual" has been definitively located in the body-brain. But it does make sense for him to cite Libet's experiment if he is a certain kind of materialist bent on privileging the role of body-brain over that of mind and intentionality in human life, culture, and behavior. For the claim that Deleuze has been misunderstood by certain affect theorists because he in fact liberated affectivity and the virtual from the body, see Richard Rushton, "Response to Mark B. N. Hansen's 'Affect as Medium, or the "Digital-Facial-Image,"'" *Journal of Visual Culture* 3 (December 2004): 353–57.

replacement by what theorists such as Michael Hardt and Antonio Negri, in their book *Empire* (2000), "thinking to follow Foucault" call the "biopolitical," is what characterizes the special political contribution of postmodernism and posthistoricism. Noting that for Hardt and Negri biopolitical struggles are "'struggles over the form of life,'" Michaels observes that

> struggles over the form of life are "ontological" rather than ideological; they have nothing to do with the question of what is believed and everything to do with the question of what is . . . So if political conflict may be imagined as conflict between two competing commitments as to what's right, biopolitical conflict appears as a conflict between what is and what isn't, or (in its more forward-looking mode) between what is and what will be.

And he cites Hardt and Negri as stating: "'Those who are against . . . must also continually attempt to construct a new body and a new life.'"[61]

Just so, we might put it that what is at stake for the theorists whose turn to affect I have been analyzing is a "logic" according to which attention to ideology or belief is replaced by a focus on bodily affects that are understood to be the outcome of subliminal, autonomic corporeal processes. Stressing bodies over ideas, affect over reason, the new affect theorists claim that what is crucial is not your beliefs and intentions but the affective processes that are said to produce them, with the result that political change becomes a matter not of persuading others of the truth of your ideas but of producing new ontologies or "becomings," new bodies, and new lives. Hunt's suggestion that historians relocate their focus of analysis from a study of human intentions to the individual "experience," defined in Damasio's neurobiological terms as a matter of bodily feelings that precede consciousness, is a case in point. So is Connolly's advocacy of self-transformation by everyday "techniques of the self" or "tactics" of the body—for example, listening to music, reading, exercise, taking Prozac, meditation, and so forth—that are said to alter or remap body-brain connections, below the threshold of consciousness, and thereby produce new political identities (*N*, 41–49, 100–104).[62]

61. Walter Benn Michaels, *The Shape of the Signifier: From 1967 to the End of History* (Princeton and Oxford, 2004), 173; hereafter abbreviated as *SS*. It follows from Michaels's critique of the logic of posthistoricism that ideas about the biopolitics of subjectivity and the agency of matter mystify the realities of globalization and provide convenient support for, not a critique of, global capitalism and neoliberalism.

62. Similarly, Massumi states that resistance to capitalism "would define itself less as an oppositional practice than as a pragmatics of intensified *ontogenesis*: at life's edge. This is the countercapitalist principle of *vitalist metaconstructivism*. This prin-

Hence, too, Connolly's defense of pluralism, defined in terms of the valorization of personal experience and feeling over argument and debate. Michaels proposes that ideological disputes, or conflicts over beliefs and meanings, are inherently universalizing, because whereas we cannot disagree about what we feel, we just feel different things, we can and do disagree about what is true, regardless of what we feel or who we are. Indeed, he suggests that it is only the idea that something that is true must be true for everyone that gives sense to disagreement, since the belief that one political system is better than another, or that a certain social arrangement is unjust, is intrinsically universal. As he maintains, the alternative to a universalism defined as involving disagreement over beliefs is a pluralism defined as involving differences of personal feeling, identity, or subject position, a position that, as Michaels puts it, "disarticulates difference from disagreement" (*SS*, 16). This is a pluralism that Connolly advocates in explicit opposition to Michaels's views, a pluralism that emphasizes political transformation at the level of the affective body, not ideas: according to Connolly, what matters politically is how something makes us feel, not what it means to us (*N*, 41–49).[63]

---

ciple can only be fully theorized . . . through experimentation" (Massumi, "Requiem for Our Prospective Dead," 60); Connolly treats the working class as a new form of identity whose status as such deserves respect rather than transformation (for a critique, see Walter Benn Michaels, "Homo Sacher-Moser: Agamben's American Dream," in *States of Emergency—States of Crisis, Yearbook of Research in English and American Literature*, eds. Winfried Fluck, Katharina Motyl, Donald Pease, and Cristoph Raetzsch, 27 [2011]: 25–36); Marco Abel suggests that a "masocritical" practice of literary criticism based on affect is engaged in a "perpetual production of subjectivity" (Marco Abel, *Violent Affect: Literature, Cinema, and Critique After Representation* [Lincoln, NE, 2007], 23); Thrift characterizes the politics of his nonrepresentationalist-affective approach as "a politics of the creation of the open dimension of being" (Nigel Thrift, "Summoning Life," in *Envisioning Human Geographies*, eds. Paul Cloke, Philip Craig, and Mark Goodwin [Oxford, 2004], 92). The French philosopher Catherine Malabou has recently put Damasio's analysis of the case of Phineas Gage at the center of her discussion of the loss of affect in traumatized subjects of all kinds. In the process, she shifts the focus away from issues of agency, intention, and belief to ontological questions concerning why not only victims of trauma but also agents of violence such as terrorists are the affectless "new wounded" in their identities as members of the class of the twenty-first century's alleged new form of being (Malabou, *The New Wounded*, 155, 212). For a critical evaluation of Malabou's arguments, see Leys, "Post-Psychoanalysis and Post-Totalitarianism," in *Psychoanalysis in the Age of Totalitarianism*, eds. Matt ffytche and Daniel Pick (London and New York, 2016), 239–51.

63. Of course, it could be argued that the new affect theory creates problems for a progressive politics in that it is not at all clear how one might go about deliberately

From Michaels's perspective, such an appeal to affect eliminates disagreement because to see or feel things from a different perspective is to see or feel the same thing differently but without contradiction. Michaels's aim is to dismantle such a logic by showing, through readings of critical, literary, philosophical, fictional, and other texts, how the replacement of ideological disagreements, or conflicts over belief and meaning, with differences in our feelings or bodies produces an indifference to political or ethical dispute. Another way of putting this is to say that the posthistoricist logic Michaels anatomizes is an attempt to get rid of the notion of meaning or belief or intention or interpretation altogether, a collective tendency he regards as a mistake, to say the least (*SS*, 31).[64] The

---

influencing what in oneself and others is beyond conscious control: the emphasis on the importance of the subliminal visceral register in people's responses makes it difficult to imagine how a political activist might intervene strategically in a particular situation (a point raised by David Campbell in "An Interview with William Connolly," 325–29). Nevertheless, Connolly and the new affect theorists operate on the assumption that when politics becomes a question of distinguishing "good affects" from "bad affects," deliberately performed manipulations operating below the level of ideology and consciousness can be countered only by manipulations of a similar kind. As Connolly himself suggests, an effective counterpolitics must somehow draw on the same resources of image control in order to challenge the sound-media campaigns of the opposition (see Connolly, "The Evangelical-Capitalist Resonance Machine," *Political Theory* 33 [December 2005]: 885, n. 15). Likewise, in reference to 9/11 and the Homeland Security Administration's color-coded alert system's manipulations of the country's fears, Massumi has commented that "The Bush administration's fear-in-action is a tactic as enormously reckless as it is politically powerful. Confusingly, it is likely that it can only be fought on the same affective, ontogenetic ground on which it itself operates" (Massumi, "Fear [The Spectrum Said]," *positions* 13 [1] [2005]: 47). The commitment to notions of "emergent causality" makes the outcome of such tactics inherently unpredictable, and this unpredictability then becomes the basis for a posthistoricist and post-Marxist "hopefulness," or "faith" in the possibilities of change. On this point, see for example Massumi, "Navigating Movements," in *Hope: New Philosophies for Change*, ed. Mary Zournazi (New York, 2002), 210–44.

64. As Michaels emphasizes, his position does not involve a commitment to rationality on the model of what Habermas calls "good reasons." Michaels argues that our reasons for believing something always seem good to us; that's what makes them our reasons. He suggests that the important difference is between those things (beliefs and interpretations) that seem to us true or false and for which we can give some reasons, and those things that seem to us to require no justification (*SS*, 188, n. 16). In a similar gesture, he answers those who complain that his characterization of what people can and can't coherently care about misses the "'crucial point that we don't live in a world in which people exercise reason all the time,'" by suggesting that such complaints depend on a mistaken idealization of "reason over reasons." "The fact

new affect theorists' contribution to this development is to try to show that with the help of the neurosciences, affect theory can be made to line up with the same posthistoricist logic because affect can be defined (erroneously in my view) as inherently non-intentionalist and indeed materialist in nature.

Critics have argued in response to Michaels's views that in emphasizing the importance of reasons and arguments in human affairs, Michaels neglects the affective dimension. But there is nothing in Michaels's position that suggests he is opposed to the idea that we are embodied, affective creatures, or indeed that our emotions and affects play a role in influencing our opinions. What I take it he does oppose, however, is the notion that this undeniable fact has any implications whatsoever for determining whether or not particular beliefs or opinions are true, or even worth taking seriously.

We can see how these issues play out by considering political theorist Jane Bennett's book *Vibrant Matter: A Political Ecology of Things* (2010). In close alliance with Connolly's views and the new materialism movement, Bennett argues that political theory needs to recognize the active participation of nonhuman things and forces, or what she calls "vital materiality." She proposes that recognition of the role of such a vital materiality will lead to a more responsible politics, one attuned to the web of impersonal, active forces that influence and affect political life. In its more philosophical aspect, Bennett's aim is to counter the tendency of modern thought to characterize matter as dull, inert, passive, and mechanical by encouraging us instead to consider the liveliness and spontaneity of material objects and things. She believes that ordinary matter and material things are so full of life that they can be said to possess agentic powers comparable in their effects to those of human intentionality and in important ways to be more decisive for the *polis* than the latter. In its more political-ethical aspect, Bennett's project is to challenge the importance traditionally given to human agency, intentionality, and reason by showing that a lively, material, and effective agency is distributed throughout the world of not just humans but also objects and things, a material agency that profoundly affects situations and events beyond the scope of human wishes and desires. She proposes that recognition of this fact

---

that people have lots of false beliefs and do lots of stupid things based on those false beliefs doesn't mean that they don't have any reasons—it just means they have bad reasons. And having bad reasons is one way (having good reasons is another) of exercising reason" (*SS*, 193, n. 39).

entails an acknowledgment of the need for a new politics defined in ecological terms as a politics in which things play a crucial role.[65]

Moreover—and this is a crucial move on her part—Bennett argues that the life or liveliness she views as immanent to matter is what she means by the term "affect." Affect on this model is defined in "Spinozist" terms as the potential for movement and reaction. As such, it is not specific to humans or even nonhuman animals but is conceptualized by her as an "actant" in Bruno Latour's sense of the term, that is, as a form of agency that can be attributed to metals, minerals, and other material things. In one extraordinary chapter, Bennett goes so far as to suggest that the massive failure of the North American electric grid in 2003 was due as much to the agency and unpredictable "strivings" of the electric current itself as it was to human agency. On her model, electricity, with its unstable electron flows, is defined as an agent in its own right, an agent that will have to "do its part" (*VM*, 30) in any plan conceived by Congress to prevent future blackouts. She thus proposes a theory of "distributive agency" that calls into question our usual understanding of human responsibility. This is a position that, as Bennett recognizes, would please the company owners, energy traders, and deregulators whose blame for the catastrophe would be attenuated. But it is a position she thinks is required if we are to stop playing the "blame game" and do justice to the role of vital affects in political life.[66]

65. Jane Bennett, *Vibrant Matter: A Political Ecology of Things* (Durham and London, 2010); hereafter abbreviated as *VM*. For another statement of her position applied to the phenomenon of hoarding, where the emphasis falls on the activity of hoarded things viewed as actants in their own right and on the hoarders viewed, not as mentally ill people, but as "differently-abled bodies that have special sensory access to the call of things," see Bennett, "Powers of the Hoard: Further Notes on Material Agency," in *Animal, Vegetable, Mineral: Ethics and Objects*, ed. Jeffrey Cohen (Washington, DC, 2012), 237–69. It is worth noting that one of the motivations for anti-intentionalism among affect theorists such as Bennett is their desire to overcome the divide between humans and other animals based on the idea that cognition and intentionality are tied to the human capacity for language and making propositions. But, as I have argued in this book, I think this is a mistake. There is nothing about the cognitivist or intentionalist position that limits intentionality and cognition to human animals. Nor is there any intrinsic link between defending the idea of intentionality and believing it is acceptable to treat nonhuman animals badly.

66. Bennett observes in this regard: "Though it is unlikely that the energy traders shared my vital materialism, I, too, find it hard to assign the strongest or most punitive version of moral responsibility to them. Autonomy and strong responsibility seem to me to be empirically false, and thus their invocation seems tinged with injus-

Bennett's arguments commit her to the view that, as agents in their own right, nonlinguistic things should be considered members of the political public. One result of this idea is an expanded sense of what constitutes the political domain, because it means conceiving nonhuman things as part of a less hierarchically organized political "ecosystem" made up of an interconnected series of parts or an impersonal swarm of vital affects and energies. A further consequence of Bennett's views is that they raise questions about how such material things might participate in the political process when they are unable to speak or represent themselves. She seems somewhat perplexed by the "many practical and conceptual obstacles" facing such a materialist vision of the *polis*. As she herself asks: "How can communication proceed when many members are nonlinguistic? Can we theorize more closely the various forms of such communicative energies? How can humans learn to hear or enhance our receptivity for 'propositions' not expressed in words? How to translate between them? What kinds of institutions and rituals of democracy would be appropriate? Latour suggests that we convene a 'parliament of things,' an idea that is as provocative as it is elusive" (*VM*, 104).[67]

---

tice" (*VM*, 37). She does not deny that sometimes moral outrage is indispensable to a democratic and just politics, but suggests that a politics devoted exclusively to outrage can do little good, especially as it encourages vengeance (*VM*, 38).

67. Of course, as Bennett herself notes, on her approach every public may be an ecosystem, but not every ecosystem is a democracy of the kind she prefers. It is an interesting question in this regard whether Clive Barnett is right when, in his valuable overview and critique of the turn to affect in political theory, he charges Connolly and other affect theorists with "cryptonormativism" on the grounds that these authors implicitly espouse certain political beliefs and norms, such as the value of democracy, without providing reasons for their beliefs because their theoretical position precludes them from doing so. One can see the appeal of Barnett's argument, for it can indeed appear that in advocating various techniques to counter the affective manipulations of the political right, Connolly invokes the persuasive force of a progressive politics that he characterizes in normative terms as more generous than that of the right (Barnett, "Political Affects in Public Space: Normative Blind-Spots in Non-Representational Ontologies," *Transactions of the Institute for British Geography* n.s. 33 [2008]: 186–200).

Yet it could be argued that Connolly's position is not contradictory but consistent because, according to his approach, political views are nothing but the expression of purely personal preferences, so that preferring democracy to despotism is like preferring tea to coffee. Connolly, Thrift, Bennett, and other like-minded affect theorists can thus be seen as replacing a concern with disagreement over political beliefs with an appeal to affective differences that they take to be independent of belief or meaning. The result is that when people have different affective responses, they don't disagree, they just are different. From this (to my mind untenable) pluralist point of view, democracy is not a normative value at all but just a personal taste, and what the political

At first sight, Bennett's quandaries would seem to be insurmountable. But in fact, she herself has suggested the answer to her concerns. For, according to her own vital materialist ideas, objects and things already communicate with us because they transmit their vital affects to us, or rather to our nervous systems, directly, without the intervention of language or representation. If communication is understood on the materialist model of a vibration of water, or of an electric current sent through a wire or neural network, then nonhuman things and objects can and do pass on their affects without recourse to representation or signification. Bennett's affect theory therefore explains how things can be included in a parliament of things by imagining that they don't have to speak or represent themselves at all, because they necessarily transmit materially the vital affects that are immanent in them. And if conceiving politics in this way makes it hard to imagine what parliamentary debate would look like, this would appear to be precisely her point: for Bennett, debate over ideology is irrelevant. Her fantasy is rather of a language of nature that isn't itself a representation because it consists of material vibrations or neural currents by which the affects are inevitably passed on, though not exactly understood. In her political utopia, we are invited to imagine a pluralism of vital, affective energies impinging on each other and producing new, if unpredictable, lives and effects.

activist is seeking to do is subliminally influence or manipulate others, through the use of images and other tactics, into sharing his or her likings while remaining pluralistically open to the idea that different persons may simply have different inclinations.

[EPILOGUE]

WHERE WE ARE NOW

In 2015, Andrea Scarantino, recently appointed editor of the newsletter of the International Society for Research on Emotion, invited Dacher Keltner and Daniel Cordaro, two researchers in the basic emotions tradition, and Fridlund and Russell, two of the most prominent critics of that tradition, to publish manifestos stating their respective positions and to engage in a subsequent debate. Keltner and Cordaro presented a revised version of the Basic Emotion Theory (referred to throughout the discussions as BET), while Fridlund and Russell summarized their reasons for rejecting BET and defended their by-now-well-established alternative approaches to the emotions. For Scarantino, the aim of the exercise was to see whether it was possible to forge a consensus among the participants in order to show that progress was being made in the field of affective science. Although his job was to serve as a neutral editor and moderator of the debate, we have seen in chapter 6 that he was on record as defending a variant of the Basic Emotion Theory. In the event, Fridlund and Russell challenged several of the questions he posed to them on the grounds that those questions assumed the existence of discrete or basic emotions, the main point at issue in the dispute.[1]

One of the most interesting developments in the proceedings was the explicit acknowledgment by Keltner and Cordaro of the failure of Ekman's original formulation of the Basic Emotion Theory. As they admitted in their manifesto: "Subsequent critiques have raised questions about the degree of universality in the recognition of these emotion expressions (Russell, 1994), about what such expressions signal (Fridlund, 1991), about the response formats in the studies (Russell, 1994), and about the eco-

1. Dacher Keltner and Daniel Cordaro, "Understanding Multimodal Emotional Expressions: Recent Advances in Basic Emotion Theory"; Alan J. Fridlund, "The Behavioral Ecology View of Facial Displays, 25 Years Later"; James A. Russell, "Moving On from the Basic Emotion Theory of Facial Expressions," in "Facial Expressions," *Emotion Researcher*, ISRE's Sourcebook for Research on Emotion and Affect, ed. Andrea Scarantino, http://emotionresearcher.com, accessed October 15, 2015. The debate that followed was titled "Debate: Keltner and Cordaro vs. Fridlund vs. Russell," in *ibid*. Unless otherwise indicated, all quotations cited in the epilogue are taken from the manifestos or the debate.

logical validity of such exaggerated, prototypical expressions. These productive debates have inspired a next wave of research, which advances Basic Emotion Theory in fundamental ways." Keltner and Cordaro's main innovation in response to such criticisms was to suggest that discrete emotions were not "unimodal" reactions, involving chiefly the face and autonomic systems, as Ekman had originally claimed, but "multimodal, dynamic patterns of behavior, involving facial action, vocalization, bodily movement, gaze, gesture, head movements, touch, autonomic responses, and even scent."

In other words, they proposed expanding the list of emotion categories beyond the prototypical expressions of a limited number of "basic emotions" to include expressions that varied more during an emotional episode and involved more components than had previously been envisaged. Whereas in the original BET the emphasis was on momentary facial expressions that could be captured in a snapshot method by matching static photographs of posed expressions with a few putative emotion categories, Keltner and Cordaro argued that emotions were "extended and multimodal dynamic patterns of behavior, in which the signal consists of a sequence of facial and non-facial actions that only collectively and over time convey the relevant message." On the basis of these revisions, the authors claimed to have found evidence for a list of distinct, universal emotions that expanded from Ekman's original six "basic emotions" to a current list of about twenty, including embarrassment, pride, awe, gratitude, and love, many of which are said to have counterparts in nonhuman primates and other animals.[2]

2. In their manifesto, Keltner and Cordaro cite a long list of research publications that claim to prove the existence of such discrete emotions. See especially Dacher Keltner, "Signs of Appeasement: Evidence for the Distinct Displays of Embarrassment, Amusement, and Shame," *Journal of Personality and Social Psychology* 68 (3): 441–54; idem, "Evidence for the Distinctness of Embarrassment, Shame, and Guilt: A Study of Recalled Antecedents and Facial Expressions of Emotion," *Cognition and Emotion* 10 (2) (1996): 155–72; J. L. Tracy and R. W. Robins, "Show Your Pride: Evidence for a Discrete Emotion Expression," *Psychological Science* 15 (3) (2004): 194–97; and J. L. Tracy and D. Matsumoto, "The Spontaneous Expression of Pride and Shame: Evidence for Biologically Innate Nonverbal Displays," *Proceedings of the National Academy of Sciences* 105 (33) (2008): 11655–60. Some of the experiments cited here used static photos or videos of expressions, posed according to the researchers' assumptions about the correct depictions of embarrassment, pride, shame, and guilt, as stimuli involving not just facial expressions but also non-facial components, such as head tilt, upper body and arm postures, and so forth, in studies designed to test the ability of observers to recognize (that is, to judge) expressions using both forced-choice and more open-ended response formats. In other experiments, the claim is made that spontaneous,

Crucially, Keltner and Cordaro denied that any single emotional component was the *sine qua non* of a discrete emotion. Even feelings, or "*qualia*," and distinct facial expressions were ruled out as necessary characteristics, a major correction of earlier versions of BET. As a result, in the "new BET," no one-to-one correspondence was posited between any specific facial muscle actions or other behavioral-physiological components and each distinct emotion. "Instead," the authors stated, "this approach suggests probabilistic associations between the multimodal behaviors and the occurrence of the emotion." Keltner and Cordaro even included an important role for "context and intention" in emotional expression "as long suggested by Fridlund," although the authors continued to treat emotional experiences and intentions as independent factors in terms completely at odds with the views of the latter, who, as we have seen, had emphasized the intentionality of facial signaling but had rejected the concept of "emotion" on the grounds that it had become too nebulous for the purposes of scientific investigation.

In short, Keltner and Cordaro accepted the criticism of Fridlund, Russell, and others that the hypothesized link between specific emotions and signature expressions or specific patterns of autonomic responses was not as close as Ekman's original BET had predicted. Instead, they emphasized the existence of "significant variation within each category of emotion . . . in the patterns of behavior that co-vary with the co-occurrence of the emotion." In these ways, Keltner and Cordaro attempted to accommodate critiques of the basic emotions approach while clinging to a newly expanded nosology of universal emotion types. They suggested that even the term "basic" emotion posed more problems than it solved and suggested using the term "well studied" in its place. They appealed to various publications by Ekman to justify their expanded, multimodal approach to the emotions.

———

prototypical expressions of pride and shame are displayed in both sighted and blind athletes from various cultures in response to victory or defeat at the Olympic and Paralympic Games. In light of Fridlund's and Russell's criticisms of the use of posed expressions; the experimenters' presuppositions about the existence of prototypical pride and shame facial expressions; questionable assumptions about what so-called "spontaneous" expressions mean in the absence of evidence concerning the subjects' actual emotional feelings; naïve suppositions about the absence of learning in the expressive reactions of unsighted athletes; similarly naïve suppositions about the lack of audience effects or intentional-communicative motives in the case of blind athletes (as if the latter are insensitive even to the roar of the crowd); and the risk of confounds in experiments of this kind; skepticism about the validity of these claims seems warranted. But a full discussion of such studies lies beyond the scope of this epilogue.

But Fridlund would have none of this. In line with his previous, well-honed arguments, he raised a number of objections. The tactic he adopted was to show that over the years Ekman had modified his original formulation of BET in ways that radically undermined the coherence of his position, and that it was on the basis of those shaky and incoherent alterations that Keltner and Cordaro were now pinning their hopes. The significance of Ekman's revisions had gone unremarked by many emotion scientists even as they incorporated them into their research programs, so that one of Fridlund's contributions to the debate was to draw attention to those changes in order to highlight their problematic implications.

On the premise that "in recent years, BET has presented a moving target," Fridlund made the following points:

1. In 1992 Ekman had begun to expand BET's criteria for expressions of emotion by introducing the idea of "emotion families." "Each of the basic emotions is not a single affective state but a *family* of related states," Ekman had claimed, by which he meant a "'group of things related by common characteristics." Thus, according to Ekman, anger did not comprise just one but more than sixty anger expressions, and so on.[3] Fridlund suggested that Ekman had broadened his categorical view of the basic emotions in this way in order to inoculate BET from the damage done by Ortony and Turner's attack on the theory for failing to identify the posited close connections between "emotion" and "expression."[4] According to Fridlund, however, the result was that Ekman had so widened the criteria of emotional expression that its categories now had far too fuzzy boundaries and loose correlations between the component parts to permit rigorous experimental testing.

2. Also in 1992, in a related change, Ekman had begun citing with approval John Tooby and Leda Cosmides's "solving life tasks" emotion model according to which emotions are no longer considered "vestigial reflexes" but organized, strategic motivations that arose in the "Environment of Evolutionary Adaptedness."[5] On this psycho-evolutionary model, emotions are viewed as whole-body, coordinated, strategic adaptations that, as Fridlund noted, "may or may not be suited to events in the contemporary world."

3. Paul Ekman, "An Argument for Basic Emotions," *Cognition and Emotion* 6 (1992): 169–200.

4. A. Ortony and T. J. Turner, "What's Basic About Basic Emotions?," *Psychological Review* 37 (1990): 315–31.

5. Paul Ekman, "Facial Expressions," in *Handbook of Cognition and Emotion* 16 (1999): 301–20.

The new adaptationist model reinforced Ekman's tendency to identify a wider array of components and responses, including bodily responses, than had previously been included in his original affect program theory. As Fridlund noted, in their original neurocultural model of BET Ekman and Friesen had followed Tomkins in arguing that "emotion was expressed strictly by the face, 'while the body shows the adaptive efforts of the organism to cope with the affect state.'" Emotions were thus defined as behaviors we cannot help, because they are unintended, involuntary expressions of our passions. As such, they were set against motivated behavior that was defined as instrumental, goal oriented, and strategic. But, following Ekman's amendments, Keltner and Cordaro were now reversing that position by adding back what Ekman and Friesen had earlier jettisoned, namely, "the very gestures and movements that foundational nonverbal communications researchers like [Margaret] Mead and Birdwhistell always considered important and modern ones still do (e.g., Bavelas and Chovil, 2000)."[6] From Fridlund's perspective, the outcome was an approach that clung to the idea of discrete, basic emotions while offering a "laundry list" of vague qualifiers and properties that were held to characterize the emotion categories but that still lacked complementary exclusion criteria, that is, stipulations regarding the conditions under which a particular emotion was *not* occurring. "This view reduces to nothing more than hand-waving about the knottiness of the phenomena and *ad hoc* choice of stipulated emotion measures," Fridlund objected, "with the result that surveys of research and formal meta-analyses continually find disappointing links between 'emotion' and 'expression.'" In effect, Fridlund's question became: On Keltner and Cordaro's approach, what is *not* an emotion? As he observed: "Theories of emotion have metastasized, and the concept of 'emotion' has now become so malleable that it can be injection-molded to inhabit any theory, even mine."

Fridlund pointed out that another significant consequence of Ekman's embrace of Tooby and Cosmides's evolutionary approach to the emotions was that "strategic" displays had come to be viewed as part of an

6. See Janet Beavin Bavelas and Nicole Chovil, "Visible Acts of Meaning. An Integrated Message Model of Language in Face-to-Face Dialogue," *Journal of Language and Social Psychology* 19 (2) (2000): 163–94, an analysis of how meaningful communications in human face-to-face encounters involve the use of contextually sensitive, fully integrated visibly salient facial displays and words. The authors cite Fridlund's work, among that of others, for emphasizing the importance of nonverbal as well as verbal elements in social encounters. Fridlund has argued that the chief function of human facial displays is not emotional but "paralinguistic," by which he means that facial displays accompany and supplement speech.

evolved pre-prepared package of responses. They were no longer treated as strategies that conformed to learned display rule conventions for managing and regulating the basic emotions, themselves defined as automatic discharges of affect programs. Instead, strategic responses were now regarded as built-in, evolved solutions to life tasks, such that, in ways that had not been acknowledged by BET advocates, the display rule component of the original BET had lost its rationale. As Fridlund commented: "whereas the earlier version mandated display rules to govern expressions that were stipulated to be vestigial, the newer 'adaptationist' version allowed that the expressions might have evolved on their own as components of behavioral patterns that enhanced inclusive fitness, with no requirement for cultural display rules to squelch or counter them on the fly." As I observed in chapter 6, Griffiths had recently put forward a version of the same thesis when he suggested that audience effects, which is to say strategic response to the presence of real (or imagined?) other individuals, could be considered part of the evolved affect system itself, with the result that the intentional, tactical nature of such responses was denied. The incorporation of strategic facial and other displays into the repertory of evolved adapted reactions thus rendered the display rule concept—defined as a learned response for managing affective responses in conformity with the rules and norms of public behavior—an unnecessary and redundant feature of BET. That is why Fridlund contended that the acceptance of the Tooby–Cosmides approach to emotions meant forfeiting the Ekman–Tomkins Facial Affect Program: the two approaches simply were not compatible or coherent with one another.

3. As Fridlund noted, in 1997 Ekman revised his position in yet another way. It will be recalled that, in defending the claim for the sociality of facial displays, on the basis of his own and others' experimental evidence, Fridlund had contended that, contrary to Ekman's view that smiles performed in the solitary condition were authentic smiles of happiness because they were uncontaminated by the interference of conventional display rules, solitary smiles were responsive to audience effects. This finding jeopardized the basic premise of Ekman's BET by showing that the distinction between solitary, authentic, affect program expressions versus public, fake, or managed display rule expressions could not be maintained.

In 1997 Ekman and Keltner had summarily dismissed Fridlund's results and claims.[7] But, as we saw in chapter 5, and as Fridlund now pointed

7. P. Ekman and D. Keltner, "Universal Facial Expressions of Emotion: An Old Controversy and New Findings," in *Nonverbal Communication: Where Nature Meets Culture*, eds. U. Segerstrâle and P. Molnár (Mahway, NJ, 1997), 41.

out, in that very same year Ekman did a complete about-face and conceded that display rules could operate in the alone condition.[8] Although Ekman did not acknowledge it, this shift had devastating implications for his affect program theory, because it was now no longer clear how one would go about determining whether a smile in the solitary condition was authentic or not, since if it was governed by display rules it was by definition a managed and not a "genuine" emotional expression. As Fridlund commented, the admission "vitiated the rationale and conclusion of BET's only empirical claim to display rules, the original Japanese-American study which rested on the proposition that solitary expressions had to be spontaneous and 'emotional.'" This meant that Ekman's influential distinction between felt and unfelt emotions could no longer be sustained. If emotion could now be *real* but unfelt, in what sense could non-Duchenne smiles be considered false? In opposition to the Basic Emotion Theory, Fridlund commented that "For BECV [Fridlund's Behavioral Ecology View] the 'authenticity' of [a person's] smile," referring to the behavior of a car salesman, "lies not in what he feels, but in whether it predicts whether he will treat us fairly if we buy a car from him. More generally, we learn whose words and expressions are reliable indicators of their intent, and over time we bond with those individuals who prove reliable and avoid those who prove otherwise" ("BEV").

4. In yet another revision of his original BET, Ekman in 1999 offered a list of eleven characteristics of which, he stated, "'none is *sine qua non* for a basic emotion to be instantiated.'"[9] The important issue here was not only that Ekman had greatly expanded the number of variables potentially associated with distinct emotions, but that he did not regard the presence or absence of any single component—including "*qualia*" or feelings—as determining for any specific affect, a revision that Keltner and Cordaro now endorsed. But Fridlund rejected this development on two grounds. His first objection was that adding variables complicated prediction and analysis, leading to "an accumulation of error terms and the need for ever-larger sample sizes to achieve statistical power." His second, more "substantive" objection was that Ekman and, following him, Keltner and Cordaro had so loosened the criteria for defining an emotion and had hedged their position about with so many vague "slipclauses and patches"

8. P. Ekman, "Expression or Communication about Emotion," in *Uniting Biology and Psychology: Integrated Perspectives on Human Development*, eds. N. L. Segal, G. E. Weaseled, and C. C. Weisfeld (Washington, DC, 1997), 328.

9. P. Ekman, "Human Expression," *Handbook of Cognition and Emotion* 16 (1999): 301–20.

that it had no remaining coherence or explanatory value. In short, in the face of the criticisms launched by himself, Russell, and others, Fridlund charged Keltner and Cordaro with attempting to salvage BET by quietly abandoning some of its core tenets, such as the distinction between felt emotions versus phony ones or between affect programs versus display rules, while simultaneously relying on those same tenets and distinctions to fudge the actual meaning of their findings. As Fridlund forcefully put this point:

> Keltner and Cordaro want to add gestural and other bodily movements to their "expressions of emotion." By what procedures will they determine which movements, committed by whom and when, are part of the emotional expressions themselves, and which movements, committed by whom and when, reflect cultural display rules to manage those emotions or even fake them? In the original Japanese–American cultural display rules study . . . self-report data were used to verify the presence of "emotion," but what would be criterial for emotion now without *qualia* as definitive? Keltner and Cordaro, by making *qualia* optional in their "New BET" definition of emotion, would be counterfeiting the term while profiting from its cash value among non-BET researchers who haven't been let in on the secret. Now BET researchers could claim to be studying "emotion" while they are really studying "emotion*," which is emotion-without-the-feeling. Thus, in a new BET study that concluded, "happier people smile more," the study would have to be footnoted, "NOTE: in accordance with the provisions of New BET, people who are happy do not have to 'feel happy' (Keltner and Cordaro, 20XX). This will delight readers, I am sure—although they need not feel it."

5. In arguing against BET, Fridlund also rejected various "newer backstops" that attempted to shore up the emotion concept, such as Barrett et al.'s "Conceptual Act Theory" and Scarantino and Griffiths's effort to nail down the emotion concept by declaring it to be "intrinsically fuzzy."[10]

10. Fridlund cites L. F. Barrett, C. D. Wilson-Mendenhall, and L. Barsalou, "The Conceptual Act Theory: A Road Map," in *The Psychological Construction of Emotion*, eds. L. F. Barrett and J. A. Russell (New York, 2015), 83–110; and A. Scarantino and P. Griffiths, "Don't Give Up on Basic Emotions," *Emotion Review* 3 (2011): 1–11. As I observed in chapter 6, in her "Conceptual Act Theory," Barrett et al. posit that the human mind is capable of categorizing the body's physical states and inputs as emotions in a context-specific fashion, adding concepts in a top-down fashion to nonconceptual physical sensations, such as one's own interoceptive state, as well as to the movements and vocalizations of others.

(As I showed in chapter 6, Scarantino and Griffiths have tried to isolate within the fuzzy borders of the emotion concept a narrower, properly scientific version of BET.) From Fridlund's perspective, all such attempts to save "emotion" were "tendentious and wasteful," if the goal was to understand facial displays. As he observed:

> For BECV, the essential question, thankfully, is not about whether "emotions exist," but "How do our facial movements affect the trajectories of our social interactions?" This question precludes snapshots, because our social interactions are diachronic, cumulative and interwoven. Specifying contexts (our interaction contexts are cumulative, and begin to comprise relationships), and capturing and understanding the different ways that people move their faces and how their interaction partners react, requires a dynamic, functional view, not one constrained *a priori* by BET's freeze-framed boxes.

As evidence for the value of his approach, he mentioned among other developments Russell's "influential critiques" of BET and demonstrations of powerful context effects in facial expression perception; the "masterful studies" by Fernández-Dols and his colleagues showing how the Behavioral Ecology View accounted for facial behavior in naturalistic settings; and Chovil's inductive typology of communicative facial displays. He also recommended Russell's Minimal Universality Thesis as compatible with his own thinking because it presumed little and built up from the bottom.[11] What he did not accept was that Keltner and Cordaro's revised version of BET offered a useful scientific way forward.

As for Russell, in his manifesto and in the debate that followed, he too offered a skeptical view of BET's prospects on grounds very similar to Fridlund's, namely, that the correlations between the hypothesized basic

11. Russell's Minimal Universality Thesis was formulated as the conclusion to his critical analysis of the evidence offered by Ekman and his school in favor of the existence of universal discrete emotions. The Minimal Universality Thesis stated that, at a minimum, patterns of facial movement occur in all human beings; that those facial movements are correlated with a person's state; that people everywhere can infer or perceive something of another's state, including the other's emotional state from facial movements; and that people in Western cultures tend to believe that certain specific facial movements "express" specific emotions. But Russell's thesis also stated that even if people in the West can achieve a high degree of consensus about their attribution of specific emotions to certain specific facial expressions, especially stylized or stereotypical faces, the postulated associations are not necessarily valid. James A. Russell, "Facial Expressions of Emotion: What Lies Beyond Minimal Universality?," *Psychological Bulletin* 118 (1995): 398.

emotions and specific facial expressions had turned out to be far looser than Ekman's original BET had predicted. Russell observed in this regard that even LeDoux, whose findings about neural circuits for fear in rats had been cited as evidence of the validity of BET, was now "abandoning the notion of discrete, hardwired emotion-specific brain circuits," thereby depriving BET enthusiasts of claims to the existence of hardwired affect programs.[12] Appealing to these considerations, to his earlier critique of Ekman's cross-cultural judgment studies, his several demonstrations of the role of context in determining the meaning of facial displays, and related investigations by Fridlund, Fernández-Dols, and others, Russell argued that the evidence was not compatible with BET and that the theory had to be abandoned.

In its place, Russell offered his theory of Psychological Constructionism as an alternative approach. "On my proposal," he wrote, "the term 'emotion' is treated as a folk rather than a scientific term":

> Episodes called "emotional" consist of changes in various component processes (peripheral physiological changes, appraisals and attributions, expressive and instrumental behavior, subjective experiences), no one of which is itself an emotion or necessary or sufficient for an emotion to be instantiated. Emotion is not invoked as the cause of the components nor as the mechanism that coordinates the components. Each component has its own semi-independent causal process.

To which he added:

> The components are coordinated, as are all human processes, but again, not by an affect program. Although emotion is not an entity causing the components, still, a witness, or the person having the emotion might categorize the episode as a specific emotion: we see emotions in others and experience emotions in ourselves. That categorization process is a process to be studied, and is neither necessary nor sufficient for an emotion to be instantiated. Psychological

12. Russell's reference was to Joseph LeDoux, "Afterword: Emotional Construction in the Brain," in *The Psychological Construction of Emotion*, eds. L. F. Barrett and J. A. Russell, 459–63, in which the author makes a distinction between the brain mechanisms he believes gives rise to innate emotional expressions and the processes that produce conscious emotional experience. LeDoux now rejects his previous identification as a basic emotion theorist, insofar as BET assumes that even emotional feelings are hardwired. LeDoux concludes that aspects of his present point of view fit nicely with Russell and Barrett's constructionist position.

> construction abandons the assumption that emotional episodes are pre-fabricated; it proposes instead that they are assembled in the same way as in any other behavioral episode, although often with a more extreme dose of valence and arousal.

In other words, Russell envisaged the participation in any emotional episode of a number of independent components or ingredients, including "valence and arousal" (or what he called "core affect"), none of which, however, he considered absolutely essential for any given affective event. The aim of emotion science was thus to determine experimentally how the loosely associated elements were co-assembled in any particular emotional episode. Rejecting the essentialist BET idea of biologically given categories of emotion, Russell accepted the notion that emotional events have loose boundaries and considerable variability. He presupposed that the different posited basic ingredients were universally present in humans and many nonhuman animals, but that specific categories of emotions and what he called "emotional meta-experience"—the conscious experience of having an emotion linked to an intentional object—are culturally specific.

On the surface, Russell and Fridlund appeared to be close allies not only in their condemnation of BET, but in their attempts to offer alternative approaches to the emotions. Indeed, with regard to what he called the "ontology of emotion itself," Fridlund declared his sympathy with Russell's constructionist approach largely because of its humility. "I see it as a springboard," he wrote, "to anthropological investigations as to how diverse peoples may or may not 'package' their behavior and experience into entities that may, or may not, resemble our Western emotions." And much of Russell's research program was closely aligned with Fridlund's position in its emphasis on the role of dynamic cues in determining the observer's understanding of the meaning of faces, on the multiple cues involved, on the role of audience effects, on the idea that such interactions unfold over time, and so on.

But I think it is important to note that beneath the apparent agreement there were and remain several differences between Fridlund's and Russell's respective solutions to the "emotion problem." As we have seen, from the start of his critique of BET Fridlund's view has been that he does not require the concept of emotion to carry out his analysis of facial movements. He is not saying that there are no emotions: he remains agnostic on that topic. What he does argue is that for his purposes he does not need the concept of emotion, not only because in the hands of Ekman and other BET theorists emotion has become so distended and malleable a term that it has become impossible to pin it down, but also because his functionalist view of displays "neither needs not benefits from 'emotion' as an explanatory construct."

Instead, Fridlund's emphasis falls on the radically contextual and relational nature of the interactions between humans and between nonhuman animals as they negotiate their encounters and communications. "BECV does not hold that faces signal irrevocable intentions, just the possible next steps in on-going negotiation," he writes. "Both the signals and their assessments form the dynamic interchange by which interactants negotiate and determine their social trajectory." One of the important implications of Fridlund's approach is thus that *context trumps everything else*. Whereas BET presumes that under certain conditions faces outwardly express an internal emotional state regardless of context, for Fridlund the meaning of facial and other signals is entirely situational. It follows that in some contexts the face alone can convey the significance of an encounter, but in other situations it is far less important. As he comments: "If my wife comes in from the outside and I ask her about the weather and she scowls, I see from her face that it's lousy outside. At that moment, in that context, her face *expresses what the weather is like*." But in other situations, especially if the context is unambiguous, we may not care much about nonverbal facial signals. Thus, if a policeman pulls me over for doing 75 mph in a 55 mph speed zone, I don't need to pay close attention to his facial expressions because, even without looking at his face, I know the meaning of the situation. In short, from Fridlund's perspective, what is determining for the meaning of facial and all other communicative signals, including in our own case crucially what we humans say, are the conditions and contexts governing particular social encounters.

One way of characterizing Fridlund's enterprise is to say that it remains at the descriptive level of everyday interactions. He does not seek to explain "emotion" by appealing to hidden, internal mechanisms. Rather, he is interested in understanding the trajectory of social encounters as they occur in ordinary, everyday life. His position treats human and even many nonhuman encounters as cognitive through and through, which is why he can suggest that anthropological research is one way to investigate them.[13] His position also encourages ecological-ethological attempts to study experimentally the relations between humans and between non-

13. For Fridlund's recent methodological suggestions along these lines, see Carlos Crivelli, Sergio Jarillo, and Alan J. Fridlund, "A Multidisciplinary Approach to Research in Small-Scale Societies: Studying Emotions and Facial Expressions in the Field," *Frontiers in Psychology* 7 (2016): 1–12.

human animals by controlling and manipulating the various environmental and other variables likely to influence particular exchanges in more or less natural settings. But on his approach the description and analysis will stay at the level of the everyday intentional actions of intact animals and will not reductively assume that analysis must go "beneath" the surface of such interactions by assuming the existence of putatively more fundamental, hypothetical causal entities and mechanisms lodged inside the mind-brain.[14] As he writes: "For BECV, displays evolved as social tools *directly*, not as parts of underlying mechanisms for the production of displays. Natural and cultural selection do not 'care about' (specifically select for) the inner workings of traits, only the traits themselves" (his emphasis). Or, as he also observes: "For me, BET is riddled with a preconception of emotions as hermetic 'things' with inner circuits and outer manifestations." For this reason, he also rejects attempts to pin down emotions by appealing to the existence of certain alleged internal neural mechanisms.[15]

Russell's solution to the emotion problem is different. Although, like Fridlund, he abandons the concept of "emotion" as a single causal entity, he proposes that beneath the surface of emotional episodes occurring in everyday social encounters are more basic causal ingredients or components in the mind-brain out of which such episodes are constructed. For Russell, the goal of affective science is thus to identify the ways in which such loosely associated ingredients co-occur in specific interactions. On the one hand, Russell's psychological constructionism is attentive to the cultural dimensions of emotional episodes and stresses the degree to which emotional and cognitive episodes are closely intertwined. On the other hand, his is an approach that risks bypassing or otherwise slighting the level of description of the encounters of interest to Fridlund in favor of hypotheses about elementary, independent entities or hidden mechanisms in the brain or mind—entities and mechanisms that are held to be more foundational than the surface aspects of our ordinary,

14. Cf. Wittgenstein.

15. "But what if one were to localize the proximate generators of *qualia* in the brain?" Fridlund asks in his manifesto. "Could we then say that changes in *qualia*, or those generators that produced the *qualia* changes, caused the associated events in the neuromuscular centers that produced the facial expressions? How does one ever determine that event A causes event B in the brain? I invite readers new to this question to Google 'Libet's experiment' to discover the labyrinthine complexities in determining neurocausality." Fridlund's reference is, of course, to Benjamin Libet's famous experiment on the so-called "missing half second," an experiment I discuss in chapter 7.

everyday interactions because the latter are themselves the products of more fundamental causal processes. Whereas Fridlund believes that we directly see the meaning of each other's facial and other displays, Russell thinks that the meaning of such displays lies in the co-assembly of various hypothetical elementary components out of which meaning is built up. He therefore lodges many of the constituents held to participate in the construction of emotional episodes inside the subject's mind-brain from which Fridlund has sought to expel them.

The implications of this development are especially evident in the case of one of Russell's hypothesized components, "core affect." According to Russell, core affect comprises the subject's internal feelings corresponding to the body's ever-changing neurophysiological condition, although he holds that, as a general feature of consciousness, core affect is not specific to emotional episodes. As he puts it, core affect is "a part of what are commonly called emotions, feelings, and moods, but is not synonymous with any of these."[16] He admits that "calling core affect a neurophysiological state is a promissory note that has so far been left unfulfilled," which is to say that its status as a neural process has yet to be proved ("MPCP," 196). This is the case even if people can easily answer the question "How do you feel?" when asked to respond in terms of their levels of excitement, pleasure, displeasure, arousal, and so on. For Russell, those subjective experiences do not stand on their own but must be distilled into their "core affect" components if those experiences are to be understood scientifically.

Moreover, according to Russell, although changes in core affect influence not only our physiological but also our psychological states, including our cognitions, in its essential psychological aspect core affect is not itself conceptual. On the contrary, Russell defines core affect at the psychological level as a "preconceptual primitive process." As such, he suggests that core affect has many of the features of modularity, not only in the sense that its operations are fast, mandatory, and have a unique output, an evolutionary explanation, and a dedicated brain circuitry, but also in the sense that it is informationally encapsulated, which is to say that it is "encapsulated from general beliefs and desires" ("MPCP," 197). De Sousa in a commentary notes in this connection that for Russell core affect is limited to "private subjective conscious feelings," and is also prelinguistic: "'A person's Core Affect is often influenced by cognitively pro-

16. James A. Russell, "My Psychological Constructionist Perspective, with a Focus on Conscious Affective Experience," in *The Psychological Construction of Emotion*, 195; hereafter abbreviated as "MPCP."

cessed information and is often embedded within an intentional state,'" de Sousa quotes Russell as stating, but in itself it does not need any conceptualization "'to do its work.'"[17]

In other words, according to Russell, core affect is independent of cognition and intention, which in the construction of emotional events is therefore added on by the organism's independent cognitive systems.[18] He thus posits a gap between core affect and intentionality or meaning in ways that not only appear to be at odds with Fridlund's emphasis on the pervasive mindedness and intentionality of all facial and other signaling processes and encounters, but also raise awkward questions about how the gap between nonconceptual affect and conceptuality or intention is to be closed. In this regard, as I observed in chapter 6 of Barrett's rather similar proposals about core affect, Russell's attempt to establish an alternative to the basic emotion approach comes close to reinstating Ekman's two-factor model according to which cognition or conceptuality is tacked on to a biologically primitive nonconceptual affective process—a model that Fridlund refuses.

One way of putting this point is to suggest that the net effect of Russell's psychological constructionist approach is to re-individualize and re-subjectivize "emotion" in ways that are not only foreign to Fridlund's endeavor but that risk underestimating the radical relationality and contextuality of social encounters on which Fridlund insists. That Fridlund's approach is radical is something he concedes. His position is a "tough sell" in part because it foregoes the sort of appeal to hypothetical internal brain-mind mechanisms that inform so much of the cognitive neurosciences and indeed Russell's constructionist (and Feldman Barrett's) theory. The implication of Fridlund's approach is that the answer to the question of how people and nonhuman animals behave is to be found by studying their interactions in their natural-social settings without re-

17. Ronald De Sousa, "Valence, Reductionism, and the Ineffable: Philosophical Reflections on the Panksepp–Russell Debate," *in Categorical Versus Dimensional Models of Affect: A Seminar on the Theories of Panksepp and Russell* (Amsterdam and Philadelphia, 2012), 226, 239.

18. In a recent article, Deonna and Scherer criticize Schachter and Singer, Russell and Feldman Barrett, and Paul Griffiths for inadvertently or deliberately making the intentional object disappear in their emotion theories. Deonna and Scherer are rare among scientists for paying attention to this important issue, although their own understanding of the correct way to theorize the issue of intentionality is remote from mine. See Julien A. Deonna and Klaus R. Scherer, "The Case of the Disappearing Intentional Object: Constraints on a Definition of Emotion," *Emotion Review* 2 (1) (2010): 44–52.

course to explanations based on the existence of hypothetical internal causal entities, processes, or mechanisms.

It is therefore not surprising that in addition to recommending various experimental research strategies in psychology itself, Fridlund also advocates a turn to anthropology and history as ways to proceed. But staying on a descriptive level, eschewing the search for putative hidden neural and other mind-brain causal components and advocating a turn to anthropology or history are all "tough sells" to researchers who are enamored of brain imaging and related technologies, and who find the notion of basic emotions congenial because their existence seems to them so obviously true, however confused and incoherent the notion can be shown to be. Of course, Fridlund does not suggest that research on facial expressions should be abandoned. On the contrary, he argues that the battle between BET and his behavioral ecology views has liberated inquiry on facial expressions, so that "investigators can now pursue hypotheses (e.g., genetic/epigenetic diversity of facial displays, facial dialects, infant deception) that, because they transgressed BET, were previously inconceivable or taboo." As the work of Bavelas, Chovil, Fernández-Dols, Russell himself, Brian Parkinson, and a host of other scientists demonstrates, there is plenty of room for further research on the significance of such displays and the conditions under which they operate. Nevertheless, it's as if Fridlund *is* asking investigators to give up many of the tools, methods, and assumptions that for so long have been standard in their science in order to operate differently—at best, as researchers in the tradition of the study of human nonverbal communication, or as ethologists studying the interactions of nonhuman animals in the field; at worst, to make room for anthropologists who are good at describing different cultures or historians capable of depicting past emotional regimes.

By contrast, Russell offers a more ecumenical and what must also seem to many investigators a more attractive and encouraging view of the prospects for emotion science. The premise of his approach is that once we give up the idea of emotion as a natural kind with a common core that is held to explain the ensuing response, such as an affect program, we can accept the inherent heterogeneity of affective "episodes" in ways that permit the integration of quite disparate methods and approaches. He therefore presents what he calls the "Greater Constructionist Project" as a research program that is able to shelter under one large umbrella the findings of even apparently incompatible or competing scientific accounts of emotion. According to him, not only are appraisal theories and social-constructionist theories compatible with psychological constructionism, but so is a revised version of Basic Emotion Theory itself. That is, once

BET is stripped of its commitment to the existence of discrete emotions and accepts the idea that the components of emotional episodes are less tightly correlated with each other than had previously been anticipated, then psychological constructionism and BET are not so very far apart.

This is especially the case when both are viewed as offering a set of empirically testable claims. Keltner and Cordaro's multimodal approach, Russell writes in the debate, "converges with traditional accounts in the study of nonverbal communication, including my own." He even observes that in acknowledging the dynamic quality of social interactions, "Keltner and Cordaro move towards Fridlund's account, at least a little." According to Russell, so close is the possible rapprochement between his own position and that of BET theorists that he says he sees no difference between his constructionist account and that offered by Levenson, "a prominent BET theorist." By this, Russell appears to mean that although Levenson still accepts the idea of basic emotions as hardwired, organized sets of response tendencies, a view Russell rejects, the former nevertheless regards the subjective experience of emotion itself as not fixed but as varying with cultural understanding, an opinion that matches Russell's.[19] Indeed, in a recent article Russell has stated that he finds little in Scarantino's new basic emotion theory that is inconsistent with his Greater Constructionist Project.[20]

In sum, Russell proposes that even if all the incoherencies and experimental confounds of the kind mentioned by Fridlund are still present and uncontrolled for in the work of Keltner and Cordaro and so many other BET theorists, there is room for a convergence between his own psychological constructionism and all other approaches to the emotions, such that the prospects for advances in the science of emotion are still good. It's as if he thinks that once he has proposed as a shared foundation for the sciences of emotion the discovery that there *is* no core or foundation beyond the co-assembly of otherwise loosely correlated components, including the hypothetical component of "core affect," and once it has accepted that the display rules concept needs to be dispensed with (something Keltner and Cordaro do not yet appear to concede), everything and anything can be accommodated as an add-on.

19. This is the same Levenson whose 1990 experiment with Ekman and Friesen, on the generation of emotion-specific autonomic nervous system activity, cited favorably by Keltner and Cordaro in their manifesto, has been widely criticized, as I discuss in chapter 5.

20. Russell, "The Greater Constructionist Project," in *The Psychological Construction of Emotion*, 437.

Russell has played an exemplary role as a brilliant critic of Ekman's original BET and as an indefatigable researcher who has conducted impressive experiments showing the validity of many of his own and Fridlund's insights concerning the contextual and social nature of facial displays. Indeed, not only in the debate does he call Fridlund's manifesto "compelling," but he has gone so far as to commend Fridlund's views as "the most exciting new account of facial behavior since Darwin"—high praise indeed.[21] Russell's colleague Feldman Barrett, too, has published many cogently argued critiques of BET and invaluable overviews of the research literature on emotion demonstrating how weak the evidence in favor of BET really is. Both these researchers, along with colleagues such as Fernández-Dols and their teams, have played a vital role in leading the fight against a picture of the emotions that has bedeviled the research field for far too long.

The trouble only arises when Russell and Barrett attempt to theorize alternative approaches that end up so close to those of the "new BET" researchers as to be virtually indistinguishable from them, with all the ensuing confusions and uncertainties regarding what it is they think they are studying when they adopt a multicomponential approach of the kind Russell advocates. On the face of it, in spite of Fridlund's stated support for Russell's work for breaking the paradigm lock BET had on facial expression research, his criticisms of Keltner and Cordaro's "new BET" for having further complicated rather than resolved the definitional issue of "emotion" would seem to apply to Russell's psychological constructionism as well. To put this even more strongly, in this debate it's as if Russell offers his psychological constructionism as the last possible solution to the emotion problem that stops short of having to go all the way with Fridlund's more radical Behavioral Ecology View. My argument in this book, however, is that in the field of emotion research there is no intellectually viable alternative to Fridlund's position, whatever the cost may turn out to be to many of the existing "scientific" studies of emotion.

21. James A. Russell, "Facial Expressions of Emotion: What Lies Beyond Minimal Universality?," 389–90.

ACKNOWLEDGMENTS

For their support, conversations, and responses to aspects of my work during the years I have been writing this book, I thank: Isobel Armstrong, Jason Bridges, Felicity Callard, James Conant, David Finkelstein, Alan J. Fridlund, Stefanos Geroulanos, Avery Gilbert, Stephen Gross, Jennifer Hornsby, Phil Hutchinson, Meira Likierman, John McDowell, Toril Moi, Constantina Papoulias, Robert Pippin, Jan Plamper, William M. Reddy, Rainer Reisenzein, James A. Russell, and Meredith Williams.

My thanks also to Ute Frevert who, as Director of the Center for the History of Emotions at the Max Planck Institute for Human Development, in 2009 invited me to become a member of the Advisory Board and offered me a warm welcome and an intellectually stimulating environment there whenever I visited Berlin. I also thank Gert Gigerenzer, Ulman Linderberger, and the other members of the Institute for their helpful engagement with my work over the years. My thanks also to Thomas Dixon and the Center for the History of the Emotions, Queen Mary University London, for inviting me to contribute to the Center's lively activities as a Visiting Research Scholar.

A warm thanks to the participants in my various graduate seminars at Johns Hopkins University and to members of audiences at my lectures and presentations at universities, colleges, and other institutions in the United States, Europe, and Australia for their constructive comments and contributions to my understanding of the issues I am concerned with in this book. The Humanities Center at Johns Hopkins has been my professional home for more than two decades, and I thank my colleagues and friends on the faculty there for providing me with such an intellectually supportive environment.

To Jennifer Ashton and Walter Benn Michaels, I offer heartfelt thanks for their friendship, conversations, and insights over many years. I dedicate this book to them.

I regard this book as the third in a trilogy of studies dealing with related topics in the recent history of the human sciences. The first two are *Trauma: A Genealogy* (Chicago, 2000) and *From Guilt to Shame: Auschwitz and After* (Princeton, 2007). In the latter, I first laid out a critique of the Basic Emotion Theory.

Portions of the present book were previously published in somewhat different forms. Chapter 2 appeared as "How Did Fear Become a Scientific

Object and What Kind of Object Is It?," in *Representations* 110 (Berkeley, CA, Spring 2010): 66–104, subsequently reprinted in *Fear: Across the Disciplines*, eds. Jan Plamper and Benjamin Lazier (Pittsburgh, 2012), 51–77; chapter 5 appeared as "A World Without Pretense? Honest and Dishonest Signaling in Social Life," in *Philosophy of Education 2013*, ed. Cris Mayo (Urbana, IL, 2013), 25–42; and chapter 7 appeared as "The Turn to Affect: A Critique," *Critical Inquiry* 37 (3) (Chicago, Spring 2011): 434–72.

Finally, I wish to acknowledge with gratitude the support of Michael Fried, who over the years when this book was germinating made himself available at any and all hours to discuss my ideas, and who when the time came read constructively and generously every word of my completed manuscript.

[APPENDIX 1]

ANIMAL SIGNALING, THE SMILE, AND THE HANDICAP PRINCIPLE

In their original paper on the role of manipulation in animal signaling, Dawkins and Krebs criticized the "informational" approach in ethology with which W. John Smith's work was associated (the term "information" was not used in the cybernetic sense) and rejected cooperation as an evolutionarily improbably event. But in his reply to their criticisms, "An 'Informational' Perspective on Manipulation" (1986), Smith complained that Dawkins and Krebs had misrepresented the informational approach and accordingly had underestimated the coevolution and coadaptation of signaling responses, especially the coevolutionary possibilities of co-operation (understood in nonmoral, behavioral ecology terms). In their subsequent response to Smith's criticisms, Krebs and Dawkins conceded this latter point.

More generally, Smith reproached Dawkins and Krebs for proposing too inflexible an account of the relationship between signaler and recipient. He even accused them of reverting to the "outmoded releaser postulate" of classical ethology, a postulate that the latter had ostensibly rejected. That is, Smith accused Dawkins and Krebs of assuming that animals respond to signals "as automatically as locks open to keys," thereby making the receiver a passive recipient of the signaler's message, independently of the selection pressures on recipients to avoid being manipulated. By contrast, Smith's information approach to animal communication held that the evolution of behavior is influenced by the advantages achieved by both signalers and recipients. As he observed in this regard:

> On the one hand, signalers should make information available through signaling only if this availability elicits responses that are, on average, beneficial to them; on the other hand, recipients should evolve predispositions to respond to this information only insofar as these, on average, benefit them. Far from being "incidental," the latter effect causes signaling and responding to signals to co-evolve, and signalers and responders are coadapted. This coadaptation does not prevent manipulative signaling. Through mimicry, signalers can in effect parasitize otherwise useful response predispositions

> of recipients, as long as these predispositions remain, on average, beneficial—although the increased costs/benefits ratio for the reactors does engender evolutionary instability. Thus conflicts of interest so crucial to the manipulation perspective are encompassed by "informational" approaches. (Smith, 1986, 72–73)

As this statement suggests, Smith emphasized the probabilistic and conditional nature of animal signaling: "[P]redictions fostered by animal signals are conditional . . . on the responding individual's behavior . . . [B]ehavioral predictions are not only conditional, but also probabilistic. They do not provide the information that a particular kind of behavior *will* occur, if conditions are right. Instead, they enable assessments of the *probability* that it will—usually relative to the probability of one or more other kinds of behavior, with none of the alternatives having a probability of one" (75). On this basis, Timo Maran has contrasted Smith's probabilistic approach to signaling with the more genetically determined views of figures such as Dawkins and Krebs and especially the related opinions of the influential ethologist John Maynard Smith.

At the same time, Smith did not exclude the possibility that "handicap" signals might evolve as a solution to the problem of how signalers might certify the reliability of their signals. As famously described by Amotz Zahavi, handicaps are signals whose costly performance provides information about some feature of the signaler, such as size, robustness, or stamina. Handicaps are thus hard-to-fake displays that depend on a restrictive relationship between the properties of the signal and the quality of the sender because of some physical necessity. For example, the roar of the deer stag, tied to the animal's strength, is tiring and hence costly to maintain; it therefore constitutes a reliable signal of the deer's reproductive fitness in mating contests.

The literature on handicap signals is very large, and questions concerning their importance and role in animal behavior remain unresolved. Smith in 1986 mentioned Ekman's studies of "leakage" and "deception" clues in human nonverbal facial and bodily signaling as demonstrating procedures of "self-certification" which, like Zahavi handicaps, serve to guarantee the inherent reliability or "honesty" of certain emotional signals (Smith, 1986, 82). More recently we have seen that, following Robert Frank's commitment model of emotion, William Flesch has likewise suggested that palpable emotional expressions may be considered costly "handicaps" that are difficult to fake or hide, handicaps that as such indicate the true affective commitment of the individual.

But, given the critiques of Ekman's views documented in the present book, it is highly doubtful that facial expressions can be considered handicaps. It is worth noting in this regard that the theoretical biologist Alan Grafen, whose brilliant game-theoretical analyses of signaling have been cited by Flesch as proving the validity of the handicap concept, did not regard certain kinds of handicaps as signals at all. For example, he denied that "revealing handicaps," as these were originally described by Maynard Smith in 1985, are signals properly defined, because the content of such signals is directly revealed or observed. Thus, according to Grafen, on the revealing handicap model (or in what Maynard Smith later identified as "indexical" signals, which depend on a fixed, causal relationship between some trait of the signaler and its expression and are therefore impossible to fake), nothing can make a low-quality male appear to be a high-quality male: indeed, on the revealing handicap model to which, according to Grafen, Zahavi himself sometimes appeared to subscribe, even the cost of the handicap is not essential; rather, what matters is that the handicap makes it possible to disclose information in an uncheatable way.

But for Grafen, signals, properly defined, vary continuously in some quantity of the signaler that is of interest to the recipient. They therefore operate on a graded continuum rather than in an either/or way. In other words, they involve strategic choices in which communication is never error free. It is in these terms that Grafen defended the Zahavi handicap principle. "To anthropomorphize," he observed, "if you observe my quality directly, that is no signal—it is the real thing," whereas in reasonable models of signaling, "the element of information is voluntarily provided." In other words, Grafen adopted a definition of a handicap as signals that are strategic and flexible and continuous with those of cheaters, and not inflexible signs of inner motivational states (or genotypes).

But the latter model is the one adopted by Frank, Flesch, and more recently by Owren and Bachorowski, in the latter's 2001 attempt to apply the handicap principle to the evolution of emotional expressions, specifically smiling and laughter. Influenced by the selfish gene approach to animal signaling, Owren and Bachorowski argue that the genuine smile must have evolved as a reliable indicator of positive emotional state in a form that has safeguarded its reliability in the face of selection pressures favoring dishonest versions, with the result that in a feedback loop, receivers can also afford to show positive, cooperative behavior. The authors suggest that dishonest signaling was a later evolutionary development that arose when selective pressures produced voluntary emotional expressions controlled by separate motor systems in the brain. Nevertheless, according to them, in spite of this countervailing selection pressure, the strategy for expressing positive affect must have been protected from exploitation.

Owren and Bachorowski link their claims to Ekman's Basic Emotion Theory by suggesting that the genuine, involuntary so-called "Duchenne" smile that results from positive affect has been shown by Ekman and colleagues to be distinguishable from the false, voluntary so-called "non-Duchenne" smile that a person produces upon request. In short, Owren and Bachorowski argue against the possibility of deception in smiling owing to the built-in guarantee of the Duchenne smile as a signal of authentic or felt happiness. On the basis of these and related considerations, the authors explicitly reject Fridlund's Behavioral Ecology critique of Ekman's views.

To my mind, Owren and Bachorowski not only show a lamentable misunderstanding of Fridlund's position but make many questionable moves. For instance, in the spirit of Zahavi, the authors suggest that "dishonesty is prevented by the inherent cost of signaling" (165). This primes the reader to hear some kind of argument about the cost to the *signaler* of expressing different kinds of smiles. But the next mention of costs is about the cost of having positive affect aroused in *receivers*: "As the volitional signal could now be used in circumstances in which it might be costly to experience smile-induced positive affect toward a particular sender, receivers would do better if able to discriminate emotion-dependent from volitional smiling." What the cost to receivers of an experience of a smile-induced positive affect might be, the authors do not say.

Above all, it is important to emphasize that Owren and Bachorowski's speculations as to the origins of the smile are so closely tied to Ekman's Basic Emotion Theory that they stand or fall with that theory. Their entire argument depends on the Basic Emotion Theory that central emotional states are either represented "honestly" or "dishonestly" in the form of genuine or false facial expressions. But, as chapters 5 and 6 document, according to Fridlund and many other critics, the validity of Ekman's Basic Emotion Theory has been decisively called into question.

In a large literature, I have found the following sources to be especially relevant and/or helpful: Richard Dawkins and John R. Krebs, "Animal Signals: Information or Manipulation?" (1978); John R. Krebs and Richard Dawkins, "Animal Signals: Mind-Reading and Manipulation," in *Behavioral Ecology: An Evolutionary Approach,* 2nd ed., eds. J. R. Krebs and N. B. Davies (Oxford, 1984), 380–402; W. John Smith, "An 'Informational' Perspective on Manipulation," in *Deception: Perspectives on Human and Nonhuman Deceit*, eds. Robert W. Mitchell and Nicholas S. Thompson (Albany, NY, 1986), 71–86, with editorial commentary on the dispute between Smith, Dawkins, and Krebs by Mitchell and Thomson, 67–69; J. Maynard Smith, "Mini Review: Sexual Selection, Handicaps and True Fitness," *Jour-*

nal of Theoretical Biology 115 (1985): 1–8; Alan Grafen, "Biological Signals as Handicaps," in ibid., 144 (1990): 517–46; idem, "Sexual Selection Unhandicapped by the Fisher Process," in ibid., 473–516; Rufus A. Johnstone and Alan Grafen, "Error-Prone Signalling," *Philosophical Proceedings of the Royal Society of London B* 248 (1992): 229–33; idem, "Dishonesty and the Handicap Principle," *Animal Behavior* 46 (1993): 759–64; Amotz and Avishag Zahavi, *The Handicap Principle: A Missing Piece of Darwin's Puzzle* (Oxford, 1997); M. J. Owren and J. A. Bachorowski, "The Evolution of Emotional Expression: A 'Selfish-Gene' Account of Smiling and Laughter in Early Hominids and Humans," in *Emotions: Current Issues and Future Directions*, eds. T. Mayne and G. A. Bonnano (New York, 2001), 152–91; J. Maynard Smith and David Harper, *Animal Signals* (Oxford, 2003); Searcy and Nowicki, *The Evolution of Animal Communication: Reliability and Deception in Signaling Systems*; Ulrich E. Stegman, "John Maynard Smith's Notion of Animal Signals," *Biology and Philosophy* 20 (2005): 1011–25; Manfred D. Laubichler, Edward H. Hagen, and Peter Hammerstein, "The Strategy Concept and John Maynard Smith's Influence on Theoretical Biology," in ibid., 1041–50; Jonathan T. Rowell, Stephen P. Ellner, and H. Kern Reeve, "Why Animals Lie: How Dishonesty and Belief Can Coexist in a Signaling System," *American Naturalist* 168 (16) (2006): E180–E204; Timo Moran, "John Maynard Smith's Typology of Animal Signals: A View from Semiotics," *Sign Systems Studies* 37 (3–4) (2009): 477–95; and Kim Sterelny, *The Evolved Apprentice: How Evolution Made Humans Unique* (Cambridge, MA, and London, 2012).

[APPENDIX 2]

DAMASIO'S SOMATIC MARKER HYPOTHESIS (SMH)

It is understandable that the new affect theorists such as Connolly are drawn to the Somatic Marker Hypothesis because Damasio's insistence on the link between rational action and emotion and feeling is attractive to them. According to that hypothesis, decision-making is the result of a combination of "high reason," which is capable of unemotional cost-benefit analyses of a given action, and somatic signals or body state profiles from the emotional body. Damasio's suggestion is that those somatic signals are transmitted to the higher brain centers, where they help screen out certain choices, thereby establishing constraints on decision and action. He therefore explains the emotional and everyday decision-making deficits displayed by patients suffering from damage to the ventromedial sector of the frontal lobes by suggesting that these patients lack the relevant nonconscious "somatic markers" or emotional "hunches" that normally influence abstract reasoning and thinking. On Damasio's model, somatic markers of emotional reactions serve as automated, corporeal alarm signals to protect individuals from making bad decisions by reducing the number of options from which they can choose. According to Damasio, persons with damage to the frontal lobe of the kind that occurred in the famous nineteenth-century case of the railway worker Phineas Gage, whose skull and brain were accidentally pieced through by a tamping iron, are capable of a certain kind of detached reasoning but lack the somatic-emotional indications on which practical reasoning and daily life depend.[1]

1. A well-known instance of brain damage, the nineteenth-century case of the American railroad worker Phineas Gage, whose brain was pierced by a tamping iron but who survived the accident, is at the center of Damasio's somatic marker hypothesis. Although remarkably, Gage did not appear to suffer subsequent deficits in intelligence, memory, speech, or movement, according to Damasio he did suffer from certain emotional changes. Based on a reconstruction of the location of Gage's wound, Damasio has suggested that what happened to Gage when he was so gravely injured was that he lost the bodily indications by which the reasoning part of the mind is normally helped to make sound decisions. The result was that his decision-making processes became radically defective, which is why after the accident he could no longer hold on to a job. Gage is the first in a series of cases that Damasio has used to illustrate

In effect, Damasio stresses the limits of "pure reason" and instead foregrounds the constitutive role in our thoughts of affects said to operate quickly, below the threshold of reflection and argument. Since somatic markers signal the inmixing of innate and learned components of our affective responses, the somatic marker hypothesis suggests a mechanism for conceptualizing how culture and the body interact. Somatic markers are thus said to be culturally influenced "gut reactions" that provide guidelines for decision-making. These ideas are appealing to those who contest theories of "deliberative democracy" and the role of rational choice in ordinary life.

But it would not be difficult to show that Damasio's account of the influence of emotion and feelings in deliberative reason is theoretically confused and empirically problematic. From my perspective, his fundamental error is to claim that all emotions, including the "secondary" ones such as shame and guilt, are built up out of the basic or "primary" emotions, which are then defined, according to the Ekman paradigm, as hardwired physiological states triggered by emotionally competent stimuli to discharge involuntarily in stereotypical ways. Damasio conceptualizes the basic emotions in non-intentionalist terms as inherently independent of cognition, knowledge, and belief. His ideas are therefore vulnerable to many of the same criticisms that can be launched against Ekman's paradigm of the emotions.[2]

---

his "Somatic Marker Hypothesis." See Antonio Damasio, *Descartes' Error: Emotion, Reason, and the Human Brain* (London, 1994).

2. For useful critical assessments of Damasio's somatic marker hypothesis, see Bennett and Hacker, *Philosophical Foundations of Neuroscience*, 210–16; Barnaby D. Dunn, Tim Dalgleish, and Andrew D. Lawrence, "The Somatic Marker Hypothesis: A Critical Evaluation," *Neuroscience and Behavioral Reviews* 30 (6) (2006): 239–71; John Cromby, "Integrating Social Science with Neuroscience: Potentials and Problems," *BioSocieties* 2 (June 2007): 149–69; G. Colombetti, "The Somatic Marker Hypothesis, and What the Iowa Gambling Task Does and Does Not Show," *British Journal for the Philosophy of Science* 59 (2008): 51–71; and Stefan Linquist and Jordan Bartol, "Two Myths About Somatic Markers," *British Journal for the Philosophy of Science* 64 (2013): 455–84. For a critique of Damasio's anti-intentionalism, see also Colin McGinn, "Fear Factor," a review of Damasio's *Looking for Spinoza: Joy, Sorrow, and the Feeling Brain* (2003), *New York Times*, February 23, 2003.

INDEX

Note: Page numbers in *italics* indicate figures.